Book Packaging & Marketing
3 Murswell Lane
Silverstone
Towcester
Northants NN12 8UT
Tel & Fax: 01327 858380

Fields of Battle

The GeoJournal Library

Volume 64

Managing Editor: Max Barlow, Concordia University,
                    Montreal, Canada

Founding Series Editor:
                    Wolf Tietze, Helmstedt, Germany

Editorial Board: Paul Claval, France
                    R.G. Crane, U.S.A.
                    Yehuda Gradus, Israel
                    Risto Laulajainen, Sweden
                    Gerd Lüttig, Germany
                    Walther Manshard, Germany
                    Osamu Nishikawa, Japan
                    Peter Tyson, South Africa
                    Herman van der Wusten, The Netherlands

The titles published in this series are listed at the end of this volume.

Fields of Battle
Terrain in Military History

edited by

PETER DOYLE

and

MATTHEW R. BENNETT
Department of Earth and Environmental Sciences,
University of Greenwich, U.K.

KLUWER ACADEMIC PUBLISHERS
DORDRECHT / BOSTON / LONDON

A C.I.P. Catalogue record for this book is available from the Library of Congress

ISBN 1-4020-0433-8

Published by Kluwer Academic Publishers,
P.O. Box 17, 3300 AA Dordrecht, The Netherlands.

Sold and distributed in North, Central and South America
by Kluwer Academic Publishers,
101 Philip Drive, Norwell, MA 02061, U.S.A.

In all other countries, sold and distributed
by Kluwer Academic Publishers,
P.O. Box 322, 3300 AH Dordrecht, The Netherlands.

Printed on acid-free paper

All Rights Reserved
© 2002 Kluwer Academic Publishers
No part of the material protected by this copyright notice may be reproduced or
utilized in any form or by any means, electronic or mechanical,
including photocopying, recording or by any information storage and
retrieval system, without written permission from the copyright owner.

Printed in the Netherlands.

Contents

Contributors	vii
Preface	ix
Terrain in Military History: An Introduction *Peter Doyle & Matthew R. Bennett*	1
'Markings on the Land' and Early Medieval Warfare in the British Isles *Kerry Cathers*	9
Geology and Warfare in England and Wales 1450-1660 *Trevor Halsall*	19
Battles on Chalk: the Geology of Battle in Southern England during the First Civil War 1643-1644 *Trevor Halsall*	33
Terrain and Guerrilla Warfare in Navarre, 1808-1814 *Shell Kimble & Patrick O'Sullivan*	51
Effective use of Terrain in the American Civil War: The Battle of Fredericksburg, December 1862 *Judy Ehlen & R. Abrahart*	63
Tullahoma: Terrain and Tactics in the American Civil War *Walter Earl Pittman*	99
The Mountain is their Monument: An Archaeological Approach to the Landscape of the Anglo-Zulu War of 1879 *Tony Pollard*	117
Maps and Decisions: Buller South and North of the Tugela, 1899-1900 *Martin Marix Evans*	137
The Thirty-Years War, 1914-1945: Mapping the Battlefields of the Past for the Construction of the European Future *Edmon Castell & Sonia Roura*	143

Terrain and the Gallipoli Campaign, 1915
Peter Doyle & Matthew R. Bennett 149

British, French, German Mapping and Survey on the Western Front in the First World War
Peter Chasseaud 171

Terrain and the Messines Ridge, Belgium, 1914-1918
Peter Doyle, Matthew R. Bennett, Roy Macleod & Louise Mackay 205

Zero Hour: Historical Note on the British Underground War in Flanders, 1915-1917
Franky Bostyn 225

Mud Blood, and Wood: BEF Operational and Combat Logistico-Engineering during the Battle of Third Ypres, 1917
Rob Thompson 237

Across the River: Interpreting the Battle of Ebro *or* Battlefields as a Didactic Resource
Edmon Castell & Lluís Falcó 257

Fortification of Island Terrain: Second World War German Military Engineering on the Channel Island of Jersey, a Classic Area of British Geology
Edward P.F. Rose, W. Michael Ginns & John T. Renouf 265

Piracy on the High Desert: the Long-Range Desert Group 1940-1943
James Underwood, Jr. & Robert F. Giegengack 311

The Geology of the Battle of Monte Cassino, 1944
John A. Ciciarelli 325

Terrain as a Factor in the Battle of Normandy, 1944
Stephen Badsey 345

Airfield Country: Terrain, Land-Use and the Air Defence of Britain, 1939-1945
Ron N.E. Blake 365

Subject Index 385

Contributors

Robert J. Abrahart
School of Geography, University of Nottingham, Nottingham, NG7 2RD, UK

Stephen Badsey
Department of War Studies, Royal Military Academy Sandhurst, Camberley, Surrey GU15 4PQ, UK

Matthew R. Bennett
School of Earth & Environmental Sciences, University of Greenwich, Pembroke, Chatham Maritime, Kent ME4 4TB, UK

Ron N.E. Blake
Department of Building & Environmental Health, Nottingham Trent University, Burton Street, Nottingham, NG1 4BU, UK

Franky Bostyn
Astridlaan 6, Zonnebeke B-8980, Vlaanderen, Belgium

Edmon Castell
Departament de Didactica en les Ciencias Socials, Universidad de Barcelona, Edificio de Llevant, despatx 121, 08035 Barcelona, Catalunya, Spain

Kerry Cathers
University of Reading, PO Box 227, Whiteknights, Reading, RG6 6AL, UK

Peter Chasseaud
School of Earth & Environmental Sciences, University of Greenwich, Pembroke, Chatham Maritime, Kent ME4 4TB, UK

John A. Ciciarelli
Pennsylvania State University-Beaver Campus, 100 University Drive, Monaca, Pennsylvania 15061, USA

Peter Doyle
School of Earth & Environmental Sciences, University of Greenwich, Pembroke, Chatham Maritime, Kent ME4 4TB, UK

Judy Ehlen
US Army Topographic Engineering Center, 7701 Telegraph Road, Alexandria VA 22315-3864, USA

Lluís Falcó
Departament de Didactica en les Ciencias Socials, Universidad de Barcelona, Edificio de Llevant, despatx 121, 08035 Barcelona, Catalunya, Spain

Robert F. Giegengack
Department of Geology, University of Pennsylvania, Philadelphia, PA 19104-6316, USA

W. Michael Ginns
Channel Islands Occupation Society (Jersey), 'Les Geonnais de Bas', Rue des Geonnais, St. Ouen, Jersey, JE3 2BS, UK

Trevor J. Halsall
: *Postgraduate Research Institute for Sedimentology, University of Reading, PO Box 227, Whiteknights, Reading, RG6 6AL, UK*

Shell Kimble
: *National Geographic Society, 1145 17th St NW, Washington DC 20036-4688, USA*

Louise Mackay
: *School of Earth & Environmental Sciences, University of Greenwich, Pembroke, Chatham Maritime, Kent ME4 4TB, UK*

Roy M. Macleod
: *Department of History, University of Sydney, Sydney, New South Wales, Australia 2006*

Martin Marix Evans
: *3, Murswell Lane, Silverstone, Towcester, Northants NN12 8UT, UK*

Walter E. Pittman
: *Department of History & Social Science, University of West Alabama, Livingston AL 35470, USA*

Tony Pollard
: *Department of Archaeology, University of Glasgow, Gregory Building, Lillybank Gardens, Glasgow, G12 8QQ, UK*

John T. Renouf
: *Le Côtil des Pelles, Petit Port, St. Brelade, Jersey, JE3 8HH, UK*

Edward P.F. Rose
: *Department of Geology, Royal Holloway, University of London, Egham, Surrey, TW20 0EX, UK*

Sonia Roura
: *Departament de Didactica en les Ciencias Socials, Universidad de Barcelona, Edificio de Llevant, despatx 121, 08035 Barcelona, Catalunya, Spain*

Patrick O'Sullivan
: *Department of Geography, Florida State University, Tallahassee, Florida, 32306-2190, USA*

Rob Thompson
: *Centre for Contemporary History and Politics, University of Salford, The Crescent, Salford, M5 4WT, UK*

James L. Underwood
: *Department of Geology, Kansas State University, Manhattan, KF 66506-3201, USA*

Preface

Terrain has a profound effect upon the strategy and tactics of any military engagement and has consequently played an important role in determining history. In addition, the landscapes of battle, and the geology which underlies them, has helped shape the cultural iconography of battle certainly within the 20th century. In the last few years this has become a fertile topic of scientific and historical exploration and has given rise to a number of conferences and books.

The current volume stems from the international *Terrain in Military History* conference held in association with the Imperial War Museum, London and the Royal Engineers Museum, Chatham, at the University of Greenwich in January 2000. This conference brought together historians, geologists, military enthusiasts and terrain analysts from military, academic and amateur backgrounds with the aim of exploring the application of modern tools of landscape visualisation to understanding historical battlefields. This theme was the subject of a Leverhulme Trust grant (F/345/E) awarded to the University of Greenwich and administered by us in 1998, which aimed to use the tools of modern landscape visualisation in understanding the influence of terrain in the First World War. This volume forms part of the output from this grant and is part of our wider exploration of the role of terrain in military history.

Many individuals contributed to the organisation of the original conference and to the production of this volume. In particular credit is due to Linda Murr, our conference administrator, both for helping us shape the original concept and doing much to organise the conference. In this context Richard Waller and Jason Wood also deserve mention for their invaluable assistance. We are grateful for the support of our partners, particularly Peter Simkins, Mark Seaman and Brad King at the Imperial War Museum; and John Nowers and John Rhoads at the Royal Engineers Museum for their enthusiastic support. During the production of the volume Marilyn Croucher has provided secretarial support and Ben Holmes provided IT expertise at a critical moment. We would also like to thank all those who reviewed papers within the volume, often while revising their own contributions; to these individual we simply say thank you. Finally, it is to those who have contributed to this volume and to the delegates of the original conference that we owe the greatest debt. Our last word is in memory of Chris Campbell, delegate and enthusiastic expert on the Korean War, who died shortly after the conference.

Peter Doyle
Matthew R. Bennett
Chatham
July 2001

Terrain in Military History: an Introduction

Peter Doyle & Matthew R. Bennett

According to the *Concise Oxford English Dictionary*, terrain is defined as a 'tract of land as regarded by the physical geographer or the military tactician'. Military considerations are therefore at heart of any definition or exploration of terrain, and it is therefore unsurprising that most of the methods of terrain evaluation are born from military needs (Whitmore, 1960; Beckett & Webster, 1969; Parry, 1984; Mitchell, 1991). As a concept terrain is, therefore, something that encompasses both the physical aspects of the earth's surface, as well as the human interaction with them. Consequently, the study of terrain is by necessity multidisciplinary in nature, and can involve geology, geomorphology, hydrology, meteorology, agriculture, and civil engineering.

Terrain underpins military engagement, and therefore the study of historic battles. In any military action there are two basic levels of engagement: strategic, and tactical. Strategic considerations ultimately influences decisions to engage in warfare, and underpin war aims. Strategic assessments of terrain concern the disposition of large-scale geographic features, the location of urban centres, resources — minerals, oil, water, for example — transport systems, lowlands, uplands, rivers and oceans (Falls, 1948; Mitchell & Gavish, 1980; Rassam, 1980; Nathanail, 2000). Tactical assessments of terrain are associated with the prosecution of battle in the pursuance of strategy. Clearly, once a battle is entered into, then all aspects of the terrain may be employed by astute commanders, and many examples exist where geology, geomorphology or meteorology have combined to defeat an attacker or help a defender (Winter, 1998). Despite the widespread recognition of the importance of terrain within military action, it is has rarely been used as an historical tool to help deconstruct events, actions and outcomes of military engagements, yet clearly its potential to impact on our understanding of such actions is considerable. In recent years, however, the relevance of terrain as a tool in the analysis of historical engagements has gained some momentum (e.g. Doyle & Bennett, 1997, 1999; Rose & Nathanail, 2000) and it is hoped that this volume will continue this trend.

Historical and cultural resonances from contested landscapes are also of importance, having anthropological significance in addition to providing an archaeological presence. These aspects may have actively controlled the site of engagement, for example according to ancient rite or custom; may play a part in the creation of a sacred or revered site of action, a process that continues through memorialisation to the present day; and ultimately may

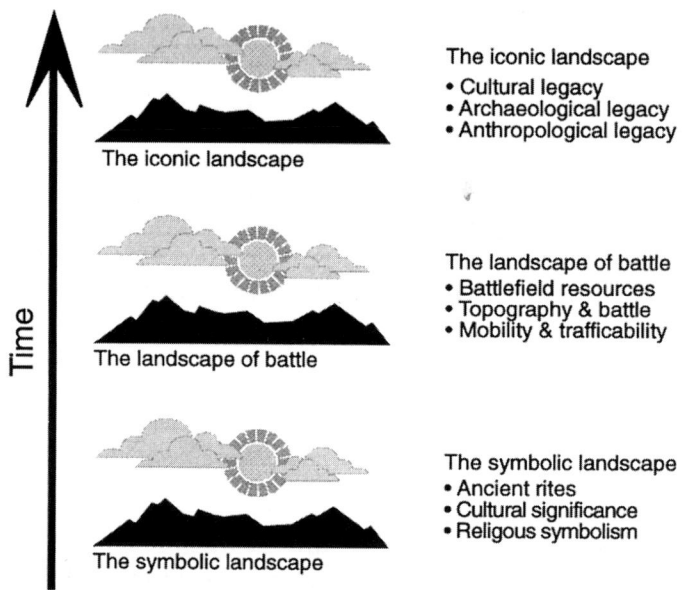

Figure 1: Conceptual model of terrain and military history

have had a role in the restoration of a devastated landscape after battle (e.g. Clout, 1997, 1999; Childs, 1998; Saunders, 2000; Freeman & Pollard, 2001). These aspects are as significant as any consideration of the terrain 'through a commander's lens' and are also explored within these pages.

The papers in this volume explore terrain in military actions both from the viewpoint of historical commanders as well as from its cultural influence, and collectively they illustrate the historical, cultural and archaeological importance of battlefield landscapes. The purpose of this introduction is to provide a framework with which to view the collective message provided by the component parts within this volume.

1. Aspects of terrain and military action

Terrain, the landscape of battle, forms a backdrop to any military action. The relevance of terrain to these actions, and to the interpretations that we place upon them, may change with time. Figure 1 presents a conceptual model which charts the association of military action with three levels, or layers, of landscapes: the symbolic, the practical, and the iconic. This model, effective from ancient times through to the present day, creates a mechanism whereby the symbolism, physical attributes and cultural aspects of terrain may be related and this model is elaborated below.

2. The symbolic landscape

Symbolism ascribed to landscapes or terrain elements, may control, through reference to ancient rites or customs, the placement of traditional battlegrounds for warring nations or communities. Such symbolism is ascribed a greater significance in ancient warfare **(Cathers, this volume)**, with a decreasing relevance in the industrialised warfare of the late 19th and 20th centuries, although ancient significance may resonate through centuries **(Pollard, this volume)**. However, the association of national identity with components of the landscape — such as the 'white cliffs of Dover' as an icon of 'beleaguered' Britain in the Second World War — is one way that new symbolism of value to a nation may be created. These aspects are further explored in the recent volume edited by Freeman & Pollard (2001).

3. The landscape of battle

This involves the way that military actions were and are controlled or influenced by terrain. The majority of the papers in this volume are concerned with this aspect of our understanding of terrain. The process of military engagement may in itself add to or develop the terrain, creating a layer of landscape with parallels to that of the earlier symbolic one — the iconic landscape. In any consideration of military actions, the landscape of battle, or more properly the sum total of its physical attributes, is of greatest significance in helping determine the outcome of battle (Mitchell & Gavish, 1980; Parry, 1984). Two aspects are necessary in the consideration of terrain: firstly a commanders grasp and understanding of terrain, in part dependent on the resources available to them, which we refer to as terrain intelligence; and secondly terrain assessment and utilisation. These are discussed below.

Terrain intelligence
Both strategic and tactical assessments of terrain require there to be adequate terrain intelligence, provided in advance of the operation. These should ideally represent a databank of information available to commanders in considering the implications of operations in prosecution of the strategic aims of their nation or group. Such intelligence is gathered by terrain specialists and is presented to military staffs in published form, most appropriately in the form of maps, plans and photographs (Whitmore, 1960; Beckett & Webster, 1969; Parkinson, 1997). In the Great War alone over 850 million military maps of all scales were produced by the main protagonists **(Chasseaud, this volume)**. The adequacy of this intelligence, and its appropriate use by commanders, is perhaps one of the most strongly debated

aspects of all considerations of military operations. This is typified by aspects of the Boer War of 1899-1902 **(Marix Evans, this volume)**, but more specifically by the raging debate over the adequacy of mapping, staff work and intelligence gathering before the allied landings at Gallipoli in 1915 **(Doyle & Bennett, this volume)**. Often, however, it is difficult to assess the extent to which commanders had detailed information at their fingertips, as with aspects of the Napoleonic wars **(Kimble & O'Sullivan, this volume)** or the American Civil War **(Pittman, this volume)**.

Terrain assessment
Typically, there are five aspects to be considered in the tactical assessment of terrain (Parry, 1984):

1. *Position*. Position is everything in a battle where the possession of high ground means that a defender is able to command the lowlands surrounding it. This is a characteristic of the battles for position in the Flanders lowlands during the Great War **(Doyle et al., this volume)**, as well as in the Battle for Monte Cassino in 1944 **(Cicarelli, this volume)**. Modern technologies allow us to re-examine the battlefield and recreate lines of sight, vantage and concealment thereby aiding historical analysis **(Ehlen & Abrahart, this volume)**.

2. *Mobility*. Adequate mobility of troops, animals and machinery requires an understanding of the ground conditions. Correlations may be made between ancient battle sites and geology which appear to suggest that certain geological terrain were favoured as battle sites **(Halsall, this volumea,b)**. However, it is in considering 20th century warfare that the importance of terrain to mobility becomes most apparent. The war on the Western Front during 1914-1918 demonstrated the need for appropriate assessment of 'going surfaces', that is surfaces over which the men and materials could be transported. The creation of rapidly driven plank roads to combat the unsuitable, soft wet ground of the Ypres Salient, for example, was an attempt to address this issue **(Thompson, this volume)**. The physical barriers provided by the bocage or hedgerow country of Normandy **(Badsey, this volume)**, the sands of the Western Desert **(Underwood & Giegengack, this volume)**, and the rivers and slopes of Italy **(Cicarelli, this volume)** during the Second World War are also excellent examples of the difficulties to be addressed in any tactical assessment of terrain.

3. *Ground conditions*. Any consideration of ground requires an assessment of the geology and the ability of a body of soldiers to dig into it to create permanent emplacements, defensive positions, and airfields. Adequate terrain intelligence is required, and the fortification of the Channel Island of Jersey during the Second World War is a particularly important example **(Rose et al., this volume)**, as is the exploitation of appropriate sites in Britain — airfield country — for the siting of air bases **(Blake, this volume)**. Offensive military engineering, such as the mining carried

out during throughout the history of siege warfare, and resumed during the Great War **(Bostyn, this volume)**, also requires similar levels of terrain intelligence.
4. *Resource provision.* The provision of adequate supply line and communications is of great importance to the supply of troops on the ground and this is relevant in the extended and vulnerable supply lines of the Western Desert preyed upon by roving 'private armies' such as the Long Range Desert Group during the Second World War **(Underwood & Giegengack, this volume)**. The local terrain itself will provide some of the necessary supplies, derived from agriculture, for example, and natural resources such as aggregates **(Rose et al., this volume)** and particularly water supply **(Doyle & Bennett, this volume)**, requires adequate intelligence.
5. *Hazard mitigation.* Natural hazards, such as weather conditions, can cause difficulties, such as the excessive rainfall during the Third Ypres campaign of 1917 in Flanders (Griffiths, 1989), or the changes in wind direction that hampered the first offensive use of poison gas by the British on the Western Front in September 1915, or the trajectory of shells (Watt, 1918). Floods, mass movements and other natural hazards similarly need to be considered.

4. The iconic landscape

The significance of battle sites as scenes of slaughter are such that they have become national symbols, icons of the ideals ascribed to by the dead of their protagonists **(Castell & Roura, this volume)**. Many examples can be drawn from the ancient battle sites of Europe, but perhaps the greatest examples may lie in the wars of the 20th century, where the significance of the landing beaches of Gallipoli and the chalk upland of Artois have strong cultural associations with nation building for Australia, New Zealand and Canada. The development of a mythology of the Great War as a war of trenches and mud — both ultimately a factor of the underlying geology and terrain — is another potent example of this (Doyle, 2001). Such significance ascribed to the iconic landscape has led to the development of the relatively new consideration of battlefield archaeology, particularly for ancient battle sites, where myth and legend can significantly cloud the historical accuracy of our understanding of the prosecution of battle. For the wars of the last two centuries, this is less of an issue, and sufficient resource exists to compare archaeological and documentary resources to establish the progress of a particular battle. But for far off wars — the English Wars of the Roses of the 15th century, for example — the received wisdom about a battle, such as the Battle of Towton in 1461 where 28,000 soldiers were supposed to have perished, archaeological evidence is at odds with local knowledge, folklore and accepted fact (Tim Sutherland, pers comm.). Ultimately, the creation of modern myths and legends, and the memorialisation of this layer of terrain are the new impetus for a burgeoning industry, with the publication of magazines such as the UK's *Battlefields Review*, and the development of a major battlefield tourism industry. These cultural aspects are currently under scrutiny by a wide range of researchers, enabling detailed aspects, forgotten or suppressed — as in the Battle of the Ebro in 1938 **(Castell & Falco, this volume)** — to be examined.

5. Conclusion

We hope therefore that this volume collectively makes and illustrates a case for the importance of terrain as a perspective in viewing historical military actions, both at a practical level — 'through the eye of the commander' — and in unravelling the anthropological and cultural resonance of battle.

References

Beckett, P.H.T. & Webster, R. 1969. *A Review of Studies on Terrain Evaluation by the Oxford-MEXE-Cambridge Group, 1960-69.* Military Engineering Experimental Establishment [MEXE] Report 1123, Christchurch.

Childs, J. 1998. *The Military Use of Land. The History of the Defence Estate.* Peter Lang, Bern.

Clout, H. 1997. War and recovery in the countryside of north-eastern France: the example of Meurthe-et-Moselle. *Journal of Historical Geography* 23, 164-186.

Clout, H. 1999. Destruction and revival: the example of Calvados and Caen, 1940-1965. *Landscape Research* 24, 117-139.

Doyle, P. 2001. Geology as an interpreter of Great War battle sites. *In:* Freeman, P. & Pollard, T. (Eds), *Fields of Conflict: Progress and Prospect in Battlefield Archaeology.* BAR Int Series, 958 Oxford.

Doyle, P. & Bennett, M.R. 1997. Military geography: terrain evaluation and the British Western Front, 1914-18. *Geographical Journal* 163, 1-24.

Doyle, P. & Bennett, M.R. 1999. Military geography: the influence ofterrain on the outcome ofthe Gallipoli Campaign, 1915. *Geographical Journal* 165, 12-36..

Falls, C. 1948. Geography and war strategy. *Geographical Journal* 112, 4-18.

Freeman, P.W.M. & Pollard, T. (Eds) 2001. *Fields of Conflict: Progress and Prospect in Battlefield Archaeology.* BAR Int Series, 958 Oxford.

Griffiths, P. 1989. *The Effects of Weather Conditions on the Third Battle of Ypres, 1917.* School of Geography, University of Birmingham, Working Paper 51, Birmngham.

Mitchell, C.W. 1991. *Terrain evaluation.* Second Edition. Longman, Harlow

Mitchell, C.W. & Gavish, D. 1980. Land on which battles are lost or won. *Geographical Magazine* 52, 838-840.

Nathanail, C.P. 2000. Geoenvironmental security — the challenge for tomorrow's geologists? *In:* Rose, E.P.F. & Nathanail, C.P. (Eds) *Geology and Warfare, Examples of the Influence of Terrain and Geologists on Military Operations.* Geological Society, London, 359-380.

Parkinson, P. 1997. Geographic support to the armed forces - military Survey's 250-year history. *Royal Engineers Journal* 111, 210-215.

Parry, J.T. 1984. Terrain evaluation, military purposes. *In:* Finkl, C.W. (Ed.) *The Encyclopedia of Applied Geology.* Encyclopedia of earth sciences, volume 13. Van Nostrand Rheinhold, New York.

Rassam, G. 1980. How geology affects war in Iran, Iraq. *Geotimes* 25, 20-22.

Saunders, N.J. 2000. Matter and memory in the landscapes of conflict: the Western Front 1914-1999. *In:* Bender, B. & Winer, M. (Eds) *Contested Landscapes: Movement, Exile and Place.* Berg, Oxford, 37-53.

Watt, A. 1918. Meteorology and the war. *Scottish Geographical Magazine* 34, 263-265.

Whitmore, F.C. 1960. Terrain intelligence and current military concepts. *American Journal of Science* 258A 375-387.

Winters, H.A. 1998. *Battling the Elements, Weather and Terrain in the Conduct of War.* Johns Hopkins University Press, Baltimore.

Peter Doyle & Matthew R. Bennett
School of Earth & Environmental Sciences
University of Greenwich
Chatham Maritime
Kent, ME4 4TB

'Markings on the Land' and Early Medieval Warfare in the British Isles

Kerry Cathers

ABSTRACT: Rather than examining how terrain affected the strategies of battles and how it was a factor to be exploited, this paper focuses upon its symbolic function. Early medieval warfare in Britain was characterised by numerous, swift raids into enemy territory for the purpose of inflicting damage and capturing booty, not for the acquisition of land or political domination. These raids were carried out by relatively small groups which could not be considered armies. The best examples of this style of warfare are the early Viking raids. Warfare was not, however, limited to raids; campaigns for the purpose of conquest were also fought. These were more complex and infrequent. Codes of behaviour surrounded the process of going to war in these situations. The absence of standing armies meant that troops had to be summoned from the surrounding area before they were mustered and then moved to what would be the battlefield. Such a process was time consuming and, as a result, invading armies did not carry out their campaign with the swiftness later armies would. They entered their opponent's territory, made camp and awaited the arrival of the defending army. It was essential that not only their presence be known, but also their location. For this purpose, well-known and easily identified locations were chosen. This paper examines this practice. Was it to be found throughout the British Isles? In Ireland, many battles were fought by waterways or other territorial borders. Was this for utilitarian purposes or was there symbolic significance involved? How late did the practice survive and what happened if enemy did not respond to the challenge? These questions are explored.

1. Introduction

Early medieval battlefields have been used to determine various things such as cultural expansion, reasons for a particular war and what kingdoms or political groups participated. Though attention has been given to locating the sites, little has gone into examining the uses of terrain in warfare (Wood, 1980) and it is the purpose of this paper to examine its role. As with other periods of history, terrain was a factor in battles of the early Middle Ages and contemporary warleaders were aware of its importance and took it into consideration in their strategies and in the deployment of their troops. For this period the terrain or 'markings on the land' served another purpose; one which could be interpreted as symbolic as it was not a factor in battles themselves, but was part of the process of warfare. In an era when an army could remain elusive and avoid a pitched battle simply by remaining mobile, mechanisms had to be devised to bring an army to war. This device was setting up camp and staying put. It is at this point that the location of the eventual conflict and its characteristics play their first role.

2. Defining the location of battles

Before this function is discussed, it is appropriate to talk first about what we do and do not know about the battlefields in the British Isles in the early Middle Ages. As a whole, evidence available to us about warfare during this period is scarce. Contemporary sources are not as detailed as historians would like them to be. Information regarding the battlefields is no exception. In many, if not most, instances before a battlefield can be discussed, it must first be located; a task not easily done. Writing in 1950, Lieut.-Col. Alfred Burne stated that for no battlefield from the early medieval period 'can the exact site be pointed to with absolute certainty' (Burne, 1950). Even though battles, recorded in the contemporary sources, are provided with names their locations remain obscure and for many, completely unknown. However, this has not prevented attempts being made to locate them.

For most searches, the starting point has been with the place names themselves. But difficulties arise with this type of evidence which in many cases prevents certainty. Place names do not remain constant through time: *Eoforwic* is now York; *Dineiddyn* is now Edinburgh. For some, an element of the original name can be found in its present-day form — *eiddyn*, for example forms part of Edinburgh — and because of this it can be traced through a number of documents and forms. This occurs for only a few of the larger, better-known and well-recorded sites. Other names do not allow for this for a number of reasons. Contemporary sources provide different spellings for the same name resulting in confusion over its original meaning and its metamorphosis into its final form. Names were incorrectly copied: a confusion of *hëdh* 'hill', with *heidhr*, 'heath' can have significant impact upon a search based on a correlation between the typography of a proposed site and details contained in the name of the battlefield. This is even more complicated if we consider the mixing of languages. Is the root Scandinavian, Anglo-Saxon or Celtic? Another problem with the actual recording of these sites is that different sources give different names. In the *Anglo-Saxon Chronicle* it is the Battle of Hatfield while in the *Annales Cambriae* it is the Battle of Meigen[1]. Other names simply disappear from written sources. Some names are distinct to the period of Scandinavian dominance in the northeast only to be abandoned (Pearson, 1995). The small-scale migration of people led to the moving of settlements (which could supersede an existing name) or a complete abandonment of them.

The search for the battlefield of Brunanburh illustrates the difficulties encountered when using placenames as identifiers. The name itself is recorded only in a poem and could be a false one given to the site for alliteration purposes. Its closest, contemporary equivalent is found in the *Annales Cambriae* where it is referred to as Brune[2]. As Pearson explains in her article about battlesites around Bramham Moor, the etymology and origin of Brunanburh are uncertain (Pearson, 1995). This increases the number of present-day places which could have had it as their original form, but does not allow for absolute identification. Until recently, the general locality of the battle was uncertain which has resulted in over thirty sites being proposed, ranging from Scotland to the south of England. It is accepted as having been fought in northern England, but no placenames resemble Brunanburh exists in the regions suggested. As has been stated, the vagueness of the name allows for numerous possibilities to be argued — though with much manipulation. The search for the battle site continues.

Place names are not the only evidence available. For a few battles we have had some description of the battlefield, or at least clues to the surroundings recorded as well as their name. This information can be used as Pearson has done with Brunanburh; to eliminate locations with similar placenames in support of Bramham Moor (Pearson, 1995). Other methods involve examination of archaeological evidence to establish a zone of settlement or penetration and thereby limit the geographical region of the search. From this a borderland is established within which battles were most likely to have taken place. Names and topographical features are then examined to find the battlefields in question. This approach has been used for battles of the migration period to isolate a region along a proposed borderland between the advancing Anglo-Saxons and the Britons. It has resulted in three sites being proposed for Badon within a certain zone (Burkitt, 1990). Caution must be taken when using this method and an awareness that its usefulness is limited to the early battles between the Britons and the invading Angles and Saxons, or the movement north by the Angles into southern Pictland.

For other periods and conflicts where there is no invasion by a cultural group, the establishment of borders has been used to narrow the possibilities. This is based on the assumption that battles were fought in border regions, in so-called buffer zones. Michael Wood uses this assumption as the basis for his search for Brunanburh. He identifies a border zone along a series of fortified sites running between Dore and Whitwell Gap just south of the Don River. He has isolated this area based on the political situation in the north (the animosity between the southern English — those under the rule of Wessex — and the Northumbrians and Danes of that region). From this, and a few hints from the sources, Brinsworth is proposed as the possible site (Wood, 1980). It should be remembered that Pearson identifies Bramham Moor as the site for Brunanburh — a location further to the north.

Leslie Alcock warns against the co-ordination of borderlands and battle sites and vice versa and their use in locating the 'spearhead' of an expansion or invasion (Alcock, 1977). Though battles could take place in borderland regions or frontier zones, they could also be fought deep within an adversary's territory. When he fought the Battle of Chester in 613, Æthelfrith penetrated far into Celtic territory[3]. His victory did not result in the immediate establishment of Anglian settlements that far west; they did not exist before the middle of the 7th century (Chadwick, 1963). Nora Chadwick comments that: 'It is not easy to see any general historical importance in the battle itself' as no territory was gained as a result (Chadwick, 1963). But this did not have to be the purpose of the battle and its occurrence so far into Celtic territory need not have any significance on our understanding of Anglian influence in this area.

3. Methods of war

It is appropriate at this point to interject a brief comment upon on early medieval warfare as its purpose and methods are perhaps different from later periods and are part of the factors affecting the choice of battlefields. For simplicity's sake and for the purpose of this paper, two categories have been established into which warfare has been placed: campaigns whose purpose was the acquisition of wealth through plunder and to inflict damage through the ravaging of a territory; and, those which had a political purpose. Plunder was as much a part, and aim, of warfare as the conquest of territory. Monetary gain and not political victory or the conquest of land acted as the motives for many expeditions (Reuter, 1985). The *Annales Cambriae* record twelve references to such attacks (Nom's, 1980). The *Anglo-Saxon Chronicle* tells of various episodes where raids were launched for the purpose of plunder and devastation. There are a number of references to plundering in Ireland in the *Cogadh Gaedhel re Gallaibh*. Osraigh and Caimhghen were ravaged and plundered by the Laighin[4]. This style of warfare was characterised by small groups of men moving swiftly into foreign territory, inflicting damage, gathering booty and withdrawing before a force could be mustered to challenge them.

The second category of warfare occurred on a larger scale, over a longer period of time and its purpose was to draw the opponent into battle. These were risky undertakings and often were avoided whenever possible — Maldon and Hastings illustrate what could happen if the battle was lost (Hooper, 1989; *cf.* Halsall, 1989). Even when the armies faced each other, battle was necessarily the end result. The purpose of these campaigns was to force one or other of the protagonists into submission which was indicated through the giving of hostages and tribute and the promise to provide military service. If failure seemed imminent and the amount of the tribute could be agreed upon, a battle could be avoided with victory attained. In 655 Oswy gave battle to Penda only after he was unable to pay tribute[5]. The majority of Brian Boru's campaigns in the 10th and early 11th centuries consisted of Brian offering his opponent the opportunity to submit to him without a battle; a surprising number accepted. Though there were regions along the borders which moved back and forth between kingdoms, victory on the battlefield did not necessarily mean that territory was obtained. The example of the Battle of Chester above illustrates this.

4. 'Marking' a battlefield

It is this second category of warfare that the 'markings' on the geography are significant. As stated above, if a force remained mobile then it could avoid battle by making itself difficult to find. Thorkel's army was warned that an English force was being collected against them in London. It 'in turn crossed at Staines' and was able to make it to the ships and escape without battle[6]. It is one reason which has been proposed as to why Bryhtnoth stood for battle at Maldon. Not because it was his chosen battlefield, but if he did not call the Danish force to battle at that time and they left, he might never find them again.

How was the intention to stand and do battle communicated? The element overlooked or perhaps diminished by historians is the difficulty of locating an army. If an 'invading' force wished to bring what would be the defending force to battle it had to make its presence known and allow itself to be found. Two criteria had to be met. First, it had to be somewhere where it would be seen and second, it had to be somewhere where it could be located easily. The best way of achieving the latter was to make camp by a landmark either natural or man-made.

Can this premise be supported? As stated above, battles were given a name in the sources or were described in association with something: a fortress; a settlement; a river; or, an island. The explanation for this could be one of convenience for the chronicler by allowing a location rather than simply saying that in a certain year 'x' fought 'y'. Asser tells of one of the many battles between Alfred and the Danes as having been fought at the place called Aclea, translated as 'oak field'[7]. This may be interpreted as an non-descript place where there were many oak trees. But placenames follow this formulae. For example, Whin Moor ('moor where whin (i.e. furze) bushes grow') has a translation just as ambiguous, but is a defined place in the north of England (Pearson, 1995). Such descriptive forms were how many places were identified and, thereby, named and they do not diminish the relevance of their association with a battle. If the location of a battle were unknown, there does not seem to be anything which would compel the chronicler to falsify a name or location. It is possible, therefore, that the presence of named locations with the battles is because battles took place close to identifiable locations.

For the battles which can be located within reason if not with absolute certainty, almost all of them are by river crossings, intersection of Roman roads, ancient monuments or fortifications. Of the 28 battles Guy Halsall discusses in his survey of Anglo-Saxon warfare, all but one could be associated with a monument or other markings (Halsall, 1989). Alcock has found that the majority of early battles between Celt and invading Saxon occurred in proximity to a river-crossing.

By locating a battle by a ford or at the intersection of Roman roads, detection was guaranteed. They were more than likely major routes of travel and communications. An invading force would not have to wait long before it was detected and message was sent to the king or thegn. Such a location was not chosen for defensive purposes, as Halsall has queried. His concern arises from an army commander permitting his opponent to cross a ford safely (Halsall, 1989). If guarantees were not made that they would be able to cross safely it is unlikely that a warleader would divided and weaken his troops to such an extent that their annihilation would be effortless. Halsall proposes that 'holding of the ford' was the initial phase of fighting, before dismissing it. He does so for the wrong reasons. It would be senseless, if not bordering on insane, to invest time and wealth into mustering an army, marching it across what could be a great distance, succeed in calling an opponent to battle and then not allowing them to take up the field. If the purpose of the expedition is understood as the contest between the two armies, allowing an opponent to cross a ford does not seem ridiculous. Byrhtnoth wanted to do battle with the Danes at Maldon and the only way he could do this was to allow them to cross over onto the 'mainland'. If he did not do so, the Danes would have retreated to their ships and departed.

Ancient monuments include such features as barrows, dykes or earthworks; Halsall includes Roman structures in this category. These appear infrequently in the sources with regard to battles. With the uncertainty of and inability to identify names of battle sites, it is possible that more of the names refer to such land markers, but this is only conjecture. Locations such as Adam's grave and Egbert's Stone are referred to as battles or mustering locations. These were visible and well known sites. Whether their association with supernatural elements had any influence on their use as sites is difficult to determine. It is unlikely that military commanders would prefer a site with religious or supernatural connections if it were not a suitable battlesite, over another location which was more suited to battle.

In none of these circumstances is it being suggested that the actual battle took place around these items. While an opponent was crossing the ford, a commander could be deploying his troops close by according to the lay out of the surrounding area. To suggest that a battle took place over the Roman road or a river is to do an injustice to early medieval 'generals'. They had a knowledge of how terrain could act for or against an army and how to use it to their advantage. Asser draws his audience's attention to the fact that the field at Ashdown was not a fair one as the Danes had arrayed their forces on the higher and more advantageous ground before the arrival of the West Saxons. These markings of roads, fords, monuments were used for the sole purpose of declaring war. This is not to say they were not taken into account as possible defences, obstacles or routes of escape, but to suggest that the battlefield included them during the fighting is arguing contrary to common sense.

A number of the battlefields are associated with fortifications, but this does not necessarily mean that they were sieges, or that the fortifications played any part in the battles themselves. Sieges were not unheard of. The *Annales Cambriae* record the besieging of Cadwallon, as well as the destruction of the fortresses at Degannwy by the Saxons and Alt Clud by the Scandinavians which might indicate some military activity centred around the fortifications[8]. Fortified sites could be found in the British Isles during this period, but they do not seem to have been designed to serve a military function. The forts in Wales were used to consolidate control of the land and were located in high locations — often above the 300 m contour — overlooking an approach along a river valley. Most were constructed to withstand raids but not a siege. They were not close to water supplies nor were they furnished with buildings for food storage to withstand a prolonged siege (Marshall-Cornwall, 1957). In northern Anglo-Saxon England and in what is now Scotland the building of fortifications was an integral part of kingship. They acted as displays of authority and wealth and housed the monarchs and the members of their household including officials and warriors. These enclosed residences were hierarchical, both in their construction and their function (Alcock, 1977).

Alcock argues extensively that the fortifications did not play a part in the battles which bear their name. There is nothing within the sources to indicate that forces — either Anglo-Saxon or British — had taken up position inside the named fortifications (Alcock, 1988). He argues that the translation of *'in - ære stowe - e is genemned æt....'* as 'in' that place rather than 'at' that place, gives the impression that one of the two armies occupied a place was inside the fortification at the time of the battle (Alcock, 1977). By translating it as 'at that place' a more accurate impression is given neither army was behind walls. If not a siege, why then were these fortifications being mentioned in association with battles, especially if some of them had been abandoned? The forts were used as markers and the battles took place at a suitable site within their vicinity. Many of the fortifications were constructed close to routes of transportation and communication (by fords or Roman roads).

As a final point in support of the use of markers in warfare is the mention made of the mustering of Alfred's army. Not only did a contesting army have to allow itself to be found, so to the king or thegn who was mustering his army had to be located. An army could not be constructed if the soldiers did not know where to meet. In 878 Alfred rode to Egbert's Stone where 'all the inhabitants of Somerset and Wiltshire and all the inhabitants of Hampshire...joined up with him'[9].

By making camp and remaining in a place which was both visible and public, it made it difficult, if not impossible, for a king or a thegn not only to be unaware of an army's presence, but where to find it. The response was not expected to be immediate, but it was expected. The absence of standing armies meant that troops had to be summoned from the surrounding area (from those who owed service) before they were mustered and then moved to what would be the battlefield. Such a process was time consuming and, as a result, invading armies — or defending ones — did not carry out their campaigns with the swiftness later armies would. Consensus stand that 11th century armies could be mustered in two to three weeks (Hooper, 1989), depending on time and distance travelled. For this reason, guarantees could be demanded that raiding and ravaging would not be carried out by the foreign army until after a battle was had.

This is not to claim that these rules were always adhered to or that battles under other circumstances did not occur. In his campaign against Æthelfrith, Redwald moved with such swiftness that his adversary was unable to muster his army. Rather than taking up position, Redwald forced Æthelfrith into battle before he was able to prepare his entire force[10]. Armies were intercepted and battle was done at that location. In 1009 the Anglo-Saxons intercepted the Danes on route to their ships and forced them to battle. Brian mustered an army and marched toward Ath Cliath (Dublin) to lay siege. En route he came across the opposing army at Glen Mama — where they were hiding their cattle and families — in a crook of the land there and forced them into battle[11].

Logistics of the time, especially the difficulty in locating an enemy who wished to remain hidden, imposed a regime upon the armies and their leaders to allow battles to occur. The solutions was simple. Locate an area which was easily identifiable, set up camp and wait for the defending army to approach. Terrain in the early Middle Ages, not only performed a role in how battles were fought, but also functioned as a declaration of war.

References

Alcock, L. 1977. Aspects of warfare of Saxons and Britons: Her...gefeaht wip Wealas. *Bulletin Board of Celtic Studies* 27, 413-424.
Alcock, L. 1988. An heroic age: war and society in northern Britain, AD 450-850. *Proceedings of the Society of Aniquarians of Scotland* 118, 327-334.
Burkitt, T. & Burkitt, A. 1990. The frontier zone and the siege of Mount Badon: a review of the evidence for their frontier location. *Somersetshire Archaeological and Natural History Society* 134, 81-94.
Burne, A. 1950. Ancient Wiltshire battlefields. *The Wiltshire Archaeological and Natural History Magazine* 53, 397-412.
Chadwick, N.K. 1963. The Battle of Chester. *In*: Anon. (Ed.) *Celt and Saxon: Studies in the Early British Border*. Cambridge University Press, Cambridge, 167-185.
Gillingham, J. 1984. Richard I and the science of war in the Middle Ages. *In*: Holt, J. & Gillingham, J. (Eds), *War and Government in the Middle Ages*. Woodbridge, London, 194-207.

Halsall, G. 1989. Anthropology and the study of pre-Conquest warfare and society. *In:* Hawkes, S.C. (Ed.), *Weapons and Warfare in Anglo-Saxon England.* Oxford University Committee for Archaeology, Oxford, 155-178.

Hooper, N. 1989. Anglo-Saxons at war. *In:* Hawkes, S.C. (Ed.), *Weapons and Warfare in Anglo-Saxon England.* Oxford University Committee for Archaeology, Oxford, 155-178.

Marshall-Cornwall, J. 1957. The military geography of the Welsh Marches. *Geographical Magazine* 30, 1-12.

Morris, J. (Ed.) 1980. *Nennius' British History and the Welsh Annals.* Phillimore. London.

Pearson, W. 1995. Bramham Moor and the red, brown and white battles. *Yorkshire Archaeological Journal* 67, 23-50.

Reuter, T. 1985. Plunder and tribute in the Carolingian Empire. *Transactions of the Royal Historical Society* 35, 74-95.

Wood, M. 1980. Brunaburh revisited. *Saga-Book of the Viking Society* 20, 200-217.

Notes

1. *Anglo Saxon Chronicle,* 633; *Annales Cambriae* 630.
2. *Annales Cambriae* 938, p. 49 and 91: 'an. Bellum Brune'.
3. Bede, *Ecclesiastical History of the English People,* ii.2.
4. *Cogadh Gaedhel re Gallaibh,* p. 151.
5. *Anglo Saxon Chronicle,* 655
6. *Anglo Saxon Chronicle,* 1009
7. Asser, *Life of Alfred,* c. 5.
8. *Annales Cambriae* 629 (note that Cadwallon was on the island of Glannuac); Degannwy 822; alt Clud, 870.
9. Asser, c. 55.
10. Bede, *Ecclesiastical History of the English People,* ii.12. translation taken from Judith McClure and Roger Collins (trans), Oxford University Press, Oxford (1994).
11. *Cogadh Gaedhel re Gallaibh,* p. 111. It is possible that on or around this physical feature there was a fortress — Ascall Gall — as the plundering of a fortress after the battle is referred to in a poem about the time.

Kerry Cathers
University of Reading
PO Box 227
Whiteknights
Reading, RG6 6AL

Geology and Warfare in England and Wales 1450-1660

Trevor J. Halsall

ABSTRACT: Arguably, geology had an important influence on battlefield tactics in the Middle Ages and 17th century in Britain, but did it also constrain patterns of warfare at the strategic level in those times? Between 1450 to 1660 AD there were several periods of active warfare in England and Wales, in the course of which some 75 significant field engagements have been identified. An investigation of the geographical distribution of these battle-sites in relation to the solid (bedrock) geology depicted on the 1:625,000 geological maps of Britain, reveals that some chronostratigraphical units sufficiently widespread to be depicted at this scale (notably the Permo-Trias and Upper Carboniferous) are associated with more battlefields than might be predicted from the relative extent of their area of outcrop, whereas others (notably the Cambrian, Ordovician, Silurian, Lower Carboniferous, Upper Jurassic and Lower Cretaceous) exhibit a below-average number of battlefields per unit of outcrop area. The application of a Chi-squared test confirms at a >99.5% confidence level that the relationship between bedrock geology and these battlefields is non-random, which strongly suggests that geology was an important influencing factor on the conduct of the campaigns in question. Other than the avoidance of militarily unsuitable, mountainous or rugged terrain underlain by relatively resistant strata, the precise nature of this relationship remains to established but the following outline is proposed as the basis for further investigation: lithology, structure and geological history of an area constrain relief, topography and the distribution of mineral resources, affecting drainage and overlying soil type and hence vegetation patterns and the 'going', which in turn determine agricultural productivity and the routing of lines of communication. Areas of resource production, substrate properties and drainage and lines of communication locate centres of population. Tactics on the battlefield are determined by topography and going, while strategy is dictated by the need to control centres of population, resource production and lines of communication, objectives whose location is further removed from, but none the less in part attributable to, the influences of the underlying geology.

1. Introduction

Arguably, geology is an important factor constraining military activity on the battlefield at the tactical level (e.g. Halsall, 2000a, this volume), and the design and construction of fortifications (Halsall, 2000b). However, it remains to be tested whether geology also plays a significant and more fundamental role in shaping patterns of warfare at campaign level, i.e. in determining strategy and grand-tactics. One test is the study by area of the distribution of battle sites in relation to the geology, during specific periods of history. As such, this paper presents a preliminary analysis of the distribution of battlefields in England and Wales for the period 1450 to 1660 AD, with respect to geology.

The period of British history most associated with active if intermittent warfare in England and Wales was between 1450-1660 AD. This relates, in particular, to three main episodes of conflict: (1) the Wars of the Roses 1450-1487, including the precursor action at Sevenoaks (1450) in the course of Jack Cade's Rebellion (Weir, 1995) and the battles of Bosworth (1485) and Stoke Field (1487) which finally brought this conflict to a close (Haigh, 1995); (2) the Anglo-Scots Border Wars which rumbled on intermittently through the first half of the 16th century and which featured engagements at Flodden (1513) and Solway Moss (1542) (Phillips, 1999); and (3), the 17th century Civil Wars between King and Parliament, commencing with the Battle of Newburn Ford in the Bishop's War of 1640 (Newman, 1985; Emberton, 1997) and concluding with the major battle at Worcester which brought the Third Civil War to a close (Rogers, 1968), together with George Booth's Rising of 1659 which preceded the Restoration (Newman, 1985). This 210 year time span was a period of relatively rapid development in military technology and practice (Roberts, 1967; Boardman,1998; Haythornwaite, 1998; Phillips, 1999), so a hypothetical relationship between bedrock geology and warfare can be tested against a background involving different styles of warfare.

The armies involved during this time period range considerably in size from the 20,000–30,000 or so per side at Towton (1461), Flodden (1513) and Marston Moor (1644) (Smurthwaite, 1993), to a mere 1,000 on each side at Powick Bridge (1642) (Guest & Guest, 1996). In the majority of cases the opposing forces were relatively evenly matched, so the strategy and tactics employed were not cumulatively biased in favour of, for example, the need for a defensive posture by smaller outnumbered armies, or, for that matter, by technologically disadvantaged forces (Rogers, 1968; Seymour, 1975; Boardman, 1998; Phillips, 1999; Barratt, 2000).

2. Methodology

Military engagements have been located geographically and placed in context by reference to various source texts, such as: Ross (1976), Lander (1990), Haigh (1995), Weir (1995) and Boardman (1998), for the Wars of the Roses;

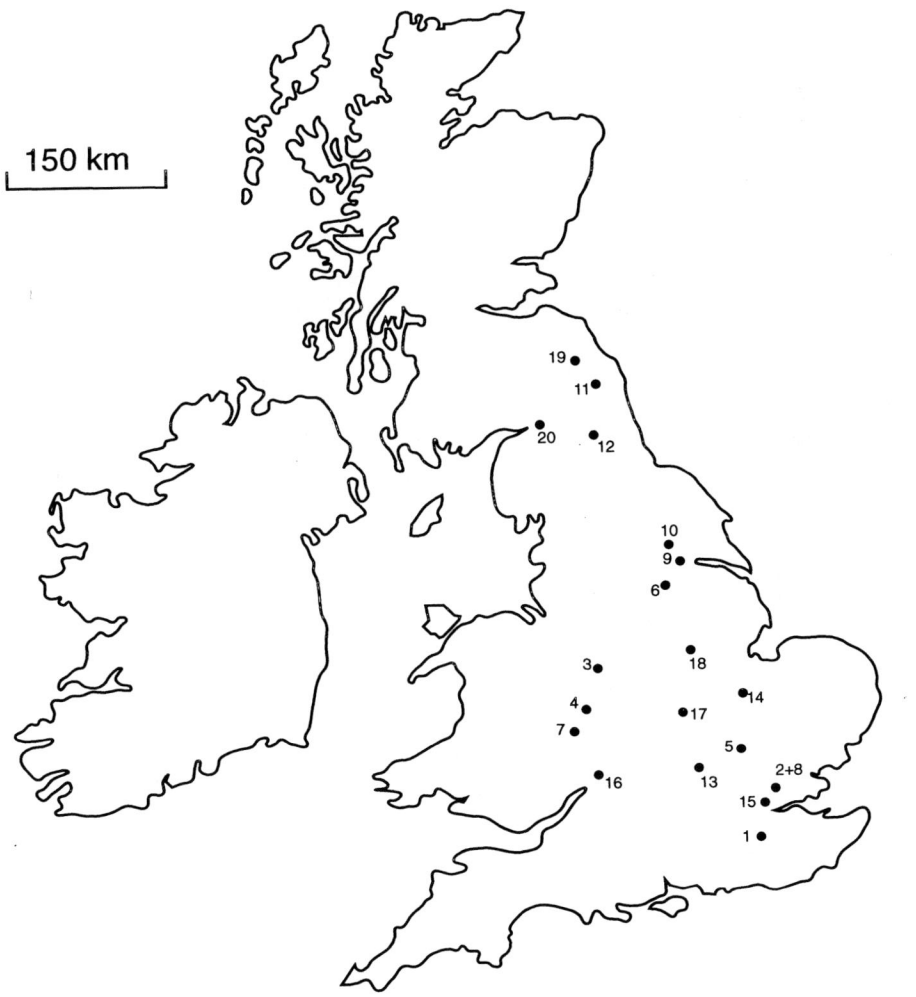

Figure 1: Battlefield sites in England and Wales: the War of the Roses and Anglo-Scots Border War. Table 1 provides a key to battlefield site numbers

Macdougal (1997), Cameron (1998), and Phillips (1999) for the Anglo-Scots War; and Burne & Young (1959), Rogers (1968), Newman (1985), Gardiner (1987a, b), Ashley (1990), Hibbert (1993), Kenyon & Ohlmeyer (1994), Emberton (1997), Haythornwaite (1998) and Reid (1998) in the case of the Civil Wars of the 17th century. A number of more general texts such as Green (1983), Clark (1993, 1995), Burne (1996), Young & Adair (1996), Kinross (1998) have been consulted, and, Seymour (1975), Smurthwaite (1993) and Guest & Guest (1996), have been used extensively. Minor skirmishes and engagements involving relatively insignificant and sometimes indeterminate forces, such as the action at Babylon Hill (1642) (Reid, 1998) and the raid on Sowerby (1644) (Newman, 1985) were excluded to avoid skewing the data with a number of such minor events. Similarly sieges and assaults on castles and fortified garrison towns were also omitted, whereas the battles at Tadcaster (1643), Wakefield (1643), Alton (1643) and Torrington (1646) were included because the defending forces were essentially field armies. Events such as the Yorkist capitulation at Ludford Bridge (1459) (Weir, 1995) and the facing down of the King at Turnham Green in 1642 (Reid, 1998) were also included because, even though there was no significant fighting, conditions for a major battle were achieved with the deployment of large forces by both sides. Following the Restoration of the Monarchy in 1660, there has been a prolonged period of internal peace in England and Wales, interrupted only occasionally by military activity, most notably at Sedgemoor (1685) and during the Jacobite Uprising of 1745-46 (Guest and Guest, 1996). Scotland was excluded from this investigation, but will in itself form a rich source of data for subsequent analysis.

The solid geology underlying the battlefields was established by reference to the 1:625,000 Geological Map of the United Kingdom, North (British Geological Survey, 1979a) and South (British Geological Survey, 1979b). In the case of battles which took place close to a geological boundary, reference has also been made, where necessary, to the relevant more detailed 1:50,000 or 1:63,360 (one inch to the mile) Geological Survey maps. When established that a battlefield actually lies astride a boundary between two underlying geological units, the battle is ascribed as occurring half on each unit. For example the battlefield of Flodden qualifies as half (0.5) on rocks of Devonian age and half (0.5) on Lower Carboniferous strata (British Geological Survey, 1979c; Halsall, 2000a). A more precise allocation is neither necessary nor appropriate. Only solid geology has been considered. Superficial drift deposits have been ignored on the basis that it is the solid geology which, in most cases, largely determines large-scale geomorphological features, and because drift deposits tend to be thin and discontinuous, and hence it is more difficult to assign battlefields to specific drift units.

The percentage aerial extent of the outcrop of the identified geological units was determined by a method similar to that described by Blake (2000). The solid substrate underlying the intersection of grid lines in the north east corner of each 10 km grid square on the Geological Maps of the United

GEOLOGY & WARFARE, 1450-1660

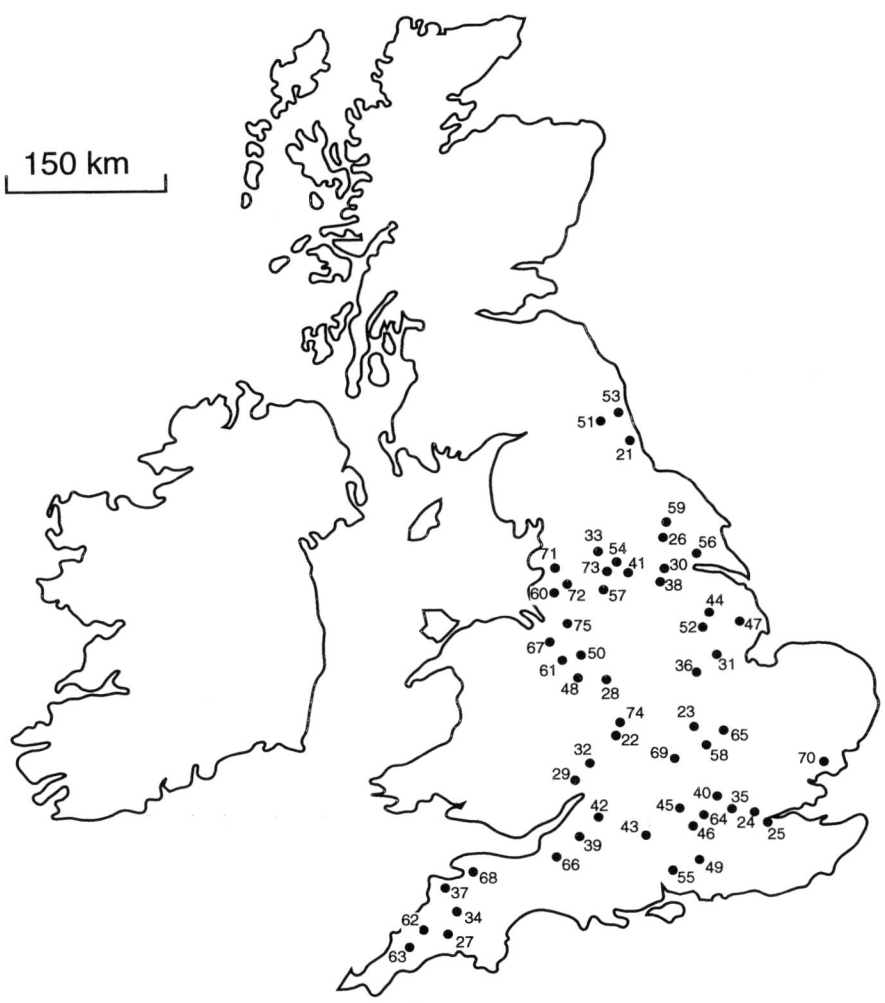

Figure 2. Battlefield sites in England and Wales: the Bishop's War and Civil Wars. Table 1 provides a key to battlefield site numbers

Kingdom (British Geological Survey, 1979a, b) was identified and recorded. The outcrop area for each unit was then calculated as a proportion of the total number of such grid square intersections (1,511) counted for the whole of England (including the Isle of White) and Wales. The geological units initially selected are the smallest units that can be readily differentiated on the geological maps, given that for some areas sections of the geological column are not subdivided. For example, on the Geological Survey map of southern Britain (British Geological Survey, 1979b) the Lower Cretaceous in Lincolnshire is undifferentiated, while elsewhere it is subdivided. For the sake of convenience at this scale of observation, the Old Red Sandstone of northern and central England is equated with the Devonian of southwest England, although not directly comparable in lithology. The Carboniferous is divided into the Upper and Lower Carboniferous, the Culm of southwest England being equated with the Upper Carboniferous elsewhere. The Permian Limestones and Marls were initially treated separately from the Permo-Triassic New Red Sandstone lithologies, which are undifferentiated. The Jurassic is divided into Lower (Lias), Middle and Upper, the Cretaceous into Lower and Upper, but the Tertiary is undifferentiated and the Pleistocene of East Anglia included with the Tertiary. Volcanic rocks have been included with their stratigraphical equivalents, while major intrusions are treated cumulatively as a separate unit, and are undifferentiated.

3. The data

The battle sites are listed chronologically alongside substrate geology in each case, in Table 1, their location is indicated in Figures 1 and 2.

Table 1. *The data set of battle sites relative to solid (bedrock) geology. The identifying number for each battlefield is that shown on Figures 1 or 2*

No.	Engagement	Year	Geological Unit	References
1.	Sevenoaks	1450	Lower Cretaceous	Smurthwaite, 1993
2.	1st St Albans	1456	Upper Cretaceous/Tertiary	Gillingham, 1981
3.	Blore Heath	1459	Permo-Trias (New Red Sandstone)	Haigh, 1995
4.	Ludford Bridge	1459	Silurian/Devonian	Gillingham, 1981
5.	Northampton	1460	Lower Jurassic (Lias)	Guest & Guest, 1996
6.	Wakefield	1460	Upper Carboniferous	Haigh, 1996
7.	Mortimer's Cross	1461	Silurian/Devonian	Smurthwaite, 1993
8.	2nd St Albans	1461	Upper Cretaceous/Tertiary	Seymour, 1975
9.	Ferrybridge	1461	Permian (Limestones & Marls)	Boardman, 1994
10.	Towton	1461	Permian (Limestones & Marls)	Halsall, 2000a
11.	Hedgeley Moor	1464	Lower Carboniferous	Haigh, 1995
12.	Hexham	1464	Upper Carboniferous (Millstone Grit)	Haigh, 1995

Table 1. Continued

No.	Engagement	Year	Geological Unit	References
13.	Edgecote	1469	Lower/Middle Jurassic	Clark, 1993
14.	Empingham	1470	Middle Jurassic	Clark, 1993
15.	Barnet	1471	Tertiary	Hammond, 1990, Seymour, 1975
16.	Tewkesbury	1471	Permo-Trias/Lower Jurassic	Hammond, 1990
17.	Bosworth	1485	Permo-Trias (New Red Sandstone)	Bennett, 1985, Holmes, 1997
18.	Stoke Field	1487	Permo-Trias (New Red Sandstone)	Bennett, 1987
19.	Flodden	1513	Devonian/Lower Carboniferous	Halsall, 2000a
20.	Solway Moss	1542	Permo-Trias (New Red Sandstone)	Guest & Guest, 1996
21.	Newburn Ford	1640	Upper Carboniferous	Guest & Guest, 1996
23.	Edgehill	1642	Lower Jurassic (Lias)	Young, 1998a
24.	Brentford	1642	Tertiary	Reid, 1998
25.	Turnham Green	1642	Tertiary	Reid, 1998
26.	Tadcaster	1642	Permian (Limestones & Marls)	Reid, 1998
27.	Braddock Down	1643	Devonian	Guest & Guest, 1996
28.	Hopton Heath	1643	Permo-Trias (New Red Sandstone)	Green, 1973
29.	Highnam	1643	Lower Jurassic	Reid, 1998
30.	Seacroft Moor	1643	Upper Carboniferous	Reid, 1998
31.	Ancaster Heath	1643	Middle Jurassic	Reid, 1998
32.	Ripple Field	1643	Permo-Trias	Reid, 1998
33.	Whalley	1643	Lower/Upper Carboniferous	Newman, 1985
34.	Sourton Down	1643	Upper Carboniferous	Reid, 1998
35.	Caversham	1643	Upper Cretaceous	Reid, 1998
36.	Grantham	1643	Middle Jurassic	Newman, 1985
37.	Stratton	1643	Upper Carboniferous	Guest & Guest, 1996
38.	Wakefield	1643	Upper Carboniferous	Reid, 1998
39.	Chewton Mendip	1643	Permo-Trias/Lower Jurassic	Reid, 1998
40.	Chalgrove	1643	Lower Cretaceous	Clark, 1993
41.	Adwalton Moor	1643	Upper Carboniferous	Guest & Guest, 1996
42.	Lansdown	1643	Lower/Middle Jurassic	Halsall, 2000a
43.	Roundway Down	1643	Upper Cretaceous	Halsall, 2001
44.	Gainsborough	1643	Permo-Trias (New Red Sandstone)	Reid, 1998
45.	Aldbourne Chase	1643	Upper Cretaceous	Reid, 1998
46.	1st Newbury	1643	Tertiary	Halsall, 2001
47.	Winceby	1643	U.Jurassic/L.Cretaceous	Clark, 1993
48.	Lee Bridge	1643	Permo-Trias (New Red Sandstone)	Newman, 1985
49.	Alton	1643	Upper Cretaceous	Reid, 1998
50.	Nantwich	1644	Permo-Trias	Reid, 1998
51.	Corbridge	1644	Upper Carboniferous	Reid, 1998
52.	Newark	1644	Permo-Trias (New Red Sandstone)	Kinross, 1998
53.	Boldon Hill	1644	Upper Carboniferous/Permian	Reid, 1998
54.	Bradford	1644	Upper Carboniferous	Newman, 1985
55.	Cheriton	1644	Upper Cretaceous	Halsall, 2001

Table 1. Continued

No.	Engagement	Year	Geological Unit	References
56.	Selby	1644	Permo-Trias (New Red Sandstone)	Rogers, 1968
57.	Bolton	1644	Upper Carboniferous	Reid, 1998
58.	Cropredy Bridge	1644	Lower Jurassic (Lias)	Smurthwaite, 1993
59.	Marston Moor	1644	Permo-Trias (New Red Sandstone)	Young, 1998b
60.	Ormskirk	1644	Permo-Trias (New Red Sandstone)	Newman, 1985
61.	Malpas	1644	Permo-Trias (New Red Sandstone)	Newman, 1985
62.	Beacon Hill	1644	Devonian	Rogers, 1968
63.	Castle Dore	1644	Devonian	Rogers, 1968
64.	2nd Newbury	1644	U. Cretaceous/Tertiary	Halsall, 2001
65.	Naseby	1645	Lower/Middle Jurassic	Halsall, 2000a
66.	Langport	1645	Permo-Trias/Lower Jurassic	Reid, 1998
67.	Rowton Heath	1645	Permo-Trias (New Red Sandstone)	Guest & Guest, 1996
68.	Torrington	1646	Upper Carboniferous	Reid, 1968
69.	Stow on Wold	1646	Middle Jurassic	Guest & Guest, 1996
70.	Colchester	1648	Tertiary	Newman, 1985
71.	Preston	1648	Permo-Trias (New Red Sandstone)	Reid, 1998
72.	Wigan	1648	Upper Carboniferous	Reid, 1998
73.	Wigan	1651	Upper Carboniferous	Reid, 1998
74.	Worcester	1651	Permo-Trias (New Red Sandstone)	Smurthwaite, 1993
75.	Winnington Bridge	1659	Permo-Trias (New Red Sandstone)	Newman, 1985

The percentage aerial outcrop of each geological unit, the number of battles occurring on each geological unit (counting battles fought astride a boundary as 0.5 to the units either side of that boundary), and the percentage of the total of 75 battles that number represents for each unit, are listed in Table 2.

In the case of the two battles of the Anglo-Scots Border Wars proximity to the mutual national border constrained their general location (Figure 1), but geology may have played a role in determining more precisely the site of both battlefields. Similar considerations do not apply to the siting of the other 73 engagements.

4. Analysis

If geology did not have a significant influence on the patterns of warfare in England and Wales between 1450 and 1660, then the distribution of battle sites, with reference to the underlying geological units, should be essentially random. It would also be expected that the percentage number of battle sites per outcrop of a given unit should approximate to the aerial extent of that unit as a proportion of the total area. However, even at a qualitative level, it is apparent that some geological units, most notably the Upper Carboniferous,

Permo-Triassic, New Red Sandstone, and Middle and Lower Jurassic, have a large number of battle sites relative to the aerial extent of their outcrops. This is balanced by the outcrop of other units such as the Upper and Lower Cretaceous, Upper Jurassic, Lower Carboniferous, Silurian and Ordovician which have relatively small numbers of battles for their outcrop area. Unsurprisingly, mountainous and rugged terrains, such as much of Wales and the Pennines, which inhibit the deployment and operation of large field armies, exhibit an absence of battlefields. These areas are underlain by the outcrop of relatively resistant rocks such as Precambrian, Cambrian, Ordovician Silurian and Lower Carboniferous strata. In these cases there was a clear and direct influence by geology on the distribution of military activity in the period. From this discussion, it remains to be demonstrated whether there is a more generally applicable relationship between geology and the battlefields.

To test this, the Chi-squared test (Gregory, 1968; Chatfield, 1983) was applied to the data to confirm the apparent non-random nature of the relationship between geology and warfare represented by sites of combat in the field. For this purpose the Quaternary and Tertiary (predominantly soft rocks) were treated as a single unit, as were the Middle and Lower Jurassic (limestones and mudrocks), the Permo-Triassic New Red Sandstones, and the Permian limestones and marls, avoiding the use of geological units whose separate outcrop is proportionally very low (< 5% of the total). The Silurian, Ordovician, Cambrian, Precambrian and major igneous intrusions were also lumped together as a single unit ('others'), representing folded and often harder rocks. The number of battle sites predicted to occur on each unit for the purpose of the test was calculated by multiplying the total number of battles by the percentage aerial outcrop of the host unit, and rounding the resulting figure to the nearest 0.5 (the smallest attributable increment recorded for the actual observed number of battles on any given substrate. The calculated value for Chi^2 is 33.13174, which when cross referenced with the number of degrees of freedom, 9 on a Chi^2 table (number of geological units in Table 3; Chatfield, 1983), indicates the relationship between geology and battle sites, is not random, with a greater than 99.5% level of confidence.

5. Conclusions

There is a very high probability that there was a constraining relationship between underlying solid geology and the geographical distribution of battlefields in England and Wales for the period 1450–1660 AD. This suggests that geology had a significant if unknown influence on the strategies of the protagonists at that time. Resistant Palaeozoic strata forming rugged uplands were avoided in military campaigns, but the more general relationship remains to be explained. Though a detailed explanation is beyond the scope of the current study, an outline model is proposed as the basis for further investigation, using data from other theatres and periods of warfare:

Table 2: Percentage of battles per geological (chronostratigraphical) unit. Points refer to the number of 10 km grid intersections in England and Wales overlying the geological unit in question. Area is the relative area of outcrop of each unit as a percentage calculated from the number of points as a proportion of the total number (1,511). Battles refers to the number of battlefields overlying each geological unit; the percentage of the total number of battles (75) it represents is given in the Percentage column. Discrepancies between the Area and Percentage values for most units indicate a non-random relationship between geology and the location of battle sites

Geological Unit	Points	Area	Battles	Percentage
Quaternary	30	1.99%	0	0%
Tertiary	109	7.2%	6.5	9.29%
Upper Cretaceous	197	13.04%	6.5	9.29%
Lower Cretaceous	86	5.69%	2.5	3.3%
Upper Jurassic	101	6.68%	0.5	0.67%
Middle Jurassic	68	4.5%	5	6.67%
Lower Jurassic	95	6.29%	8.5	11.33%
Permo-Trias (New Red Sandstone)	188	12.44%	19	25.33%
Permian limestone/marl	35	2.3%	3.5	4.67%
Upper Carboniferous	219	14.49%	16	21.33%
Lower Carboniferous	85	5.63%	1.5	2.14%
Devonian	106	7.03%	4.5	6.42%
Silurian	89	5.89%	1	1.43%
Ordovician	53	3.51%	0	0%
Cambrian	8	0.53%	0	0%
Lower Palaeozoic volcanics	6	0.4%	0	0%
Precambrian (incl. metamorphics)	10	0.66%	0	0%
Intrusive Rocks	25	1.65%	0	0%

Table 3: Chi-squared analysis of the data

Geological Unit	O-Battles: observed	E-Battles: estimated	$(O-E)^2/E$
Quaternary & Tertiary	6.5	7	0.035714
Upper Cretaceous	6.5	10	1.225
Lower Cretaceous	2.5	4.5	0.888889
Upper Jurassic	0.5	5	4.05
Middle/Lower Jurassic	13.5	8	3.78125
Permo-Triassic	22.5	11	12.02273
Upper Carboniferous	16	11	2.272727
Lower Carboniferous	1.5	4	1.5625
Devonian	4.5	5.5	0.181818
Others	1	9	7.111111
			33.13174

1. Lithology and structure are fundamental attributes of the geology of an area; in conjunction with its geological history, they are important factors constraining relief, topography and the distribution of mineral resources.
2. Lithology and relief are the principle geological constraints determining the nature of overlying soils and the drainage properties exhibited by terrain, which in turn affect vegetation patterns and, in military terms, the 'going'.

3. Topography and relief, soil type and structure, drainage patterns and properties, coupled with prevailing weather conditions (in part determined by relief), affect agricultural productivity. Topography, drainage patterns and going determine the routes adopted by lines of communication.
4. At the furthest remove, the distribution of areas of resource production, both agricultural and mineral, the patterns of interconnecting lines of communication in conjunction with relief and the local properties of substrates and their drainage, serve to locate centres of population.
5. Military operations at the level of battlefield tactics are constrained by considerations of the local topography and going and hence by factors which are more readily attributable to the underlying geology.
6. Strategy, which determines the pattern of campaigns, is dictated by the need to control and defend centres of population and resource production and the lines of communication which connect them, and to deny these to your enemy. These objectives are further removed from the underlying geology, but, none the less, their location and distribution is determined by the interplay of a number of constraints, many of them geological. Hence the geology of a region is indirectly but critically formative of the patterns of warfare which take place therein, albeit the various factors at this level are complex and difficult to unravel.

Acknowledgements

I would like to express my gratitude to Ron Blake of Nottingham Trent University for his guidance on methodology in relation to determining the proportional outcrop area of the major geological units in England and Wales, and to Howard Grubb and David Foot of the University of Reading for their kind advice with reference to the statistical analysis of the data.

References

Ashley, M. 1990. *The English Civil War*. Alan Sutton Publishing, Stroud.

Barratt, J. 2000. *Cavaliers, The Royalist Army at War 1642-1646*. Alan Sutton Publishing, Stroud.

Bennett, M. 1985. *The Battle of Bosworth*. Alan Sutton Publishing, Stroud.

Bennett, M. 1986. *Lambert Simnel and the Battle of Stoke*. Alan Sutton Publishing, Stroud.

Blake, R.N.E. 2000. Geological influences on the siting of military airfields in the United Kingdom. *In:* Rose, E.P.F. & Nathanail C.P. (Eds), *Geology and Warfare: examples of the influence of terrain and geologists on military operations*. Geological Society, London, 236-274.

Boardman, A.W. 1994. *The Battle of Towton*. Alan Sutton Publishing, Stroud.

Boardman, A.W. 1998. *The Medieval Soldier in the Wars of the Roses.* Alan Sutton Publishing, Stroud.

British Geological Survey 1979a. *Geological Map of the United Kingdom North (North of National Grid Line 500 km N)* 3rd Edition. British Geological Survey, Keyworth, Nottingham.

British Geological Survey 1979b. *Geological Map of the United Kingdom South (South of National Grid Line 500 km N)* 3rd Edition. British Geological Survey, Keyworth, Nottingham.

Burne, A.H. 1996. *The Battlefields of England.* Greenhill Books, London.

Burne, A.H. & Young, P. 1959. *The Great Civil War.* Eyre & Spottiswoode, London.

Cameron, J. 1998. *The Stewart Dynasty in Scotland, James V The Personal Rule, 1528-1542.* Tuckwell Press, East Linton.

Chatfield, C. 1983. *Statistics for Technology, Course in Applied Statistics* 3rd Edition. Chapman and Hall, London.

Clark, D. 1993. *Battlefield walks: The Midlands.* Alan Sutton Publishing, Stroud.

Clark, D. 1995. *Battlefield walks: North.* Alan Sutton Publishing, Stroud.

Dodds, G.L. 1996. *Battles in Britain.* Arms and Armour Press, London.

Emberton, W. 1997. *The English Civil War Day by Day.* Grange Books, London.

Gardiner, S.R. 1987a. *History of the Great Civil War Volume 1, 1642-1644.* Windrush Press, London.

Gardiner, S.R. 1987b. *History of the Great Civil War Volume 1I, 1644-1645.* Windrush Press, London.

Gillingham, J. 1981. *The Wars of the Roses.* Weidenfeld & Nicolson, London

Green, H., 1983. *The Battlefields of Britain and Ireland.* Constable, London.

Gregory, S. 1968. *Statistical Methods and the Geographer* 2nd Edition. Longman,London.

Guest, K. & Guest, D. 1996. *British Battles.* Harper Collins, London.

Haigh, P.A., 1995. *The Military Campaigns of the Wars of the Roses.* Alan Sutton Publishing, Stroud.

Haigh, P.A. 1996. *The Battle of Wakefield, 30 December 1460.* Alan Sutton Publishing, Stroud.

Halsall, T.J. 2000a. Geological constraints on battlefield tactics: examples in Britain from the Middle Ages to the Civil Wars. *In:* Rose, E.P.F. & Nathanail C.P. (Eds), *Geology and Warfare: Examples of the Influence of Terrain and Geologists on Military operations.* Geological Society, London, 32-59.

Halsall, T.J. 2000b. Geological constraints on the siting of fortifications: examples from Medieval Britain. *In:* Rose, E.P.F. & Nathanail C.P. (Eds), *Geology and Warfare: Examples of the Influence of Terrain and Geologists on Military Operations.* Geological Society, London, 3-31.

Halsall, T.J. 2001. Battles on Chalk: the geology of battle in southern England during the first Civil War. *In:* Doyle, P. & Bennett, M.R. (Eds), Fields of Battle, Kluwer, Amsterdam, This Volume.

Hammond, P.W. 1990. *The Battles of Barnet and Tewkesbury*. Alan Sutton Publishing, Stroud.

Haythornwaite, P.J. 1998. *The English Civil War 1642–1651, an Illustrated Military History*. Brockhampton Press, London.

Hibbert, C. 1993. *Cavaliers and Roundheads, The English at War 1642–1649*. Harper Collins, London.

Holmes, R. 1997. *Battlefield Walks 2*. BBC Books, London.

Kenyon, J.P. & Ohlemeyer, J. (Eds) 1998. *The Civil Wars of England, Scotland and Ireland, 1638-1660*. Oxford University Press, Oxford.

Kinross, J. 1998. *Discovering Battlefields of England and Scotland*. Shire Publishers, Princes Risborough.

Lander, J.R. 1990. *The Wars of the Roses*. Guild Publishing, London.

Macdougall, N. 1997. *The Stewart Dynasty in Scotland - James IV*. Tuckwell Press, East Linton, Scotland.

Newman, P. 1985. *Atlas of the English Civil War*. Croom Helm, London.

Phillips, G. 1999. *The Anglo-Scots Wars, 1513-1550, A Military History*. The Boydell Press, Woodbridge..

Reid, S., 1998. *All the King's Armies, A Military History of the English Civil War, 1642-1651*. Spellmount, Staplehurst.

Roberts, M. 1967. *Essays in Swedish History*. Weidenfeld & Nicholson, London.

Rogers, H.C.B. 1968. *Battles and Generals of the Civil Wars 1642 - 1651*. Seeley Service & Co., London.

Ross, C. 1976. *The Wars of the Roses, A Concise History*. Thames & Hudson, London.

Seymour, W. 1975. *Battles in Britain 1066 - 1746*. Sidgwick & Jackson, London.

Smurthwaite, D. 1993. *The Ordnance Survey Complete Guide to the Battlefields of Britain*, New Edition. Michael Joseph Ltd, London.

Weir, A. 1995. *Lancaster and York - The Wars of the Roses*. Jonathan Cape, London.

Young, P. 1998a. *Edgehill 1642, The Campaign and the Battle*. Windrush Press, Moreton-in-the-Marsh.

Young, P., 1998b. *Marston Moor 1644, The Campaign and the Battle*. Windrush Press, Moreton-in-the-Marsh.

Young, P. & Adair, J. 1996. *Hastings to Culloden, Battles of Britain*. Alan Sutton Publishing, Stroud.

Trevor J. Halsall
Postgraduate Research Institute for Sedimentology
University of Reading
PO Box 227
Whiteknights
Reading, RG6 6AL

Battles on Chalk: the Geology of Battle in Southern England during the First Civil War, 1643-1644

Trevor J. Halsall

ABSTRACT: The impact of geology on 17th century warfare is illustrated by four battles in the English Civil War. Actions at Roundway Down and Cheriton and the two battles at Newbury took place within a sixteen month period from 1643 to 1644. At the strategic level, the Chalk outcrop significantly constrains the route of important lines of communication in southern England. Roads between Swindon and Newbury are a case in point; the failure of the Parliamentarian Army to secure its line of march along such well drained tracks led directly to the First Battle of Newbury. The course of that engagement was steered by the geomorphology of the battlefield. Battlefield tactics and their outcome were also constrained by geology at the Second Battle of Newbury, the principal action taking place on a restricted front on the low ridge between the Lambourn and the Kennet. The Parliamentarians completed a flank march by night to avoid the tactical difficulties imposed by the terrain, yet their conduct of the battle was less successful. At Cheriton a dissected erosion surface on the Chalk provided a sound defensive position for Hopton's Royalist troops in March 1644, but unfortunately his junior officers abandoned their position to engage the enemy on lower ground. The latter proved to be a killing field on which the Royalist cavalry was destroyed. At Roundway Down the Upper Greensand escarpment to the Parliamentarian rear converted the defeat of their horse into a disaster as they retreated from the field.

1. Introduction

Terrain plays a crucial role in the conduct of warfare. Geological factors including lithology, structure, hydrogeology and erosional history are fundamental influences on key terrain attributes which include geomorphology, drainage, vegetation patterns, land use and the distribution of mineral resources. As such, it follows that geology is of fundamental importance in warfare. In recent papers, I have demonstrated how geology was significant in determining the course and outcome of a number of British battles in the 14th, 15th, 16th and 17th centuries (Halsall, 2000a), as well as constraining the design and construction of medieval fortifications (Halsall,

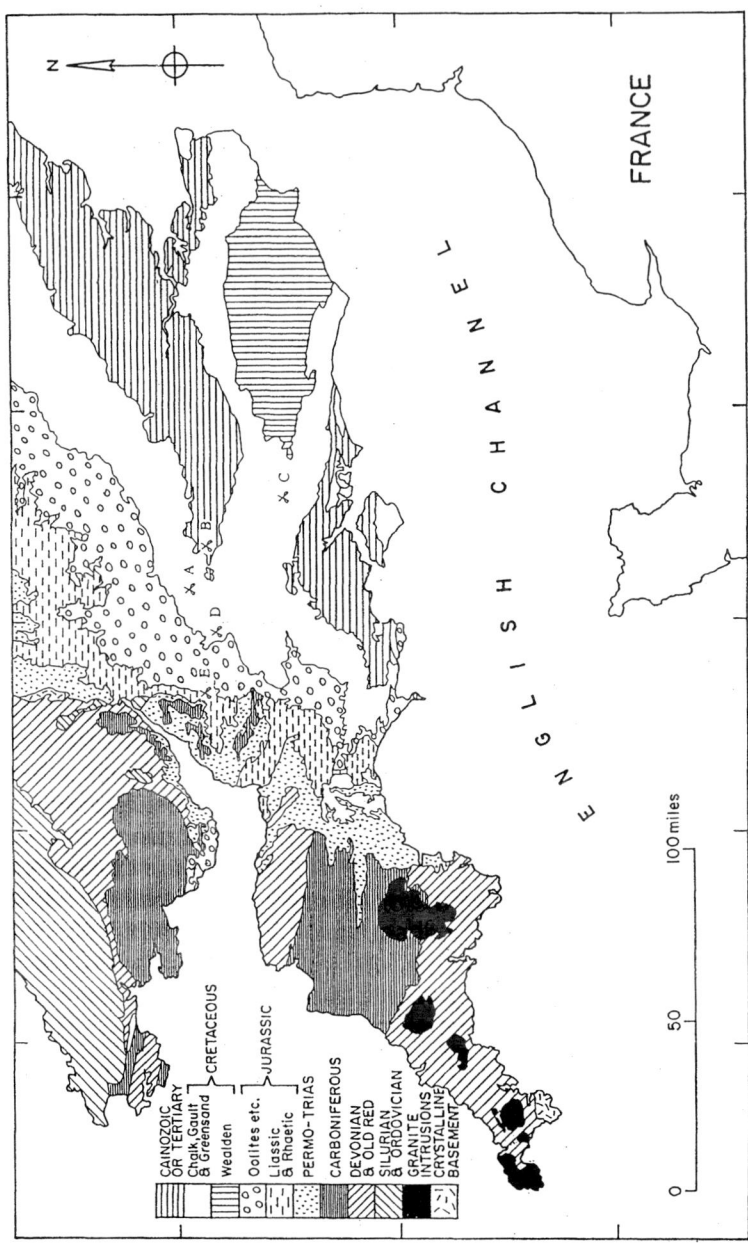

Figure 1. The location of military actions in central, southern England during the Civil Wars, in the context of the geology of southern Britain (after British Geological Survey, 1979). A, Aldbourne (skirmish, 18th Sept. 1643); B, Newbury (1st Battle, 20th Sept. 1643, 2nd Battle, 27th Oct. 1644); C, Cheriton (battle, 29th March, 1644); D, Roundway Down (battle, 13th July, 1643); E, Lansdown (battle, 5th July 1643)

2000b). The Civil Wars between King and Parliament in Britain, in the mid 17th Century, is susceptible to such analysis, and the major battles in central southern England (Figure 1) are a case in point. The contribution of terrain and its underlying geology at Lansdown (5th July 1643) has already been described (Halsall, 2000a). This article focuses on the other four important battles in the South: Roundway Down; 1st Newbury; Cheriton and 2nd Newbury.

These engagements are selected to demonstrate how, in different ways, terrain can influence the location, course and outcomes of battle, at different stages in the developing action, within a narrow time frame of history. All four battles took place between July 1643 and October 1644, and involved opposing armies using essentially similar equipment and tactical doctrine, with neither side having an overwhelming numerical advantage. They took place within a relatively small area, and, with the exception of Lansdown, were the only significant engagements in that area between the 13th century and the present day. However, the four battles are each in their own way unique. Two (1st and 2nd Newbury) resulted in the victory of the protagonist adopting an essentially defensive tactical stance. At Roundway Down the attacking side was victorious, while at Cheriton the outcome was essentially decided by insubordination. All four were influenced, in one way or another, by the outcrop of the Chalk. Though fought in the same area, they were each part of a different campaign, and each the culminating and strategically important, if not decisive, action in that campaign.

Rather than describe the four battles in chronological order, they are considered in terms of the stages of battle when terrain, and therefore geology, proved to be most critical to the ensuing action. Namely, in: (1) campaign strategy and the approach march (1st Newbury); (2) battle strategy and grand tactics (2nd Newbury); (3) tactical deployment and the outcome of insubordination (Cheriton); and (4) during the retreat of defeated troops (Roundway Down).

The four battlefields discussed are covered by three 1:50,000 sheets in the Ordnance Survey Landranger Series; Roundway Down by Sheet 173, the 1st and 2nd Battles of Newbury by Sheet 174, and Cheriton by Sheet 185. Eight character grid references are provided for key locations on each battlefield, using standard Ordnance Survey convention, which provide an accuracy to within 100 m.

2. The impact of geology on strategy and the approach march: the First Battle of Newbury, September 1644

Good roads were of great importance to commanders during the Civil War. The supply trains and horse drawn artillery of the time were heavy and cumbersome, easily becoming bogged down and immobilised in difficult going, such as that offered over poorly drained clay substrates in wet conditions. In many places the routes adopted by ancient Roman roads were still among the best available in 17th century England, as they typically followed a direct

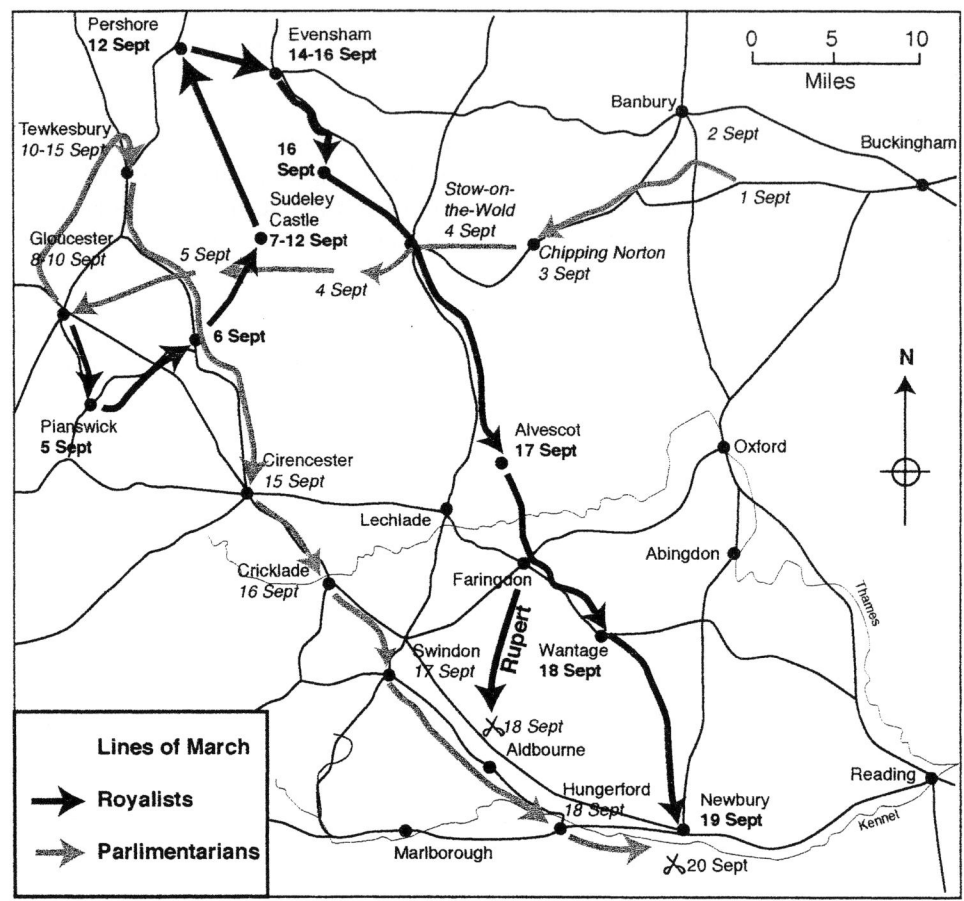

Figure 2. The campaign manoeuvres of the opposing armies in September 1643, culminating in the 1st Battle of Newbury. The location of the two armies on a day-by-day basis are indicated: Royalist in bold text; Parliamentarian in normal text. (Modified from: Ashley, 1990)

line of least resistance over well drained limestone ridges and Chalk uplands. For example, the network of roads which radiated from Roman Silchester across Berkshire, Wiltshire and Hampshire were, and in many cases still are, important routeways. In the 17th century such roads were of strategic importance, the Ermin Way between Silchester and Gloucester via Newbury and Cirencester being a good example.

The First Battle of Newbury was the culminating event in the successful campaign by the Earl of Essex' Parliamentarian Army to relieve the besieged and strategically important City of Gloucester. By conducting a flank march, initially at night, to the north of the Royalist capital Oxford (Figure 2) Essex was able to force the King to abandon his investment of Gloucester (5th September) and march to relieve the garrison (Rogers, 1968; Ashley, 1990; Emberton 1997). Having restocked the city with food and powder, Essex set out on 10th September to return to his base in London, before his army could be intercepted and cut off by the Royalists. To confuse his opponents and evade pursuit, Essex first moved north to Tewkesbury before slipping south to Cirencester on 15th September (Gardiner, 1987; Rogers 1968).

The Parliamentarians marched rapidly using the relatively good going following the line of Roman roads traversing the limestone country of the Cotswolds, via Cirencester to Swindon. From there Essex planned to climb the Chalk escarpment to the south east, and continue to utilise the direct, well-drained route to Newbury, a town which was sympathetic to the cause of Parliament (Money, 1972), and on to Reading which was already in Parliament's hands. Essex stole a march on the King, for by the time his troops reached Cricklade, he was already a whole day ahead of his pursuers. From there his escape should have been straight forward. However, the King set his troops on a parallel course, marching to Alvescot, where they arrived on 17th September. By then the King had a good idea of Essex' plans, so a large force of Royalist cavalry was sent ahead, under the command of Prince Rupert, to intercept the Parliamentarian withdrawal and delay their attempts to reach Newbury (Rogers, 1968). Rupert caught up with the Parliamentarians on Aldbourne Chase on 18th September (Figure 2). Strung out on the march, the Parliamentarian Army was in danger of destruction (Seymour, 1975), but by means of a series of hard fought rearguard actions, the Royalist horse were held at bay, enabling Essex to extricate his forces and reach Hungerford by nightfall (Burne & Young, 1959). However, the Parliamentarian commander had been driven south of the direct route over the Chalk to Newbury, and in order to improve the security of his flank in the light of the events at Aldbourne, was forced to continue his march south of the River Kennet, thereby keeping the River between his army and the threatening Royalist cavalry (Haythornthwaite, 1998). King Charles was at this point at Wantage with the Royalist infantry and artillery (Ashley, 1990).

The terrain between Hungerford and Newbury south of the Kennet is underlain by the outcrop of the Tertiary Reading Beds and London Clay, presenting much poorer going for an army on the march than the Chalk downs to the north, especially in very damp autumn conditions (Hibbert, 1993; Emberton, 1997). Progress next day was slow, and when Essex' advance guard approached Newbury, he found that the King had already occupied the town and the higher ground to the south, thereby blocking the Parliamentarian escape route to Reading and London. Essex had been intercepted and battle was now inevitable; his troops would have to fight their way through or face destruction (Seymour, 1975; Ashley 1990).

At the tactical level, the King had been successful in this second phase of the campaign. All his troops had now to do was hold their positions and await the disintegration of the opposing forces. The two armies were

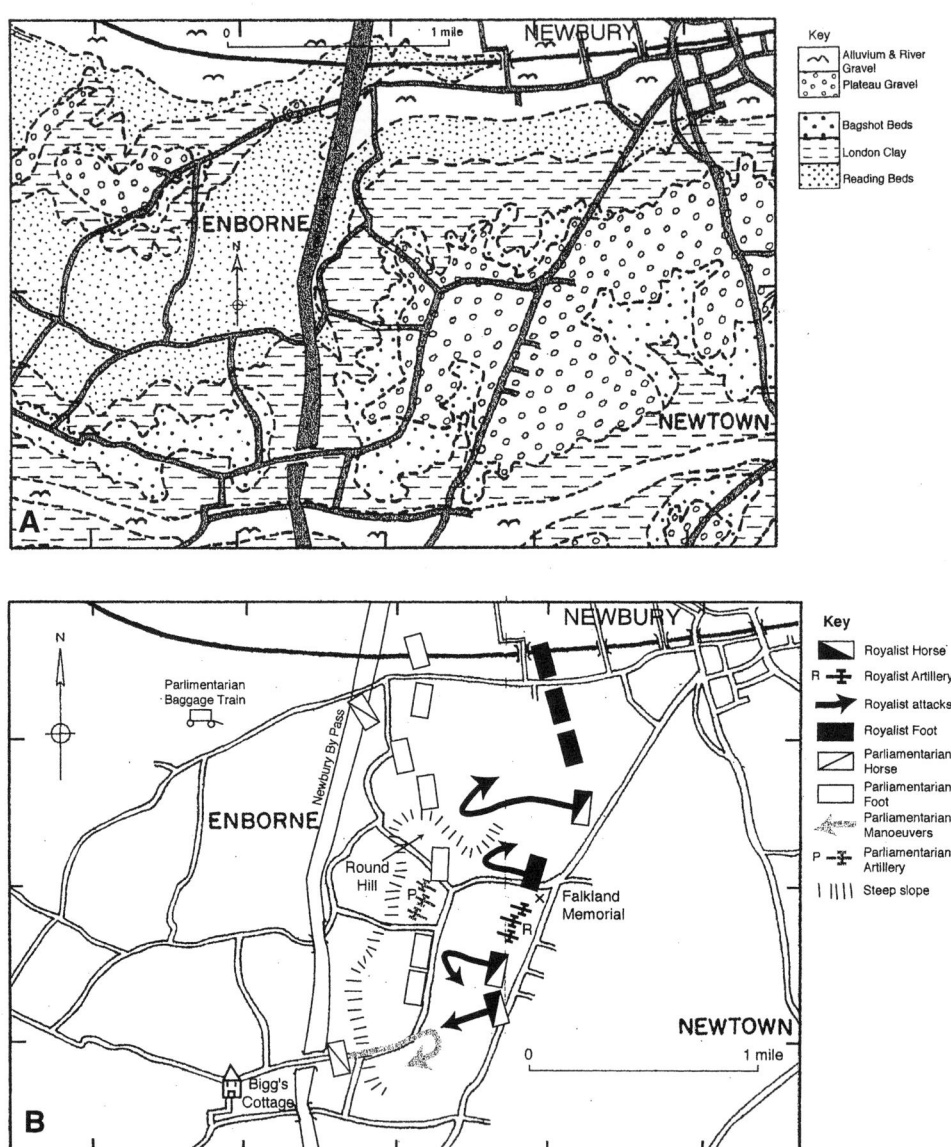

Figure 3. **A.** *The geology of the area of the battlefield of the 1st Battle of Newbury (Modified from: British Geological Survey, 1962).* **B.** *The 1st Battle of Newbury (20th September, 1643). The dispositions and principal manoeuvres of the opposing armies are indicated, together with significant terrain features*

relatively evenly matched: the Parliamentarians with some 10,000 foot, 4,000 cavalry and about 20 guns had the better infantry, while the Royalists fielding about 8,000 foot, 6,000 horse and a similar number of guns (Rogers, 1968), had the superior cavalry. They also had the advantage of defending higher ground. The plateau of Wash Common, capped by an outcrop of Bagshot Sands (Figure 3A), was occupied by the cavalry of the Royalist left wing. To the north the ground shelves gently over the London Clay to the River Kennet on which the Royalist right flank rested (Green, 1983). Newbury town and its southern outskirts lay immediately behind, securing the Royalists' right and centre (Figure 3B). The ground to the west falls away steeply to meet the rolling terrain formed by the outcrop of the Reading Beds, over which the Parliamentarian Army had advanced on 19th September. However, the Royalists had failed to secure the western edge of the plateau and the Round Hill, an almost detached outlier of Bagshot Beds overlain by Plateau Gravels (British Geological Survey, 1962), which forms a small spur at the northwest corner of Wash Common [SU452652], and is high enough to dominate the ground held by the King's troops to the east. Essex' men occupied these positions during the night and by the morning of 20th September were sufficiently well established to drive off repeated attacks by the Royalist Army (Smurthwaite, 1984; Haythornthwaite, 1998). The Royalist cavalry was unable to penetrate the hedgerows, preventing them from engaging the enemy.

The Parliamentarians established a powerful artillery battery in front of the Round Hill, and by the end of the day, the King had failed to beat them back (Rogers, 1968, Seymour, 1975). His troops withdrew that night leaving the way open for Essex to continue his march to Reading, along the ridge of Tertiary deposits forming Greenham Common and extending to Brimpton towards Aldermaston, and thereby extricating his forces from potential disaster. There was a sharp skirmish at Aldermaston as Prince Rupert's Royalist cavalry attempted to break through the Parliamentarian rearguard as they retreated, but the Royalist cavalry were driven off with loss (Money, 1972), and Essex and his men reached the safety of Theale near Reading without further losses.

3. Grand tactics and their outcomes shaped by geology: The Second Battle of Newbury

After a successful campaign in south-west England, which culminated in the defeat of the Earl of Essex' Parliamentarian forces in the region at Lostwithiel on 2nd September, 1644, the King and his army retired towards the Royalist capital, Oxford. They narrowly failed to trap and destroy separate Parliamentarian contingents near Andover and Basingstoke, before they could combine to form a significantly superior force (Rogers, 1968; Newman, 1985; Burne, 1996), at which point the Royalists fell back on Newbury (Figure 4). It was then the turn of the Parliamentarians, with their

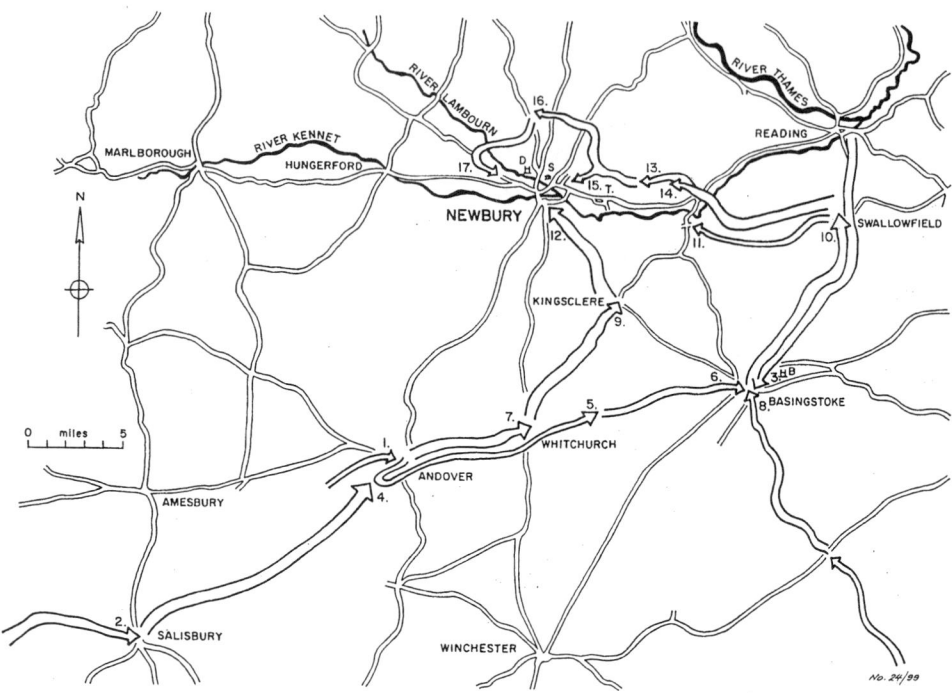

Figure 4. The campaign manoeuvres of the opposing armies in October 1644, culminating in the 2nd Battle of Newbury (Modified from: Halsall, 1999): 1, Waller moves to Andover, 15th Oct.; 2, the King moves to Salisbury, 15th Oct.; 3, Manchester advances to Basingstoke, 17th Oct.; 4, the King attempts to trap Waller, 18th Oct.; 5, Waller retreats towards Basingstoke which he reaches on 19th Oct. (6); 7, the King advances to Whitchurch, 20th Oct.; 8, Essex and his troops from Portsmouth march into Basingstoke, 20th Oct.; 9, The King moves to Kingsclere, 21st Oct.; 10, the combined Parliamentarian armies to Swallowfield, 21st Oct.; 11, Roundhead cavalry advance to Aldermaston and make contact with Royalist outposts, near Thatcham (T), 21st Oct.; 12, The Royalist Army occupies Newbury, 21st Oct.; 13 the Parliamentarian main body crosses the Kennet at Padworth, 25th Oct.; 14, Manchester advancing to Bucklebury Common and, Waller to Beenham Heath; 15, Parliamentarian troops occupy Clay Hill, 26th Oct., and, that night, their main body conduct a flank march to North Heath (16); 17, the Flank march continues and culminates in an attack on Royalist positions at Speen, 27th Oct. Other localities shown are: B, Basing House; D, Donnington Castle; and S, Shaw House

larger army, to try to bring the King to battle with the aim of inflicting a decisive defeat that could bring the war to an end.

The combined Parliamentarian armies, under joint command of Essex, Manchester, Waller and others (Ashley, 1990), began their advance from Swallowfield on 24th October, and next day crossed the Kennet at Padworth

(Figure 4). Rather than follow the direct route to Newbury along the Kennet Valley, which tends to be wet and muddy in late autumn, they chose to approach via Beenham, Bucklebury Common and Cold Ash, on the ridge of high ground north of the river formed by the outcrop of Eocene Beds overlain by a thin layer of Plateau Gravels. This may have offered better going and enabled them to threaten the Royalist lines of communication with Oxford. The Parliamentarians camped between Beenham and Bucklebury Common, on the night of 25th October (Halsall, 1996).

Reports by their cavalry patrols, which had reconnoitred the Royalist dispositions, left the Parliamentarian generals with a difficult situation. The King had deployed his troops in a strong defensive position north of Newbury, with infantry and supporting cavalry occupying the village of Shaw and Shaw House, a large Jacobean mansion with walled gardens, facing east towards their opponents (Burne, 1996). To the north west the Royalist left and rear was covered by Donnington Castle and its outworks, which they had held for over a year with a well supplied garrison and powerful batteries of artillery. In addition, the main body of the King's horse, plus significant formations of infantry, were stationed to the rear in reserve. It was clear that an attack on the positions at Shaw would be costly and have to be made on a narrow front, because of the River Lambourne which protected the Royalists southern flank, while the guns of Donnington, secured the ground to the north.

Halsall (1996, 1999) has demonstrated, that the Parliamentarian generals devised their plan to defeat the King while still in camp on the evening of 25th October (Halsall 2000a) and before Essex, who was ill, retired to Reading to recuperate, rather than during the following day as normally accepted (Rogers, 1968). The plan was for the major part of the Parliamentarian Army, under the command of Sir William Waller, to conduct a flank march by night via Cold Ash, Hermitage, North Heath and Boxford, to fall on the Royalist rear at Speen, and to defeat them there by means of a combination of surprise and overwhelming force. Meanwhile the Earl of Manchester, with some 7,000 infantry and cavalry plus the bulk of the artillery, was to occupy Clay Hill, overlooking the Shaw positions from the east, to divert the Royalists attention and pin them in their positions with synchronous diversionary attacks.

The flank march was a remarkable success, and Waller's force, totalling about 12,000 infantry and cavalry, drove a light Royalist covering force from their field defences on Speen Hill at about 3.00 PM on 27th October. Prince Maurice's Cornish Brigade, some 3,000 infantry which formed the Royalist rearguard was forced to evacuate the village of Speen [SU456682] and fall back to regroup defensively on a hedge-line further east, in the face of overwhelming odds. At about the same time Balfour's Parliamentarian Horse on the right wing, swept aside the facing Royalist cavalry and pursued them across the terraces of the Kennet, south of Speen (Rogers, 1968). However, at this point the geology and resulting geomorphology of the terrain began to dictate the course of events. Speen lies on a narrow ridge composed of Chalk

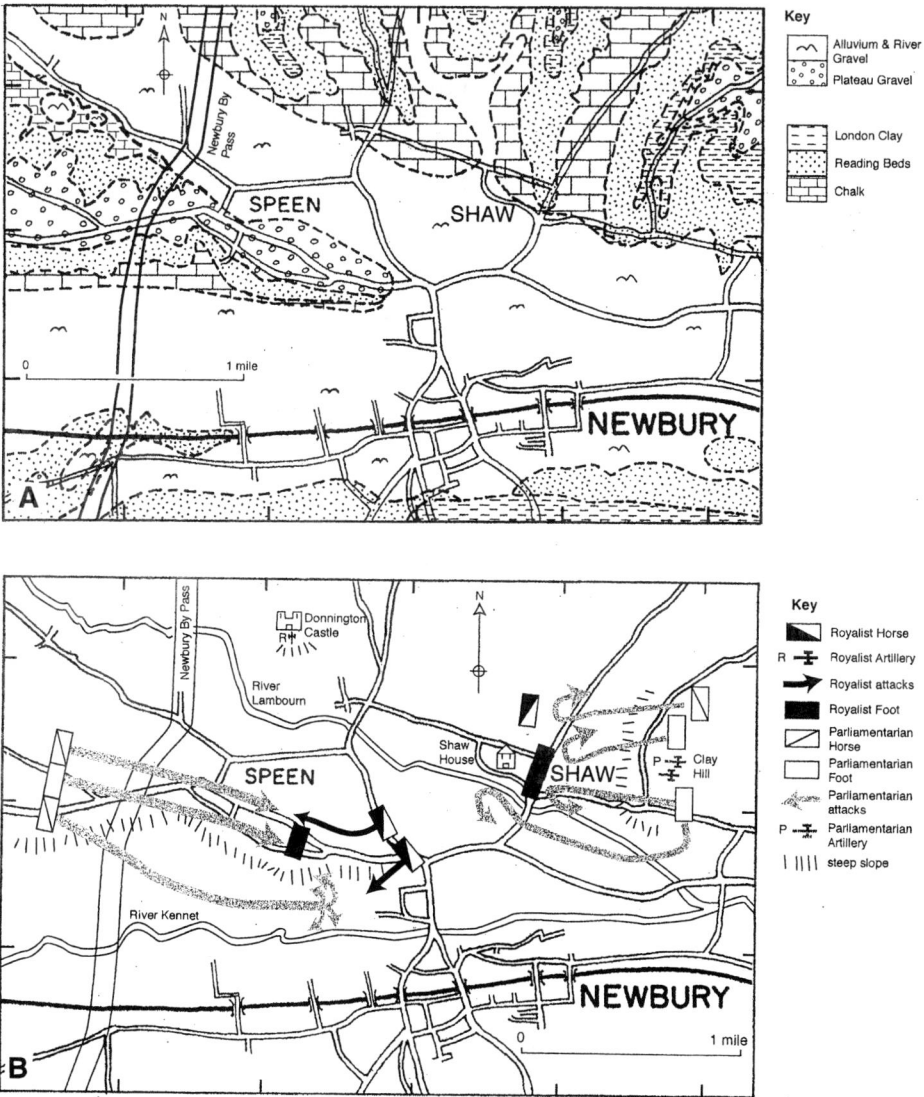

Figure 5. **A.** *The geology of the area of the battlefield of the 2nd Battle of Newbury (Modified from: British Geological Survey, 1962).* **B.** *The 2nd Battle of Newbury (20th October, 1644). The dispositions and principal manoeuvres of the opposing armies are indicated, together with significant terrain features*

and overlying Eocene Reading Beds, with a thin cover of Quaternary plateau gravels. To the south the ground slopes steeply down to the Kennet flood plain, with the River itself some 300 to 600 m beyond (Figure 5A). To the north the ground shelves into marshy ground on the flood plain of the Lambourne, beyond which rises the ridge dominated at its southern end by Donnington Castle and its guns.

The course of the action is summarised in Figure 5B. The Parliamentarians were obliged to attack on a narrow front, so their superiority in numbers did not tell. The Royalist infantry were protected by hedgerows, while their cavalry reserve was ordered forward by the King to counter the opposing horse. The Queen's Regiment of Horse charged Balfour's cavalry in flank as its advance passed them on the lower ground adjacent to the Kennet, routing the already disordered Parliamentarians and driving them back on their main body (Rogers, 1968). On the Parliamentarian left Cromwell's horse manoeuvred ineffectively, held in check by the opposing cavalry and the guns of Donnington (Newman, 1985; Kinross, 1998). The attack faltered and finally petered out by about 4.00 PM as dusk fell.

At Shaw an initial early morning attack at 7.00 AM on 27th October had been driven back, and the Earl of Manchester had spent the day ineffectively bombarding the Royalist positions. At about 4.00 PM, aware that Waller's flank marching force had arrived, but too late to influence the outcome of the battle, he renewed a series of attacks on Shaw (Smurthwaite, 1984). These assaults were defeated by Royalist defenders and Manchester's men retired to Clay Hill with significant losses.

Despite possessing a considerable numerical advantage and the tactical initiative, and having conducted a superbly effective flank march by night, the Parliamentarians had been defeated by a combination of difficult terrain and the resilience and fortitude of their opponents. The terrain had constrained their approach march, had militated against an all-out assault on the restricted and easily defensible hostile positions at Shaw, giving rise to the concept of the flank march. When it came to it in the resulting battle the principal action took place on the narrow ridge east of Speen. There the Royalists had secure flanks, and, after the initial success of their opponents, they were able to stabilise and hold a defensive line. That night the King re-provisioned the garrison of Donnington Castle and withdrew his army to safety at Oxford, unmolested by the Parliamentarians who were left in possession of the battlefield (Emberton, 1995).

The outcome of this battle had wider significance (Halsall, 1996). In order to cover up their own failings, Cromwell and others endeavoured to place the blame for their defeat at the door of the Earl of Manchester and his inability to co-ordinate his diversionary attacks with the main assault at Speen, which should have tied down the King's reserves and prevented them reinforcing the line where it really mattered. But synchronising manoeuvres on opposite sides of a battlefield remains a difficult undertaking even today with advanced methods of communication; in the 17th century it was almost impossible.

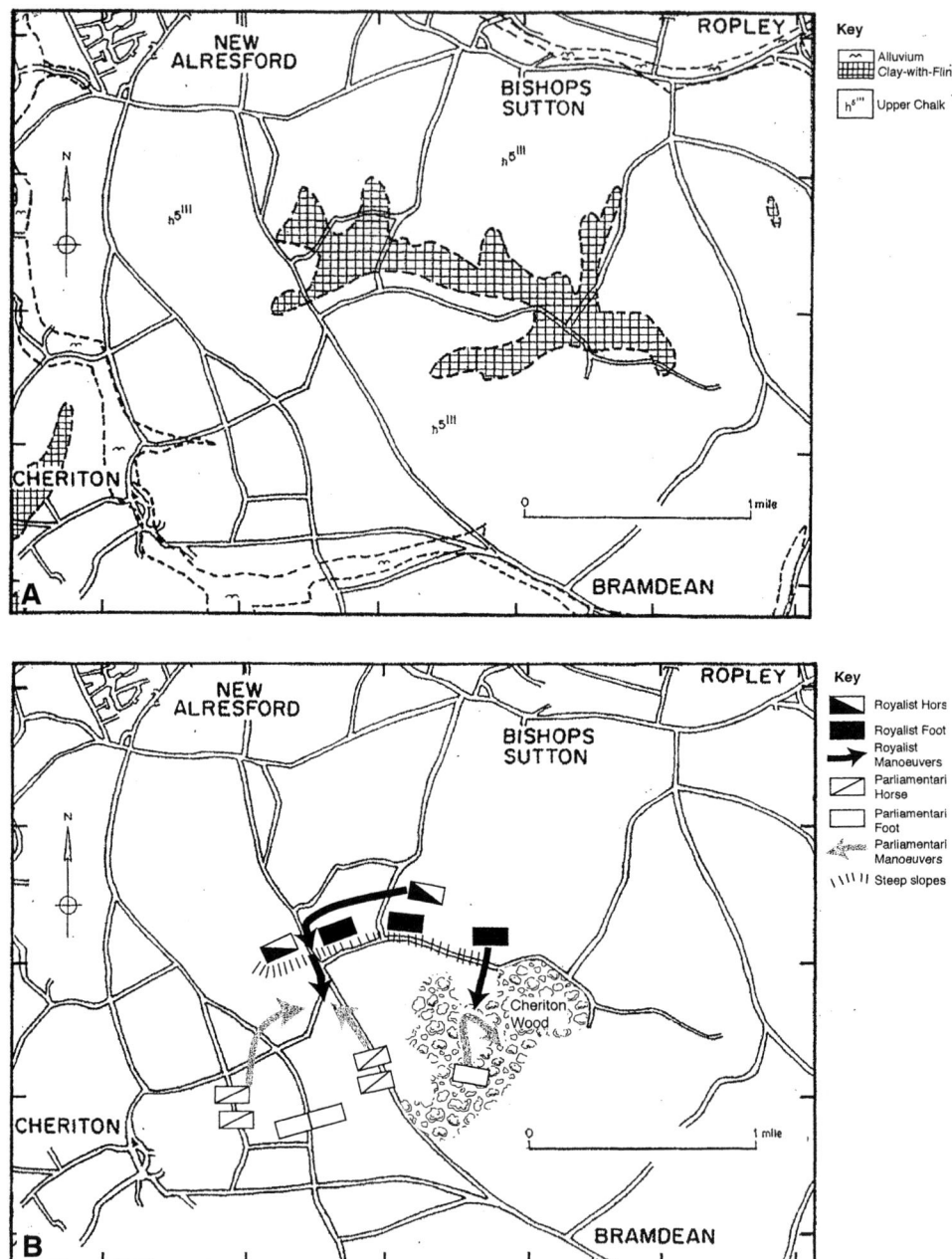

Figure 6. **A.** *The geology of the area of the battlefield of the Battle of Cheriton (Modified from: British Geological Survey, 1975).* **B.** *The Battle of Cheriton (29th March, 1644). The dispositions and principal manoeuvres of the opposing armies are indicated, together with significant terrain features*

4. A sound position on Chalk high ground lost as a result of insubordination: Cheriton, March 1644

The geomorphology of the Chalk outcrop to the east of Winchester played a significant role in shaping the outcome of the Battle of Cheriton on 29th March, 1644. The Upper Chalk outcrops over an extensive area in central Hampshire, and in the vicinity of Cheriton it is essentially flat lying (British Geological Survey, 1975). Emergence of the entire region prior to the Eocene led to the development of a sub-Palaeogene erosion surface (Jones, 1980). Subsequent uplift gave rise to later episodes of erosion and dissection of this surface, so that only remnants now remain (Wooldridge & Linton, 1955). Such remnants occur as a low plateau capped by Clay-with-Flints forming the northern and eastern part of the battlefield at Cheriton (Figure 6A). Short, steep slopes descend to the lower ground to the west and south (Figure 6B). A narrow hedged lane, which descends the plateau edge towards the south [SU595301], figured prominently in the ensuing battle. To the east of this point the plateau margin curves southwards, and its western flanks are hidden by a substantial area of woodland, extending into the valley below, known as Cheriton Wood.

In early 1644 Royalist forces, commanded by General Hopton, and a Parliamentarian army commanded by Waller were struggling for control of southern England. On 28th March, Waller with 10,000 men, camped in the vicinity of Hinton Ampner, a mile west of Bramdean (Seymour, 1975). Meanwhile Hopton had been joined by a contingent commanded by Lord Forth, whose seniority gave him nominal command. This situation gave rise to confusion in the chain of command, even though Lord Forth had agreed that Hopton should retain effective overall command of their army in the field (Seymour, 1975). The combined Royalist forces now totalled about 6,000 men, including about 2,500 cavalry. The Royalist generals decided to take the initiative and force a battle by advancing and seizing the high ground of the plateau at Cheriton. In this they were entirely successful and Waller moved north to confront them.

The Royalist position on the plateau, fronted by a significant gradient, was relatively secure, so Waller decided to achieve an advantage by occupying Cheriton Wood. This enabled his infantry to advance close enough to the opposing troops to annoy them with long range musketry from the cover of the trees, and positioned them to threaten the flank of any formations attempting to advance along the lane traversing the battlefield from north to south. The wood also provided a desirable position from which it would be possible to enfillade any enemy troops advancing into the valley below. Meanwhile the main body of Waller's men took up positions on a low east-west ridge in front of the Royalist position (Rogers, 1968).

Hopton realised that Cheriton Wood was the key to the battlefield, so he despatched 1,000 musketeers under the command of Colonel Appleyard, and supported by artillery, to secure it. This they duly did, probably in part by outflanking their opponents, who withdrew, leaving the Royalists in possession of the wood and thereby in a position to dominate the centre of the battlefield.

At this point the actions of Sir Henry Bard, a young and relatively inexperienced officer, were to influence the battle. Sir Henry commanded a regiment of Royalist cavalry on the right wing, and decided to lead his unit to attack the Parliamentarian horses on the lower ground below to his front. It is possible Bard could not see the full Parliamentarian deployment facing him because of the curvature of the slope down from the high ground on which he was stationed, as if he had he may have stood his ground as ordered. As it was, Bard led his troops down a narrow lane [SU593301], which due to the enclosure of the plateau slopes was the only accessible route to the enemy (Rogers, 1968), and advanced to meet his foes. The Parliamentarian horse closed in on his men before they could deploy their full strength from the defile, and Bard's regiment was eliminated or captured. Other Royalist cavalry units now followed in an attempt to rescue Sir Henry and his men, but each in turn was destroyed by their opponents. After a relatively short passage of time the Royalist cavalry was effectively eliminated, and without horse the remainder of the army was extremely vulnerable, so the Royalist generals were forced to order a withdrawal, which was accomplished in some disorder (Burne, 1996). The somewhat confused pattern of this battle may in part have arisen as result of misunderstandings arising from the selection of identical field signs and passwords for the day by the opposing armies (Guest & Guest, 1996), and as a result they would have difficulty in distinguishing friend from foe. The ultimate outcome, however, was a clear victory for Waller, which secured southern England for the cause of Parliament. A secure defensive position formed by a remnant of the Sub-Palaeogene erosion surface on the Chalk, had thereby been rendered of no value by the insubordination of Sir Henry Bard, leading directly to the destruction of the Royalist cavalry in this action.

5. The Greensand-Chalk escarpment turns retreat to disaster: Roundway Down, July 1643

The action at Roundway Down in Wiltshire on 13th July 1643 provides a classic example of how a terrain feature directly attributable to the underlying geology (Figure 7A) can have a major impact on the retreat of defeated troops, in this case with disastrous consequences. Waller's Parliamentarian Army, which had recently been defeated at Lansdown, was besieging the Royalists garrisoning Devizes, when a relieving force of 1800 cavalry from Oxford, under the command of Lord Wilmot, approached from the north east (Rogers, 1968; Smurthwaite, 1984). Sir William Waller deployed his 2,000 horse, 2,500 foot and 8 guns (Burne & Young, 1959) on the gentle northeastern slopes of Roundway Hill [SU023647] to block the advance of the approaching Royalists. The latter fired two cannons from Roughridge

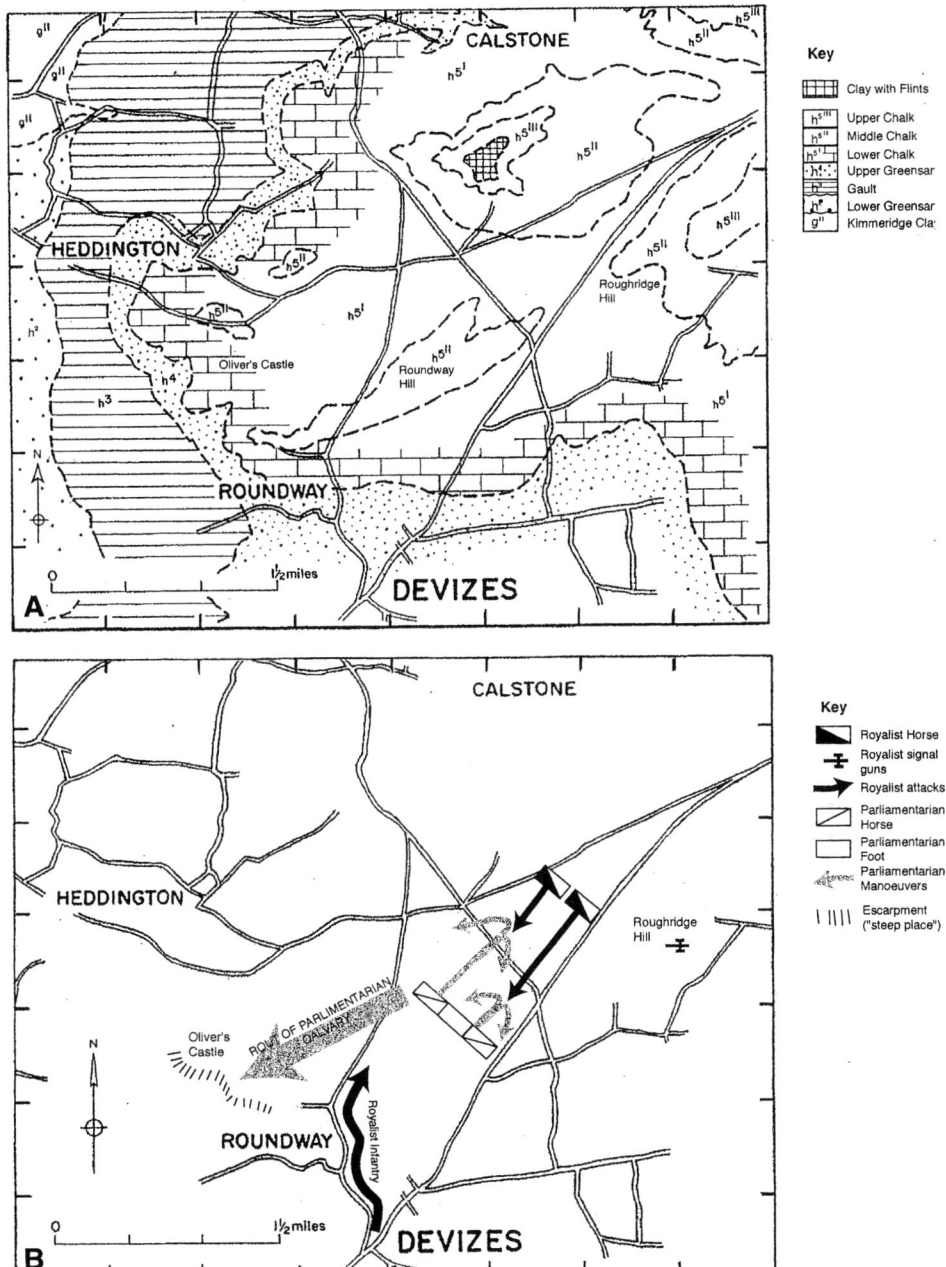

Figure 7. **A.** *The geology of the area of the battlefield of the Battle of Roundway Down (Modified from: British Geological Survey, 1975).* **B.** *The Battle of Roundway Down (13th July, 1643). The dispositions and principal manoeuvres of the opposing armies are indicated, together with significant terrain features*

Hill (Figure 7B) as a signal to notify their colleagues in Devizes of their approach (Rogers, 1968).

Roundway Hill and Roughridge Hill are outliers of Middle Chalk, in the latter case capped with Upper Chalk, resting on flat-lying Lower Chalk which forms the lower ground separating them (British Geological Survey, 1974). The hills are the eroded remnants of the Mio-Pliocene peneplain (Wooldridge & Linton, 1955). To the west of Roundway Down the land surface falls steeply over an escarpment marking the limits of the Lower Chalk outcrop and that of the underlying Upper Greensand (British Geological Survey, 1974).

The Parliamentarian horse advanced north-eastwards to meet the Royalist cavalry (Rogers, 1968). Despite the slope of the ground which favoured their opponents, the Royalists routed each wing of Waller's cavalry in turn, and swept them from the field (Guest & Guest, 1996), leaving the Parliamentarian infantry isolated. Shortly afterwards the 3,000 Royalist foot garrisoning Devizes emerged from the town. Faced by both infantry and cavalry, Waller's foot on Roundway Hill were placed in an untenable situation, and surrendered. Meanwhile the Parliamentarian horse fled towards the hidden escarpment to their rear. This steep slope caused the cavalry to be unhorsed, turning the flight into a disaster (Rogers, 1968). After this short battle and the ensuing routs and surrenders Waller's army effectively ceased to exist and the Royalists were, until Cheriton, in control in the south and south west.

6. Conclusions

The four engagements described illustrate how terrain and the geology that underlies it can influence the course of action in warfare. In the case of the 1st Battle of Newbury, the inability of Essex to secure his preferred, and direct, line of march over relatively good going on Chalk led to his entrapment by the Royalists. The tactics employed in the engagement itself also focused on the geomorphology of the battlefield, the defeat of the King's forces turning, despite their initial advantage, on lax security and lack of attention to the form of the terrain at the outset. At the 2nd Battle of Newbury the Parliamentarian commanders chose to approach the Royalist's defensive positions via higher ground to the north east, rather than along the Kennet Valley. The terrain of the battlefield itself ensured that any fighting would take place on a narrow front, and to avoid making a costly frontal assault, the Parliamentarians conducted a flank march by night, to fall on the Royalist rear. Unfortunately for the Parliamentarians, the difficulties of fighting on a narrow front, in this case on the Speen ridge between the Lambourne and the Kennet, still applied and their attack bogged down and was contained. Their difficulties were further exacerbated by the inability of Manchester's independent command to adequately co-ordinate their attacks with those of the main assault force, separated from them by more than 2 km of hostile ground.

At Cheriton, Hopton and Forth arrayed their troops to take full defensive advantage of the plateau formed by the remnants of the sub-Palaeogene erosion surface on the Chalk. The fruits of their careful deployment were undone by insubordination, in which circumstances the geomorphology of the battlefield worked to their disadvantage and the Royalist cavalry were destroyed. Finally at Roundway Down, Waller's Parliamentarian troops, although deployed on higher ground to block the advance of the attacking Royalist cavalry, were defeated by the determination of the hostile horse, who swept their cavalry from the field in rout. The presence of a steep escarpment at the base of the Chalk, to their rear, turned the retreat of the Parliamentarian cavalry into a complete disaster as their unsuspecting flight led them to plunge to destruction down the precipitous slope.

In each of these engagements the outcrop of the Chalk and its impact on geomorphology and hydrology played an important role, so that the four actions described might rightly be referred to as battles on Chalk.

Acknowledgements

The assistance of Debbie Whitmore and Alan Cross in the preparation of the diagrams is gratefully acknowledged, as is the encouragement of John E. Thomas of Reading and Ted Rose of Royal Holloway, and the many constructive comments of Ron Blake, Nottingham Trent University.

References

Ashley, M. 1990. *The English Civil War*. Sutton Publishing, Stroud.
British Geological Survey 1962. *Hungerford. England and Wales. Sheet 267. Drift*. British Geological Survey, Keyworth, Nottingham.
British Geological Survey 1974. *Marlborough. England and Wales. Sheet 266*. British Geological Survey, Keyworth, Nottingham.
British Geological Survey 1975. *Arlesford. England and Wales. Sheet 300. Drift*. British Geological Survey, Keyworth, Nottingham.
British Geological Survey 1979. *Geological Map of the United Kingdom, South* (South of National Grid Line 500km N), 3rd Edition. British Geological Survey, Keyworth, Nottingham.
Burne, A.H. 1996. *The Battlefields of England*. Greenhill Books, London.
Burne, A.H. & Young, P. 1959. *The Great Civil War*. Eyre & Spottiswoode, London.
Emberton, W. 1997. *The English Civil War Day by Day*. Grange Books, London.
Gardiner, S.R. 1987. *History of the Great Civil War (Volume 1)*. Windrush Press, London.

Green, H. 1983. *The Battlefields of Britain and Ireland*. Constable, London.

Guest, K. & Guest, D., 1996. *British Battles*. Harper Collins, London.

Halsall, T.J. 1996. The Second Battle of Newbury: a reappraisal. *Cromwelliana* 1996, 29-38.

Halsall, T.J. 1999. Beenham in the Civil Wars. *In*: Anon. (Ed.), *Beenham: a History*. History of Beenham Group, Beenham, Berkshire, 27-33.

Halsall, T.J. 2000a. Geological constraints on battlefield tactics: examples in Britain from the Middle Ages to the Civil Wars. *In*: Rose, E.P.F. & Nathanail, C.P. (Eds), *Geology and Warfare: Examples of the Influence of Terrain and Geologists on Military Operations*. The Geological Society, London, 32-59.

Halsall, T.J. 2000b. Geological constraints on the design and construction of fortifications: examples from Medieval Britain. *In*: Rose, E.P.F. & Nathanail, C.P. (Eds), *Geology and Warfare: Examples of the Influence of Terrain and Geologists on Military Operations*. The Geological Society, London, 3-31.

Haythornwaite, P.J. 1998. *The English Civil War 1642-1651: an Illustrated Military History*. Brockhampton Press, London.

Hibbert, C. 1993. *Cavaliers and Roundheads, The English at War 1642-649*. Harper Collins, London.

Jones, D.K.C. 1980. The Tertiary evolution of south-east England, with particular reference to the Weald. *In*: Jones D.K.C., (Ed.), *The Shaping of Southern England*. Academic Press, London, 13-47.

Kinross, J. 1998. *Discovering Battlefields of England and Scotland*. Shire Publishers, Princes Risborough.

Money, W. 1972. *A History of Newbury*. Newbury Bookshop and Thames Valley Press, Maidenhead.

Newman, P. 1985. *Atlas of the English Civil War*. Croom Helm, London.

Rogers, H.C.B. 1968. *Battles and Generals of the Civil Wars, 1642-1651*. Seeley Service & Co., London.

Seymour, W. 1975. *Battles in Britain 1066-1746*. Sidgwick & Jackson, London.

Smurthwaite, D. 1984. *The Ordnance Survey Complete Guide to the Battlefields of Britain*. Guild Publishing, London.

Wooldridge, S.W. & Linton, D.L. 1955. *Structure, Surface and Drainage in South-East England*. George Philip & Son Ltd., London.

Trevor J. Halsall
Postgraduate Research Institute for Sedimentology
University of Reading
PO Box 227
Whiteknights
Reading, RG6 6AL

Terrain and Guerrilla Warfare in Navarre, 1808-1814

Shell Kimble & Patrick O'Sullivan

ABSTRACT: Through recent history, the relationship between physical terrain and guerrilla success has been treated as a matter of received wisdom, but few quantitative studies on this topic have surfaced to verify this relationship. The purpose of this study is to quantitatively assess the extent to which the physical geography of the terrain is a direct factor in the implementation of guerrilla warfare and its success. This study empirically tests the validity of conventional wisdom, by investigating the statistical relationship between physical terrain and the outcome of guerrilla conflict for a sample of 35 historical battles, in which the Spanish peasantry fought against Napoleon's invading Grande Armée in the early 1800's. It was during this conflict that the term 'guerrilla' was coined. An historical reconstruction of these 35 battles has yielded information on the physical conditions in which each battle was fought and on its outcome.

1. Introduction

In 1992 the frequency of wars in the 20th century reached a peak of 34. This was not a sudden leap resulting from the end of the Cold War but a continuation of a trend since 1945 of an increasing number of small wars. These wars are mostly confined within countries in the form of civil wars or insurrections, often involving conflict between antipathetic religious or ethnic groups. Much of the fighting is low intensity in fire power and is carried on in rugged and remote areas. The name 'guerrilla' became attached to this unconventional, insurrectionist variety of warfare after the impact that Spanish rebels had on the Grande Armée two centuries ago, especially in the north along the flank of the Pyrenees. Given the irregular nature of this type of conflict and the consequent lack of records, there is continuing controversy about its efficacy as an instrument of power, and about the balance of advantage between regulars and guerrillas. Rezendes (1981) has presented evidence that for such wars since 1945, the facts run counter to the conventional wisdom that guerrillas are most successful in broken, wooded country. This was Clausewitz's (1968) position, based on his perception of Napoleonic circumstances, a view echoed by Laqueur (1970). It seems evident, however, that the received wisdom on this issue is a matter of impression rather than evidence and analysis. However, there are enough empirical data available from the engagements in the western Pyrenees to go back to the source of the concept of guerrilla warfare in Spain and test the efficacy of this warfare in differing terrain types.

The Spanish campaign against the French in the Peninsular War from 1808-1814, particularly in the Basque region of Navarre, is a classic instance of guerrilla warfare. This is perceived as one of the most successful irregular wars in history. Consequently, a detailed examination of the terrain characteristics and outcomes of a number of engagements in this region should throw light on the relationships of guerrilla success to terrain. Although it is pre-modern in weaponry and technology, pre-dating the ready availability of rifled guns and the use of cordite, it still allows for the analysis of basic tactical considerations and may offer lessons for today.

2. The war in Iberia

During the Napoleonic era, the Spanish people took up arms to purge the invading force from their lands. Their method of fighting, however, was not in compliance with the regular European warfare of the period. In fact, it was in the Iberian Peninsula during these Napoleonic conquests that the term 'guerrilla' was coined to describe the radical, undisciplined, and chaotic fighting used against the French Grande Armée by the local peasantry. Although the Spanish were not the first to utilise this method of fighting, it was during this period that its large-scale success gained international attention. This may have been the largest, unorganised guerrilla insurrection to that date. Iberia was always a major thorn in Napoleon's side and the unpacified Spanish guerrillas not only pointed out his military weaknesses but also showed that he could, indeed, be defeated.

The guerrillas were mostly unaware of how the rest of Spain fared against Napoleon. News of a few major battles, like that at Bailen, would reach them and inspire them to continue the fight, but guerrilla action within each province was generally unrelated and independent of the others. The essence of guerrilla tactics is to trade space for time. The enemy is allowed to dominate a lot of ground, but his morale and force are slowly eroded by a thousand small cuts. He is drawn to extend his supply and communication lines and spread his firepower thinly, so that his internal connections as well as his flanks may be gnawed and his resolve eroded by constant nipping.

One of the major problems with the French army, was its inability to adjust to the newer types of warfare they were encountering in Spain. Napoleon grossly underestimated the destructive influence that the guerrillas had on his empire. Iberia was already technically conquered so he never examined the different kinds of warfare that his troops were encountering, assuming that these little revolts were inconsequential and would die out in time.

The guerrillas attacked from a wide variety of unanticipated locations. They may only have caused minor damage with each attack but the uncertainty of their movements forced the Army to keep up constant guard. French communication and supply lines were difficult to keep intact, and by

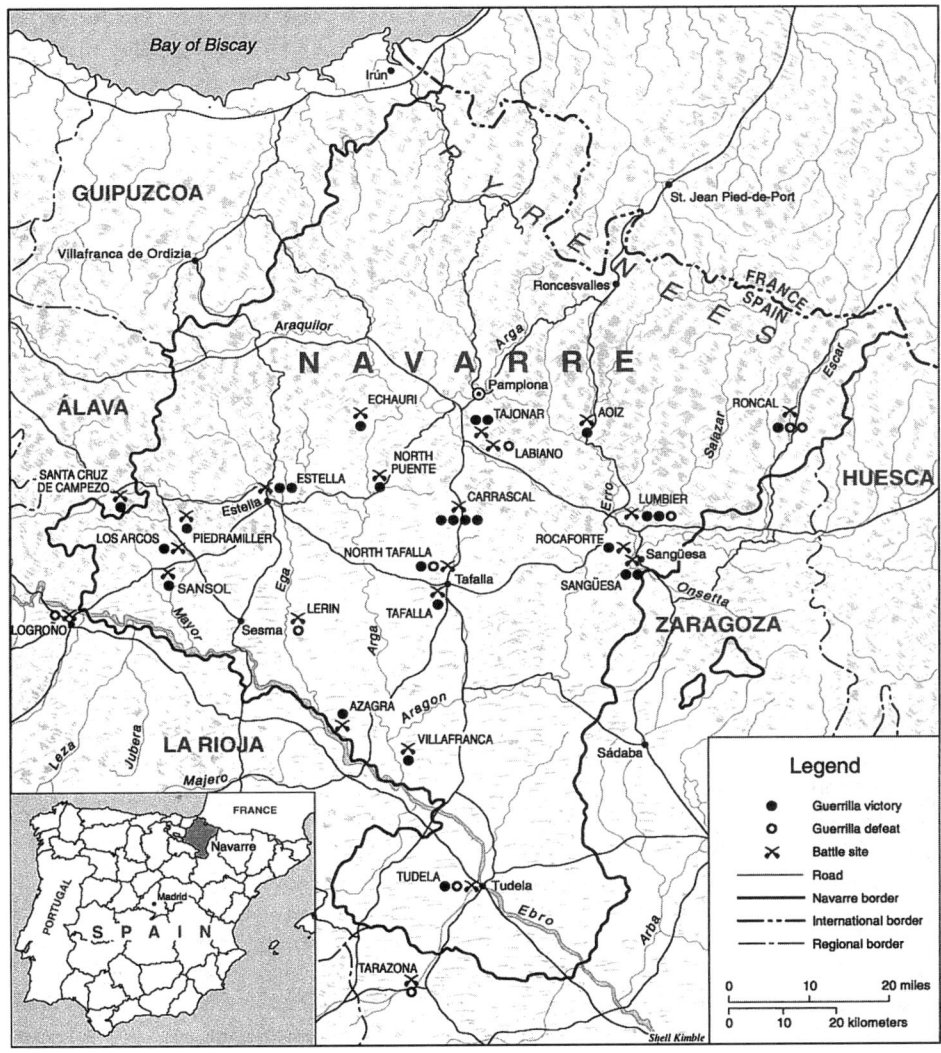

Figure 1. Battle sites within the Navarre region

1810 the French needed some 108,000 men to keep their lines of communication clear (Chaliand, 1982). Guerrilla attacks on supply wagons and couriers became a constant problem. In a letter to Ney, Soult described such problems dealing with the guerrillas. 'Despite all the measures I have taken to destroy the bands of guerrillas, we have not yet succeeded. The brigands continue to torment us, intercepting our convoys and cutting off our isolated weak detachments' (Horward, 1984). For example, guerrillas would ambush French couriers and rewrite the letters and instructions from Paris or from the other generals in Spain. They would then send one of their couriers to deliver the mail. The French were put in danger of losing several battles because of the chaos from troop deployment. As noted by Chaliand (1982) the French had a four-to-one superiority in regular soldiers on the Peninsula and it was the guerrillas' actions that prevented a mass concentration of forces against Wellington, with what well might have been fatal results.

Unfamiliarity with Iberian terrain was also a setback for the conquering army. Napoleon did not utilise superior terrain intelligence to his advantage as he had done with the Austrians and Prussians. Napoleon was known for studying the characteristics of the terrain before engaging in battle. The first French official department for cartography was started under his command and the accuracy of the maps it generated had become focal tools throughout most of his campaigns (Horward, 1984). However, this policy was not employed in Iberia under the assumption that victory was already assured. This might have been his most serious miscalculation of the campaign because it left the guerrillas with the home turf as well as the tactical knowledge for its use in their operations. Targeted points along the supply and communication lines indicate that the guerrillas did, indeed, use a form of tactical terrain analysis for much of their operations.

Navarre can be separated geographically into the rugged Montana in the north and the lower lying Ribera in the south. This division of Navarre, not only marks differences in the physical geographic make up of the region but also differences in the political and economic characteristics of both areas (Tone, 1994). The Navarrese guerrilla movement was started in the Montana, where their situation was more conducive to an insurrectionist reaction to an invading army.

The largest supply route and a major target area for the guerrillas ran from Bayonne to Irun, south through central Navarre, and then continued southward to Madrid (Figure 1). This route was of critical importance due to the difficulty of crossing the Pyrenees. This was one of the few roads conducive to artillery and supply wagon mobility. Thus, it pinpointed for the irregulars the precise location of incoming French troops, supplies, and information. This helps to explain how the Navarre region became one of the more important geographic areas for the rebellion and the Navarrese guerrillas some of the most successful insurrectionists of the war.

3. Weapons and tactics

Standard weapons for the French Grande Armée during this period were the bayonet, carbine, musket, dragoon musket, lance, pistol, and the sword. The guns were smoothbore firearms with short ranges and muzzle loading. The ammunition rarely fitted easily into the barrel and a sure fire was never guaranteed. Although rifled barrels were around, they were not used in battle because of their excessively slow loading time.

Black powder was used in all firearms which constantly gave off smoke, often enough to cloud a battlefield. Small arms would get fouled in a matter of minutes which slowed firing times considerably. Muskets could get five rounds fired per minute in the beginning. Shortly thereafter, only four rounds per every three minutes were possible. Accuracy was also a problem. The musket had a range of 1,065 m but two battalions firing 100 m apart would only hit a handful of men. Engagements at 30 to 40 m, however, could effectively eliminate half of the opponent's force. As noted by Esposito (1978), some soldiers got to be excellent shots and could pick off a guerrilla from behind a rock at 300 m.

Since weapons technology was evenly distributed between the guerrillas and the French, it was other factors which played important roles in battle outcome. An obvious factor for examination would be the tactics on the battlefield. Tactics means the employment of units in combat so that their ordered arrangement and manoeuvre will bring about maximum effectiveness and victory (Peltier, 1966). In this sense, the guerrillas had a definitive tactical system devised to successfully ambush and retreat as quickly as possible. Espoz y Mina arranged a tactical attack plan for most of his confrontations with the French. He divided his force in half, specifically reserving his second half as a back up volley, if needed, and placed his attacking force in concealed positions as close as possible around the pre-planned battle area. The guerrillas opened fire when French forces approached the designated spot for commencement. Following the initial bombardment of fire, the attacking force then rushed in for hand-to-hand combat with bayonets. Close quarter attack was brief and was followed by a swift retreat into the surrounding mountains or hills. Mina had only half his men fighting the battle and the reserve units were only used if the attacking units had difficulty in combat or during the retreat. The French never knew the numbers of guerrillas they were engaging. On several occasions, when victory seemed imminent to the French, the fresh reserves of guerrillas would join the attack and if the insurgents were unable to win, they would, at least, muster a quick retreat without leaving the potential prisoners the French had foreseen.

Napoleon's Grande Armée repeatedly used the tactics that had won them victories in the rest of Europe but in the Iberian Peninsula the regular, block formation fighting was not an effective counterinsurgent tactic.

4. Quantitative analysis of guerrilla engagements

The purpose of this study is to determine quantitatively whether or not a statistical relationship exists between the outcome of guerrilla battles and the physical and human attributes of the battlefield terrain. It is hypothesised, drawing on conventional assumptions, that guerrillas will win battles more frequently where the terrain is rugged and the vegetation heavy, the local relief is greater with more river valleys, and where the availability of roads is less, reducing the mobility of regular forces. To test this hypothesis it is necessary to develop measurements of the dependent variable, whether guerrillas lost or won, and of the independent variables, slope, vegetation, river and road density.

The military engagements examined comprise 35 battles that took place in Navarre between the years 1808 and 1812. Although guerrilla activity in this region was frequent and incidents numerous, only thirty-five of the battles researched left records which contained all of the geographic and military information needed for this study.

Many of the written records were from the French commanders and generals who tended to understate and de-emphasise the scale of the insurrectionist warfare. Any general who was unable to control his district fell into ill favour with the Emperor who was impatient with the time and resources being spent on the Iberian campaign. Therefore, reports often exaggerated the number of guerrilla casualties to reflect at least an equal number to the French losses. The guerrilla accounts, if any had been recorded, also tended to exaggerate the damage done to the enemy. These numbers had a higher level of inaccuracy since guerrilla fighters rarely stayed after a battle to count casualties. The 'hit and run' tactics of guerrilla warfare left little time to calculate the exact losses caused to the French. John Tone (1994), a scholar of the Navarrese guerrillas, estimated the casualty numbers of each side per battle. It was assumed that each side would have fairly accurate estimates for their own losses while magnifying those of the enemy. If two accounts of the battle were recorded, one from each side of the engagement, then the casualties of the French were taken from the French records and the casualties of the guerrillas were taken from the guerrilla records.

The development of the measures of overall battle outcome was determined by a variety of criteria. If either side clearly annihilated the other, the outcome was obvious. If the French were forced into retreat, either back to their previous encampment or into an already French occupied city, then the guerrillas were given the win. If the guerrillas were forced into total dispersion, without immediate reconnection or follow up confrontations, they were assigned a loss. Although they employed the split and run tactics frequently, regrouping was usually achieved within several days. After that, due to low morale and losses, the guerrillas would lose numbers of soldiers who had returned to their farms and had to be recruited again for service. The final determinant of victory over loss was the casualty count. The side with greater percentage of casualties was assigned a loss. All battles were assigned a number based on the guerrilla outcome. Guerrilla wins were assigned a 0 and losses were assigned a 1.

The independent variables, measurements of the physical and human elements of the terrain, were constructed from two sets of maps of differing scale. The *Atlas de Navarra* (1977) provided maps at a scale of 1:200,000 with 100 m contours. The slope and road characteristics were calculated from this source. Vegetation and river characteristics were based on calculations from maps at a scale of 1:50,000. Grids were placed on both map sets and circular areas drawn around the battle sites. A 2.5 cm diameter circle was drawn on the grid to indicate the battle sites for the 1:200,000 scale maps.

This represented a 5 km diameter circle on the maps. The grid was divided into 2.5 cm squares which further subdivided into tenths. Therefore, each battle site contained an average of 310 data observations. A 25 cm diameter circle, representative of about 12.5 km, specified the sites for the 1:50,000 scale maps.

Slope is the factor of major concern in the terrain analysis. There were two different methods for calculating slope for use in separate statistical tests. The 1:200,000 scale maps were used for both methods of slope calculation. The first method gave battle sites a slope value ranging from one to five, five being the most rugged and mountainous. These were found by counting the number of contour line changes from east to west and from north to south. For example, Tudela had two contour line changes from east to west and one change from north to south. It was assigned a one. Roncal had fourteen line changes from east to west and fifteen changes from north to south. It was subsequently assigned a five. Altogether, six sites warranted a one, eight sites a two, ten sites a three, seven sites a four, and four sites a five.

The second method for battle site slope calculation required the use of the elevation points off the grid. Positional, easting and northing values were assigned from the x and y axes of the grid, with the z variable being the elevation value. These x, y, z data points were used to calculate the continuous slope as a percentage for any one grid. The percentages of slope for each location were then categorised into five ranges: 0-20 = 1, 21-40 = 2, 41-60 = 3, 61-80 = 4, and 81+ = 5, each being assigned a value of one to five.

River densities and sizes per battle location were calculated according to density on the grid. One of three values was assigned to each location. Sites with no rivers or containing a few light streams were given a value of one. Sites with light streams and a small river were given a value of two and sites with a major river and streams were assigned a value of three. Sixteen sites were assigned a one. Eight sites were assigned a two and eleven were assigned a three.

Vegetation characteristics of the battle sites were assigned according to type and density. Each site was given a value of one, two, or three. These values were assigned according to the percentage of the varying types of vegetation throughout the grid of each battle site. A value of one indicated sparse vegetation, grassland, prairie, and/or cultivated land. A value of two indicated substantial vegetation and a mixing of woodland, grassland, and cultivation. A value of three indicated heavy vegetation and woodland. Eight battle sites were subsequently assigned a one. Thirteen sites were assigned a two and fourteen sites were assigned a three.

Road densities were also calculated according to a three value system. A value of one indicated a battle site with a few small roads. Locations assigned a two indicated the presence of small to medium sized roads and locations assigned a three had a major highway amongst the other road networks. Battle sites located on the road from Pamplona to Tafalla, for example, were assigned a weight of three due to the route being a major passageway between Paris and Madrid. The hypothesis for this variable states that guerrilla victory is more likely with fewer, smaller roads within the battle location. It is believed that the smaller roads in a given battle location would provide more difficulty for a regular army with heavy supplies and artillery to traverse. Based on these categories, six sites were assigned a one, twelve sites were assigned a two and seventeen were assigned a three. Figure 1 shows the battle sites and outcomes for each engagement used in the analysis.

5. Analysis and interpretation

Logistic regression models showed no significant relationship between guerrilla outcome and any of the independent variables, as the null hypothesis was not rejected (for statistical details, see Kimble, 1997). This test indicates that guerrilla victory in the Navarrese campaign against the French was not dependent on high slope, low river density, high vegetation density, and/or high road density of the battlefield.

The results of the Fisher's Exact Test could not establish a significant difference between guerrilla victories in the Montana and the Ribera. Had the null been rejected, the possible explanations for the differing outcomes by region could have been related to the tactical use of the local physical geography. Since the null could not be rejected, the indication is that terrain did not play a significant role in battle success.

The chi^2 test showed that a significant relationship did exist between guerrilla outcome and the percentage slope in the Montana but it inverted the expected direction of the relationship. The null hypothesis can then be rejected, but here guerrilla victory was significantly more likely on flat land. This result does not uphold the conventional beliefs, even though a significant relationship occurred.

6. Conclusions

The statistical analysis carried out in this study hardly constitutes a resounding confirmation of conventional wisdom about the efficacy of guerrilla warfare. Some results fail to find a significant relationship between the outcome of engagements and the terrain. The chi^2 test overturns the conventional wisdom, finding victory more likely in flatter areas. The failure to disprove the null hypothesis in the logistic regression analysis suggests that many factors other than terrain characteristics came into play in determining the outcome of battles between guerrillas and the French in Navarre between 1808-1813. It is not simply a matter of superior guerrilla exploitation of the lie of the land. Clearly the size of forces, the degree of surprise achieved, firepower amassed and deployed, and the degree of steadfastness that both sides showed all play a role.

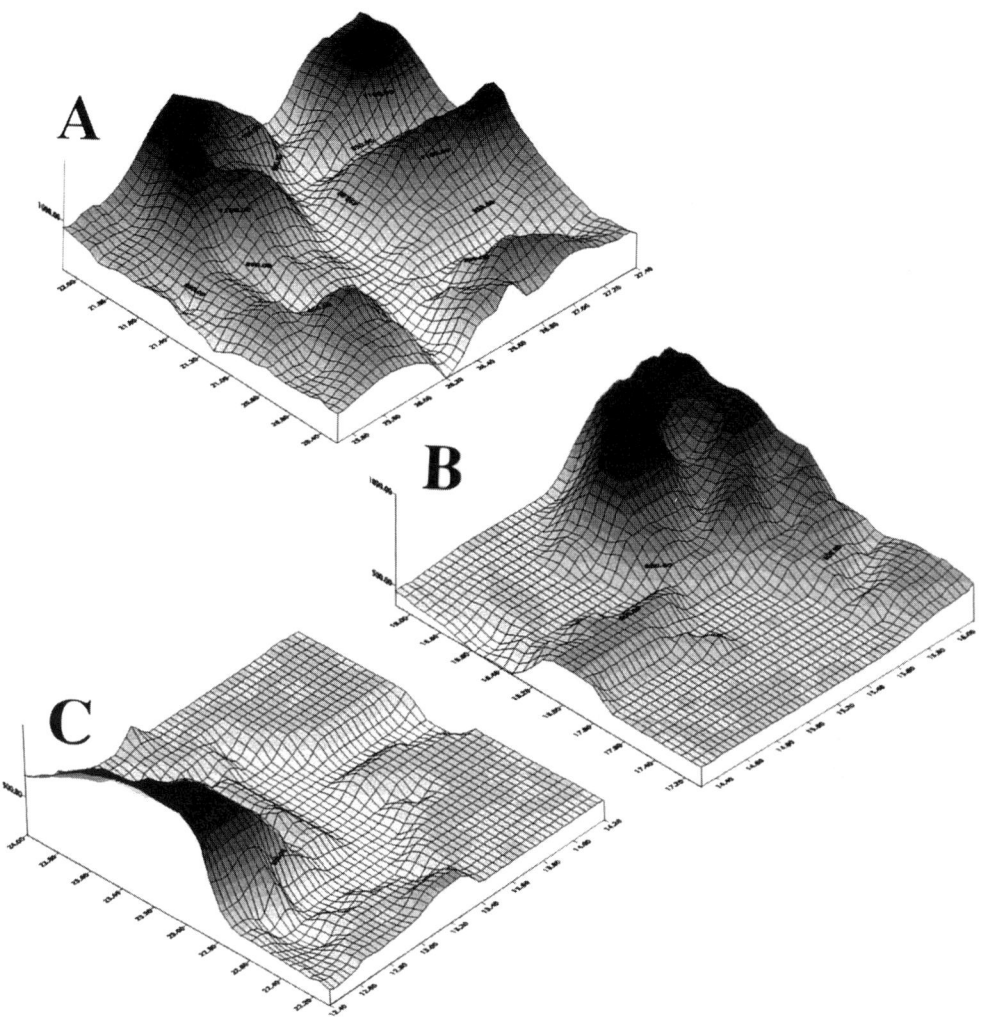

Figure 2. Digital elevation models for the Roncal (A), Carrascal (B) and Echauri (C) battle sites

There is no indication that guerrillas were more successful in more rugged terrain. They certainly succeeded in flat open areas. We cannot put all of their successes down to superior terrain analysis. Their fervour and the reduced morale of the French army may have been very important factors.

Several battle sites became popular locations for guerrilla planned confrontations, which supported the assumptions that terrain was a factor of major concern. Two of these were the Roncal and the Carrascal. The Roncal is a notably rugged area with high degrees of slope and high vegetation density

(Figure 2A). According to military doctrine concerning guerrilla success requirements, this area should have been the most suitable for their confrontations. However, of the three battles included in the analysis, two were losses to the guerrilla forces. The one victorious battle took place at the beginning of the war when the guerrilla numbers were at their lowest. The second battle in that area resulted in a defeat for the guerrillas and their complete dispersion without regrouping. The third battle was a disastrous defeat, especially due to the material losses the guerrillas suffered. French forces were able to find and take over the guerrilla supply base, including their hospitals and munitions. Guerrilla forces in the Roncal were formidable, however, and had even signed treaties with the French in the early years of the war. According to this study, as their numbers increased, so did their susceptibility to defeat. The reasons for these outcomes are not clear. Perhaps if more battles in the Roncal could have been tested, this trend would be different.

In the Carrascal (Figure 2B), an area less rugged than in the mountains of the Roncal, the guerrilla forces made several planned ambushes. Five different battles were analysed from this area and all five ended victoriously for the guerrillas. The primary reason for planning ambushes in the Carrascal was that the supply route running directly through it connected Madrid to Bayonne. The French were pretty much limited to this passageway through the mountains, consequently giving the guerrillas a predictable location for ambush. However, the roads through this area were wide and easily traversable which should have been an asset to the French troops, allowing for their use of artillery and cavalry. Repeatedly, the insurrectionists defeated the French with little losses to their own side. They were even able to chase the French troops back to the previous city they had occupied. This would sometimes involve a lengthy chase, extending 35 km or more.

The confrontation in the Echaure valley (Figure 2C) was one that typifies the expected behaviour of guerrilla warfare. Guerrilla forces were able to hit and retreat despite the overwhelming superior forces of the French, supposedly using the lie of the land to favour their escape. Technically, the regular army had surrounded the irregulars, leaving no escape route. The guerrilla force did escape and despite a two month search, was never found. Unexpectedly, false information was relayed to the French commanders, enabling the guerrillas to persuade most of the forces to abandon an important pass in the escape route, leaving only a small contingent force to be fought. Although this approach was cunning and allowed for a successful evasion of the regular forces, the degree to which the physical relief of the terrain affected the outcome is uncertain.

This study, like that of Rezendes (1986), implies that the outcomes of guerrilla warfare is not absolutely conditioned by the physical terrain. Highly mountainous areas with dense vegetation were thought to be crucial ingredients for guerrilla success. Although most military doctrines have believed this was the case, current research is finding that such dependency by the guerrillas for their physical environment is not as strong as was once believed. Even warfare prior to the 20th century, was not fully dependent on terrain. Clearly, other factors affect guerrilla outcomes.

Further study of guerrilla tactics and terrain may yet prove the originally accepted theories. Perhaps research should be conducted relating specific guerrilla tactics to certain types of terrain. Relationships may yet exist but the aspects of insurrectionist warfare may be too complex to study without subdividing its tactical variety.

The impact of such research affects the modern world as well. The accepted doctrines of insurrectionist warfare need to be reviewed. On the larger scale, insurrections since the end of the Cold War continue at an average of thirty a year. With such an abundance of activity involving this type of warfare, importance should be attached to understanding the factors that affect both it and counterinsurgency tactics. The world perspective has long been focused on mass destructive explosives and heavy firepower, but the active warfare currently being waged is smaller scaled guerrilla activity. Conflicts from the Caucasus to the Balkans are all modern insurgent battlegrounds. Guerrilla confrontations are as effective today as they were 200 years ago. This would seem to indicate that studying this type of warfare is essential for understanding what makes it so successful and how to prepare for any such future confrontations.

The implication for the armies of major powers or of smaller armies under the UN flag doing peacekeeping duties in remote settings, is that there is no simple rule of thumb for judging the dangers presented by irregulars in low intensity warfare. It is not simply enough to stay out of the trees and mountains. The US Marines discovered in Somalia that local irregulars could be very effective even in dry, lowland country. Perhaps we should give considerable credence to Walter Laqueur's (1976) statement: 'The history of guerrilla war, in brief, varied from country to country, and sometimes even from province to province. To attempt in these circumstances to formulate a definitive theory of guerrilla warfare is a vain undertaking.'

References

Anon., 1977. *Atlas de Navarra: geografico, economico, historico.* Pamplona.

Chaliand, G., ed. 1982. *Guerrilla Strategies: An Historical Anthology from the Long March to Afghanistan.* University of California Press, Berkeley.

Clausewitz, C. Von. 1968. *On War.* (translated by J.J. Graham). Routledge and Kegan Paul, London.

Esposito, V. 1978. A Military History and Atlas of the Napoleonic Wars. Arms and Armour Press, New York

Horward, D. D. 1984. *Napoleon and Iberia: The Twin Sieges of Ciudad Rodrigo and Almeida, 1810.* University Presses of Florida, Tallahassee.

Kimble, S. T. 1997. *Terrain Analysis and Guerrilla Warfare in Navarre.* Unpublished Master's Thesis, Florida State University.

Laqueur, W. 1976. *Guerrilla: A Historical and Critical Study*. Little Brown and Co., Boston.

Martin, Andres. 1953. *Historia de los sucesos militares de la Division de Navarra, y demas acontecimientos de este Reyno durante la ultima guerra contra el Tirano Napoleon*. Pamplona.

Peltier, L. C. 1966. *Military Geography*. D. Van Nostrand Company, Princeton.

Rezendes, J. S. 1986. *Geographical Aspects of Guerrilla Effectiveness*. Unpublished Master's Thesis, Florida State University.

Tone, J. L. 1994. *The Fatal Knot*. The University of North Carolina Press, Chapel Hill.

Shell Kimble
National Geographic Society
1145 17th Street NW
Washington DC 20036-4688

Patrick O'Sullivan
Department of Geography
Florida State University
Tallahassee
Florida 32306-2190

Effective use of Terrain in the American Civil War: The Battle of Fredericksburg, December 1862

Judy Ehlen & Robert J. Abrahart

ABSTRACT: The environs of Fredericksburg, Virginia, were a major theatre of operations in the American Civil War (1861-1865) beginning with the Battle of Fredericksburg in December 1862. Fredericksburg, located on the Rappahannock River at the Fall Line, was a key obstacle to Union advances on Richmond, the Confederate capital. The battlefield comprised river terraces and gentle slopes with parallel north-south ridges providing structural boundaries on either side. The Confederates exploited the natural barrier of the river and river terraces, which impeded the Union river crossing, while occupying a defensive position on high ground. Tactical advantages were also obtained from obstacles including a major road, a railroad embankment, bogs, dense woods, and stone walls and fences. The Union Army advanced uphill through these obstacles with little cover in their unsuccessful attempts to dislodge the Confederates. Effective use of terrain coupled with high quality leadership thus enabled Confederate forces to defeat their opponents at the Battle of Fredericksburg. This in turn helped to postpone Union capture of Richmond for nearly three years.

1. Introduction

It is well that war is so terrible - we should grow too fond of it!
Robert E. Lee (Freeman, 1934, p. 462)

Knowledge of the terrain is important at all levels of battle. Commanding generals need to understand the regional terrain in order to correctly place their corps for the upcoming battle. The field commander needs to understand the local terrain over which they must direct their divisions and upon which they must place artillery for most effective use. Company commanders need to know the location of obstacles that will hinder the company's forward movement as well as those that will provide protection for troops in the heat of battle. Terrain factors, such as line-of-sight, the location and effectiveness of obstacles, and state-of-the-ground, are thus important at all levels of military operation.

In the mid-19th century, at the time of the American Civil War (1861-1865), the effective use and knowledge of local terrain was of even greater consequence to military success than in modern battles. The limited ranges and reliabilities of weapons, the quality and availability of ammunition, the overall poor physical condition of soldiers, the absence of adequate maps and intelligence, and the problems of supply, among other factors, made substantial knowledge of the terrain decisive. The Battle of Fredericksburg, which was fought in December 1862, exemplifies how knowledge of the battlefield and effective use of local terrain can facilitate the defeat of superior forces — superior in terms of manpower, artillery, and supplies.

In this paper, we describe the historical background of the Civil War, and give a short description of important events that led to the Battle of Fredericksburg. We then provide a general description of the terrain over which the battle was fought. The progress of the battle is next explained in conjunction with details on the use of the terrain, and finally, we summarise the crucial effects of terrain on the outcome of the battle. The terrain is illustrated in four ways: (1) using words; (2) using sketch maps and terrain profiles; (3) using contemporary and modern photographs; and (4) using screen-grabs of a computer generated three-dimensional (3-D) virtual world[1].

2. Historical background

The immediate cause for the outbreak of hostilities that became the American Civil War was the secession of the Southern states, beginning with South Carolina in December 1860 and ending with North Carolina in May 1861, and the determination of the Northern states to maintain the Union at all costs. The reasons for secession are complex and intertwined, but can be broadly summarised under the headings economics, westward expansion, and States' Rights.

The economies of the two regions, North and South, were very different. The North was largely industrial, whereas the South was primarily agricultural. The raw materials upon which industry depended were more abundant in the North than in the South. In addition, the climate and soils of the South were more amenable to large-scale agriculture than to industry. In the early- to mid-19th century, the economy of the South was dependent upon raw cotton, exported primarily to the mills of northwestern England. Compared to the Northern states, the Southern states were sparsely populated, and there was inadequate cheap labour for large-scale, labour-intensive farming. The use of slaves solved the labour problem.

Most Northerners, with their strict Northern European Protestantism, did not support the practice or system of owning slaves, whereas a large proportion of Southerners, in particular those that owned extensive parcels of land, at least accepted that slaves formed an essential element of their agricultural practices and were needed to support the economic basis of the South. The westward expansion of the United States that occurred during this period thus led to disagreement about the adoption of such practices in the new States and Territories. The number of slave states and non-slave states was more or less equal, and both sides were concerned that changes in the

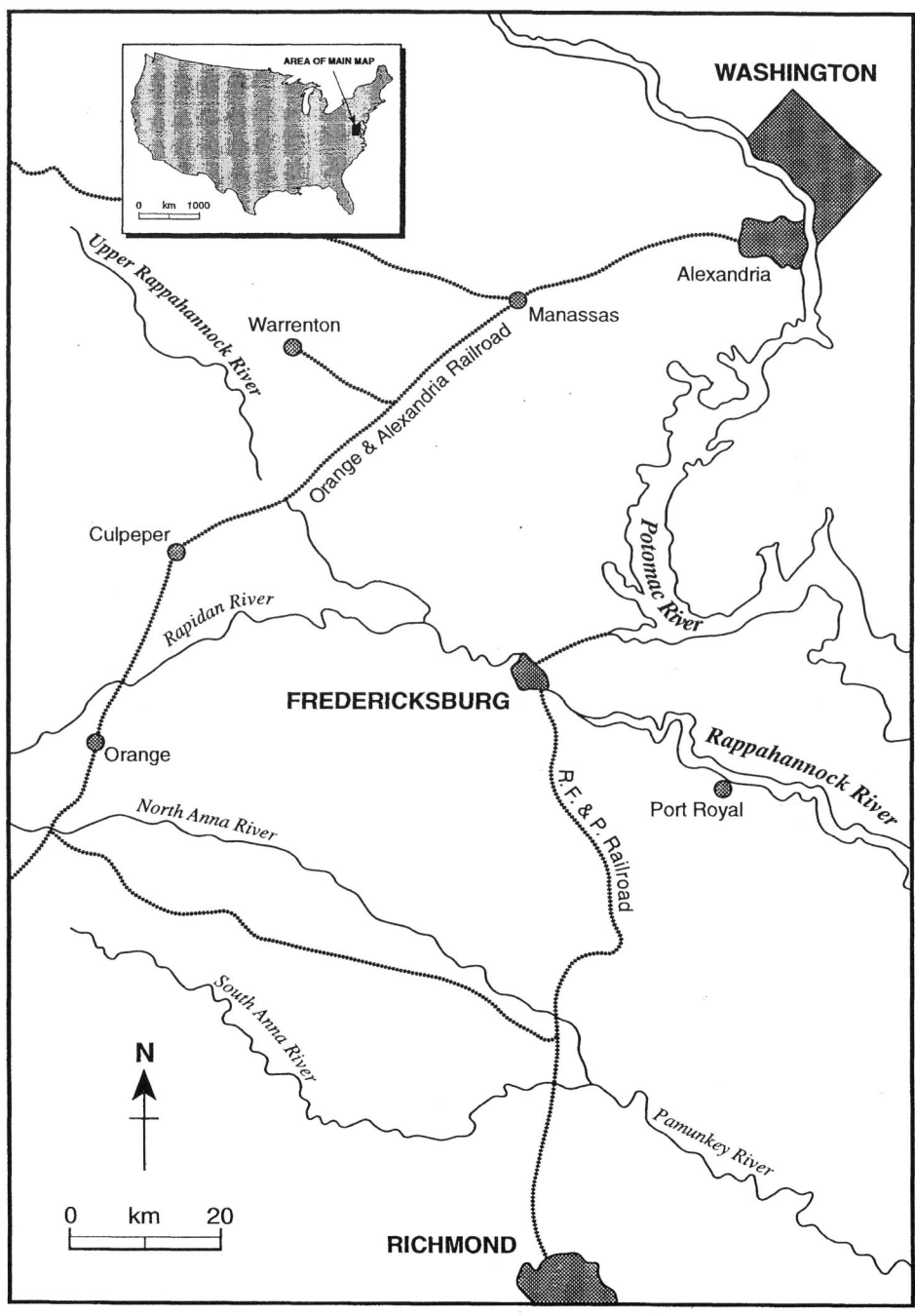

Figure 1. Regional map of the Northern Virginia theatre showing Richmond, Washington, Warrenton, Culpeper, and Port Royal. (Modified from: Whan, 1961)

balance of power would result from the admission of additional States to the Union. Between 1820 and 1854, a series of compromises addressed this problem, but did little more than postpone armed conflict. The last compromise, the Kansas-Nebraska Act of 1854, allowed the inhabitants of each Territory to determine whether or not to admit slavery prior to statehood.

To Southerners, States' Rights involved more than just the right to own slaves; it also included the right to maintain their way of life, and the right to take this lifestyle into new States and Territories. There was a fundamental disagreement between North and South with respect to the Federal Constitution. Southerners believed that States' Rights were paramount and saw the United States as a loose federation. Northerners, on the other hand, saw the federation as paramount, and believed that the rights of individual states were second to those of the nation as a whole. Thus as the Northern population swelled, due to increased immigration primarily from overseas, the balance of Congressional power moved north toward States that did not recognise slavery. Southerners feared that the South would loose control over matters that Southerners considered to be the right of individual States to decide, matters believed to be outside the jurisdiction of the Federal Government, and as defined in the Constitution. Southerners feared the loss of their lifestyle as well as their livelihoods.

This tangled web of disagreement, disagreement in practice as well as in principle, caused the Southern states to leave the Union one by one. Eleven Southern states seceded from the Union between December 1860 and May 1861. They did so individually, with no intention of forming a separate nation. They soon realised, however, that greater strength lay in banding together, so the Confederate States of America (CSA) was formed in February 1861. Montgomery, Alabama was chosen as the first capital, and Jefferson Davis was chosen as the first President of the Confederacy. The capital was moved to Richmond, Virginia (Figure 1) in May because Montgomery was considered too remote.

One of the first actions of the fledgling CSA was to negotiate with the government of the United States and its new President, Abraham Lincoln, to remove Union troops from Federal property within the CSA and to negotiate compensation for other Federal property previously seized in the South. This effort failed partly because Lincoln believed these actions were against his Constitutional duties as President. When CSA forces demanded the surrender of Fort Sumter in Charleston, South Carolina in April 1861 and the garrison commander refused, the first shots of what became the American Civil War were fired.

The first major battle of the Civil War was fought near the small railway junction of Manassas, Virginia (Figure 1) in July 1861. This battle was a tactical victory for the Confederacy. As a result, the people of the South became overconfident, whereas defeat increased Union resolve to overcome what was seen as Southern intransigence to accept the views and opinions of the North. A number of important battles, in Tennessee, Kentucky, Missouri, Maryland, and Virginia, were fought between the first victory at Manassas[2]

Figure 2. Fredericksburg from the southeast in February, 1863. (Source: National Archives Still Photo Unit, College Park, MD; Timothy H. O'Sullivan, photographer)

and the Battle of Fredericksburg in December 1862. Not one of these battles was definitive; most victories, if they can be called that, were tactical, and the number of dead and wounded was incredible (Union, almost 60,000; Confederate, almost 54,000; Davis, 1994). However, a combination of small victories and strong political persuasion was sufficient to keep wavering states within the Union and the resolution of both sides to fight to the finish was confirmed.

In 1862 the Civil War in the Virginia theatre was at stalemate. The Union Army of the Potomac had achieved no major successes, and the Confederate Army of Northern Virginia had achieved victory only at the First Battle of Manassas. After this defeat, Union forces attempted — and failed — to take Richmond beginning in April 1862. This three-month campaign was followed by a second Union defeat at Manassas in August 1862. The Army of Northern Virginia next went on the offensive and invaded the North with battle occurring near the small town of Sharpsburg, Maryland, west of Washington DC, in September 1862. This was a horrendous[3] but indecisive engagement[4]; after Sharpsburg, the Confederate forces withdrew to the south and occupied positions to the west (Shenandoah Valley) and east (near Culpeper in Virginia) of the Blue Ridge Mountains (Figure 1). The Union Army was encamped near Warrenton to the north (Figure 1).

In early November 1862, the newly appointed Union commander, Maj. Gen. Ambrose Burnside, proposed another attempt to take Richmond using the

Figure 3. The Rappahannock River above the Fall Line at Fredericksburg

shortest, most direct route. He recommended moving supplies by boat down the Potomac River from Washington to Aquia Landing, terminus of the Richmond, Fredericksburg, and Potomac Railroad (RF&P), then to continue overland using the railroad to Richmond (Whan, 1961; Figure 1). The Army was to move on foot from the west along the Rappahannock River to the Fredericksburg area and thence south to Richmond. The Rappahannock and the town of Fredericksburg were key obstacles to be overcome in this Union drive south.

3. Battlefield terrain

Fredericksburg, a small town of about 4,000 or 5,000 people in late 1862, is located on flood plain and a series of terraces about 15 m above the Rappahannock River, which is one of the major rivers in eastern Virginia (Figure 2). Fredericksburg lies about 800 m south of the Fall Line, which is the boundary between crystalline Piedmont rocks to the west and Coastal Plain sediments to the east (Mixon *et al.*, 1997), and marks the point below which the river is tidal.

Above the Fall Line, the Rappahannock River is narrow and rocky, with numerous rapids (Figure 3). There are several fords upstream, although even at these crossing points the river is prone to sudden flooding and is treacherous to cross when the water is high. Below the Fall Line near Fredericksburg, the river narrows slightly and thus runs much faster. It broadens downstream, becoming more than 800 m wide at Port Royal 29 km below Fredericksburg.

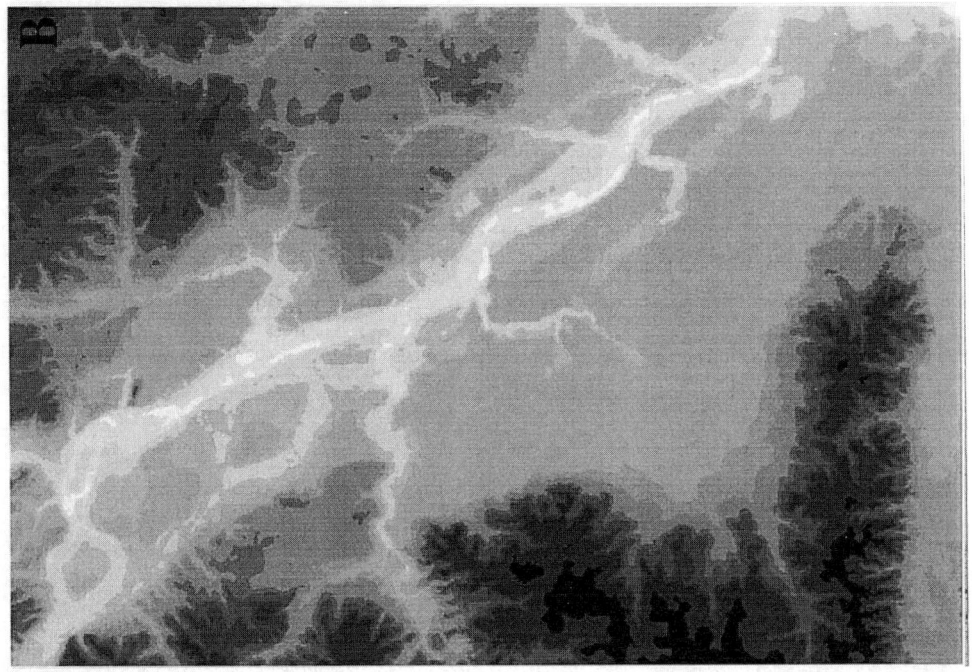

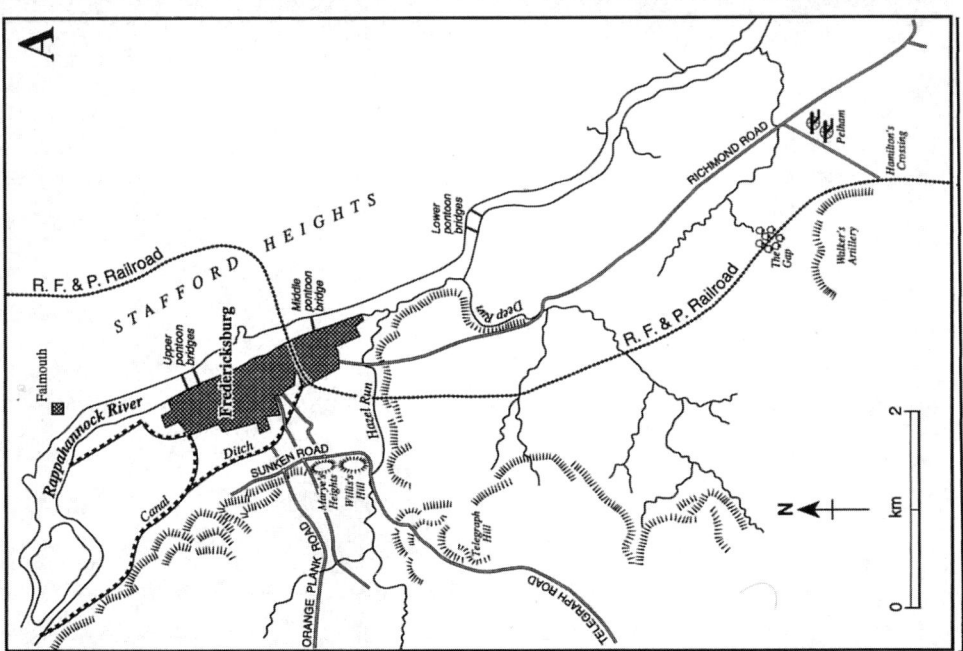

During the Civil War, there were no fords or bridges downstream from Fredericksburg.

Figures 4A and 4B show plan views of the Fredericksburg terrain and Figures 5A and 5B show cross-sections of the terrain in the two zones of heaviest fighting. The Fredericksburg area consists of two north-south ridges, one east of and parallel to the Rappahannock River (Stafford Heights), which at this point flows south, and a second, curved ridge, on the west bank. The Rappahannock flows between the two ridges near the base of Stafford Heights. Small, discontinuous flood plains are present on both banks of the river. On the west bank, there is a wide, sloping plain above the river extending to the ridge. South of Fredericksburg, the western ridge, which is further from the river than the eastern ridge, curves back to the east and approaches the river near Hamilton's Crossing about 9.5 to 10 km downstream. This ridge includes Marye's Heights, Willis Hill, Telegraph Hill[5], and Prospect Hill. The curvature produces an amphitheatre consisting of flood plain and terraces sloping from the western ridge to the river.

In the southern part of the battlefield, a low terrace slopes uphill from the river to the base of the curving western ridge (Figures 5A & 6). This gently undulating plain is up to 3.2 km wide with a grade of about 4.5% and in 1862 was covered by open fields (O'Reilly, 1993). Deep Run and Hazel Run flow across the plain to the east. Near the river both streams occupy 10 m deep, wooded ravines. The Richmond Road and the RF&P Railroad cross this area from north to south (Figure 4A). The Richmond Road, located in part on a low, north-south ridge (O'Reilly, 1993), had an earthen parapet and ditch on each side (Franklin, *in* Luvaas & Nelson, 1994) and was hedged (O'Reilly, 1993). The RF&P Railroad had high earthen banks on both sides of the track (Figure 7) and a ditch on one side (O'Reilly, 1993). During the Civil War, the land was cultivated, '... much cut up by hedges and ditches.' (Franklin, *in* Luvaas & Nelson, 1994, p. 47), and there was a substantial stone fence about 800 m long just east of the RF&P embankment below Prospect Hill (O'Reilly, 1993). Trees were present only in the ravines and near the river (O'Reilly, 1993; Nolan, 1995). The heavily wooded western ridge in this area curves to the east and reaches its highest elevation, more than 60 m above the terrace, at Prospect Hill. The terrace wraps around the east end of Prospect Hill, beyond which is the marshy valley of Massaponax Creek.

Figure 4. A. Battlefield map (Modified from: Freeman, 1943). B. Digital Elevation Model of the battlefield with identical area and identical horizontal scale to those used for the map. Interpolated elevations in this area range from -14 m [lightest shade] to +77 m [darkest shade]. The Rappahannock River crosses the battlefield from top left to bottom right and is depicted using the lightest colours. The dark shaded area west of the river in the northwest quadrant is Marye's Heights. The light and medium shaded linear feature to the immediate east of this area is Kenmore Valley. The large amphitheatre is contained in the southwest quadrant

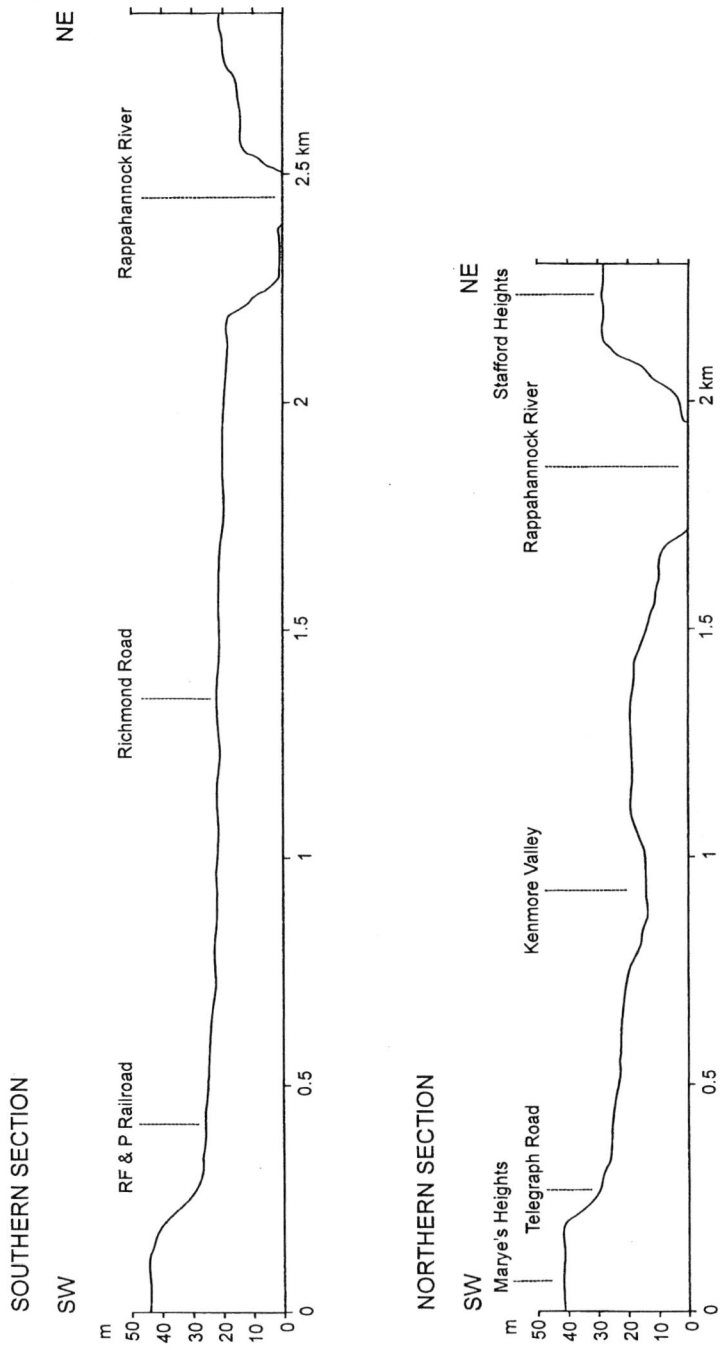

Figure 5. **A.** *SW-NE terrain profile from near Prospect Hill to Stafford Heights (Southern Section).* **B.** *SW-NE terrain profile from Marye's Heights to Stafford Heights (Northern Section)*

Stafford Heights, the eastern ridge 45 m above the river, provides a clear view of the narrow, discontinuous flood plains that are present on both banks of the Rappahannock in the northern part of the battlefield (Figs 8 & 9). The lower terrace, upon which the town of Fredericksburg is located, is some 15 m above the river. This terrace is about 150 m wide. The upper terrace, about 5 m above the lower terrace, is present only as remnants and forms a 150-200 m wide plateau west of town. It disappears beyond Hazel Run to the south (Figure 4A).

The ground on the west side of the upper terrace slopes down to the marshy Kenmore valley (Figures 5B & 10), which contained a millrace (the 'ditch'; Figure 4A) carrying waste from a canal to the north and west of town (Figure 4A). The millrace, approximately 4.5 m wide and between 1.5 and 1.8 m deep (Whan, 1961), was difficult to ford but could be crossed via the bridges that carried the main streets. A steep bluff approximately 6 m high forms the west side of the valley (Figure 10B). Marshy areas also occur along the millrace and canal to the north (Figure 11), as well as along Hazel Run to the south (Figure 4A). The embankment for an uncompleted railroad runs east-west across this slope just south of town.

Beyond and west of Kenmore valley, the land slopes gently upward to the base of Marye's Heights (the name for this part of the western ridge) with a grade of about 3% (Figures 5B, 10, & 12). In 1862 this area consisted of open fields with isolated houses and gardens (Stackpole, 1991). No woodland was present. Between the open fields and the steep wooded slope leading up to Marye's Heights, a distance of about 275 m, was Telegraph Road (Figure 4A). This road was sunk below ground level, partly by usage and partly by excavation, to even out the grade. Stone walls, about 500 m long and 1.2 m high (Kershaw, *in* Luvaas & Nelson, 1994), were present on both sides of the road (Figure 13). These walls were not visible from the grassy slope to the east (Figure 14).

4. Preparing for battle

Starting on the 15th November, Maj. Gen. Burnside, the Union commander, began his advance on Richmond by moving the first of his three 'Grand Divisions' under Maj. Gen. Edwin V. Sumner from the west near Warrenton to the Fredericksburg area (Whan, 1961; Figure 1). By the evening of 17th November, Sumner's forces had occupied Stafford Heights (Figure 4A). The other two 'Grand Divisions,' commanded by Maj. Gens William B. Franklin and Joseph Hooker, arrived in Fredericksburg on 19th and 20th November. Each 'Grand Division' comprised two corps. Franklin commanded the largest 'Grand Division' which had about 45,000 men; Hooker's 'Grand Division' had about 38,000 or 39,000 men; and Sumner's 'Grand Division' had about 30,000 men (Stackpole, 1991). Franklin was positioned on the left flank to the south; Sumner on the right flank to the north; and Hooker, in reserve, behind Sumner (Esposito, 1956).

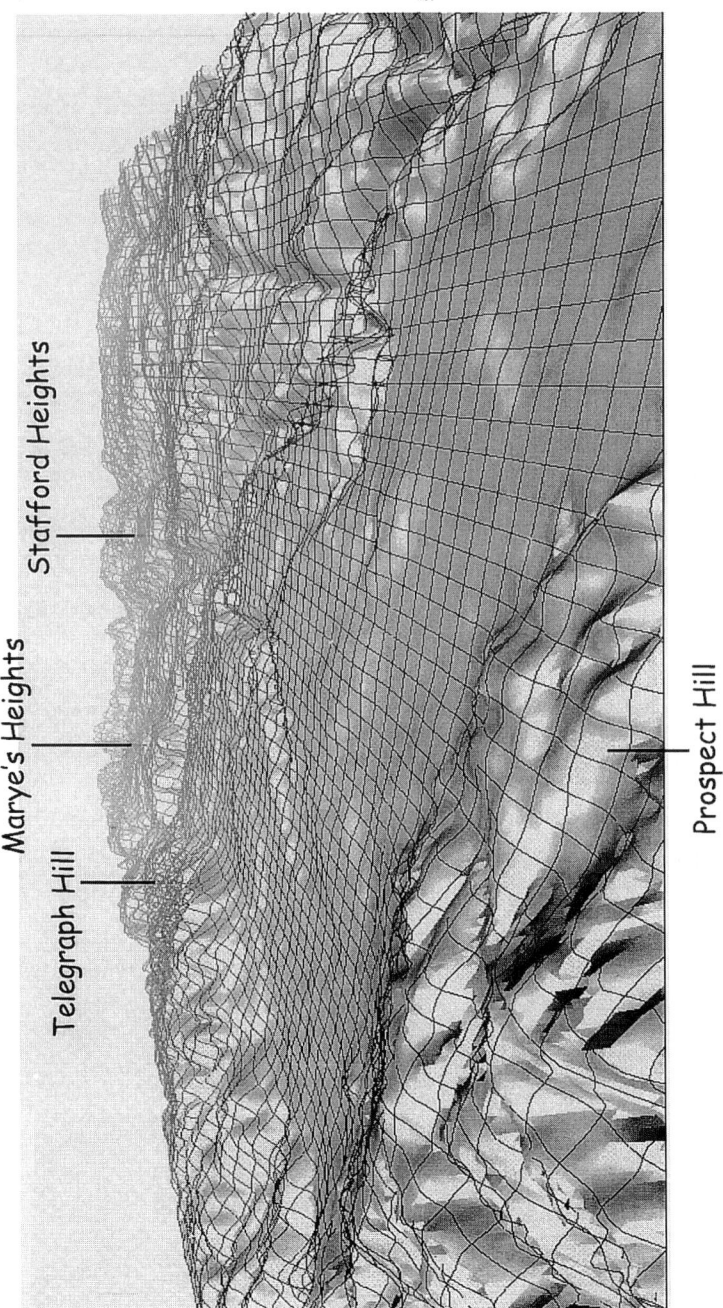

Figure 6. 3-D VRML shaded relief model looking northeast over Prospect Hill and across the amphitheatre that forms the southern sector of the battlefield. The grid spacing between each set of parallel lines is 100 m and vertical exaggeration is x8

Figure 7. The railroad embankment looking east at the southern end of Jackson's line. The trees on the left side comprise the boggy woodland penetrated by Meade's men the morning of 13th December. The stone pyramid in the background is a monument to that effort

Figure 8. Flood plain on west bank of the Rappahannock with terrace above. Location of upper pontoon bridges

In early November, General Robert E. Lee, the Confederate commander, had a small contingent of troops on Marye's Heights (Whan, 1961). Lee's main force was in western Virginia divided by the Blue Ridge: Lt. Gen. Thomas 'Stonewall' Jackson's corps was in the Shenandoah Valley west of the Blue Ridge and Lt. Gen. James Longstreet's corps was near Culpeper on the western side of the Virginia Piedmont (Figure 1). When Lee realised Burnside had dispatched the Army of the Potomac to Fredericksburg, he ordered Longstreet's corps to the Fredericksburg area. There were two reasons for this action: (1) to protect the route to Richmond; and (2) to safeguard the rich agricultural area near Fredericksburg, which could be useful in the future for provisioning his army (Stackpole, 1991). By 5th December, both Confederate corps were in position on the long, curving western ridge. Longstreet's 41,000 men occupied the northern part of the ridge west of town opposite Sumner. Jackson's 39,000 men occupied the southern end of the ridge opposite Franklin (Esposito, 1956), and his line extended southeast along the Rappahannock to Port Royal (Figure 1).

The Union plan of attack involved crossing the Rappahannock on pontoon bridges; the existing bridges had been destroyed during earlier military activity. The pontoons did not appear, however, until late November (Stackpole, 1991), due in part to inadequate transport and bad weather, but also to organisational delays caused by Maj. Gen. Halleck (General in Chief in Washington) who did not fully support Burnside's initiative (Stackpole, 1991; O'Reilly, 1993). Even after the pontoons had arrived, Burnside delayed: the positions for the river crossings, for example, were not selected until 10th December. Burnside was not willing to follow Sumner's suggestion to cross the river and occupy the lightly defended Marye's Heights prior to the arrival of Longstreet's entire corps (Stackpole, 1991). Burnside's forces could also have used upstream fords to cross the river prior to the arrival of the pontoons and of Jackson's corps; he was so advised by his subordinates (Whan, 1961). However, he preferred not to move small, isolated units across the river, since rising water could cut off troop units from both the main force and reinforcements (O'Reilly, 1993).

Lee expected Burnside to attack the right of his line, perhaps crossing some distance below Fredericksburg, so several divisions of Jackson's corps were positioned near Skinker's Neck and at Port Royal, 20 and 29 km, respectively, downstream. Jackson, who preferred to take the offence, did not like the Confederate position and predicted 'We will whip the enemy . . . but gain no fruits of victory.' (Freeman, 1934, p. 439) The terrain in front of Jackson's corps was considered more vulnerable to attack than that below Longstreet's: There was more room for Union troops to manoeuvre on this wide plain than on the narrower terraces and plain below Marye's Heights. Furthermore, Lee believed that Longstreet's position was more defendable than Jackson's to the south. Jackson's men were thus concentrated over a short distance, about 3.2 km, and Longstreet's men formed a thinner line about 9.6 km in length (O'Reilly, 1993). Jackson's men, however, were supported by cavalry, the southern part of the ridge was more heavily wooded than Marye's Heights,

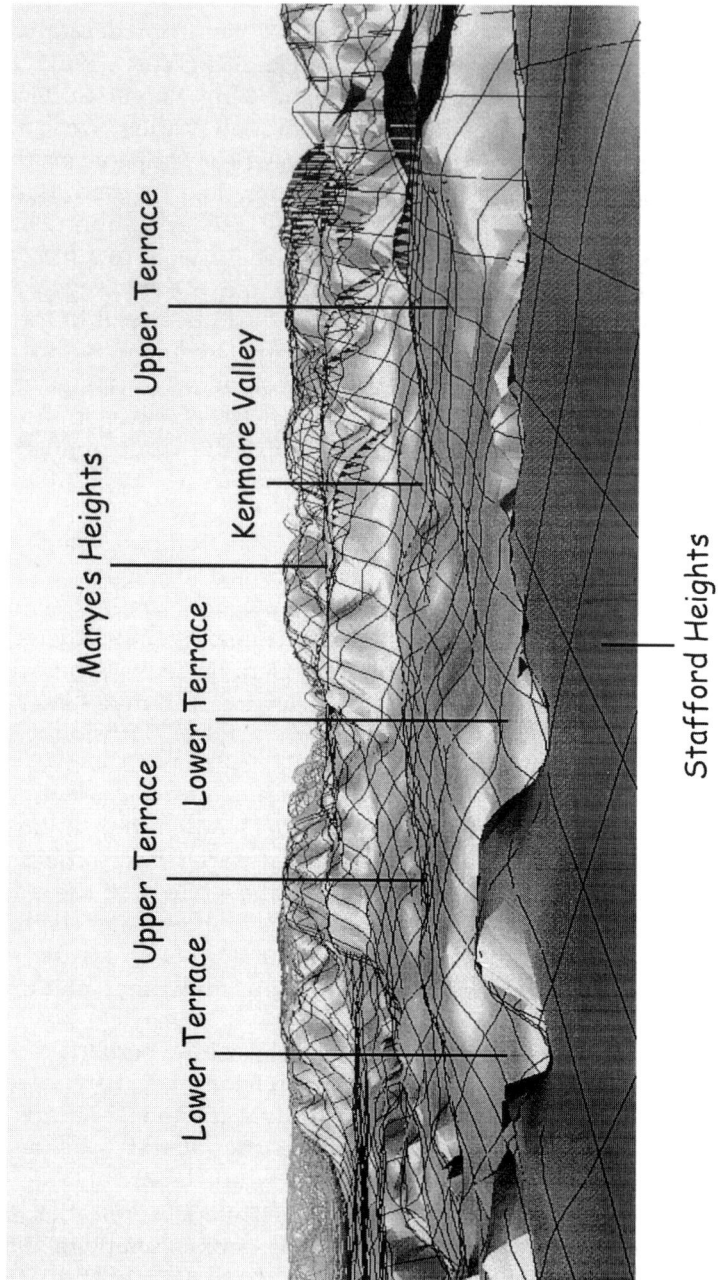

Figure 9. 3-D VRML shaded relief model looking southwest from Stafford Heights toward Marye's Heights. The Rappahannock River, which is out of sight, lies below the flat surface in the foreground. The flood plain on the west bank, also out of sight, merges upward into the lower terrace. Grid spacing between each set of parallel lines is 100 m, vertical exaggeration is x8

and although the plain was much wider there, than west of Fredericksburg, it contained more significant obstacles — the Richmond Road with its hedges, earthen parapets, and ditches; the RF&P track with its high earthen embankments (Figure 7); and two deep, wooded ravines. In addition, Lee's flanks were connected by a new road constructed along the back of the ridge for lateral communication, reinforcement, and movement of supplies.

5. The river crossing

Battle commenced on 10th December with the construction of pontoon bridges in three positions (Figure 4A). Two bridges were located toward the north end of town; one bridge, near the ferry landing at the south end; and three bridges, further south beyond the mouth of Deep Run (Figure 15). Bridge building equipment had been moved into position during the night, so Lee's forces were aware of what was to occur; the large number of bridges being built near town indicated that the attack would be made there rather than in the vicinity of Port Royal (Freeman, 1943). At this point Lee moved all but two of Jackson's divisions from Port Royal and Skinker's Neck to his main line on the ridge west of Fredericksburg.

On 11th December, construction of the upper and middle pontoon bridges (Figure 4A) was hampered by Confederate sharpshooters who were hiding in and around the buildings of the town. These riflemen made effective use of their surroundings, positioning themselves in basements and cellars of houses on the bank along the river front. Defensive rifle pits and connecting trenches were also constructed (Whan, 1961). Furthermore, these soldiers were positioned about 10 m above the river, so that when the bridges were completed, the Union infantry had to advance from the water's edge onto a narrow flood plain below the riflemen in an exposed position (Figure 16). Confederate artillery could not reach the Union forces, and the formidable Union artillery on Stafford Heights was unable to provide adequate covering fire for the engineers due to heavy fog that made accurate firing impossible (Burnside, *in* Luvaas & Nelson, 1994). The fog lifted some time after noon.

Nine attempts were made to complete the upper bridges (McLaws, 1956) before Union infantry volunteers began crossing in boats late that afternoon. Fierce hand-to-hand fighting along city streets and between buildings forced the Confederates back to the upper terrace[6]. This uphill withdrawal formed part of Lee's battle plan (Nolan, 1995) and took advantage of the bluff between the upper and lower terraces, such that after withdrawal to the upper terrace, Confederate troops again had excellent fields of fire downward into the advancing Union troops. The efforts of these riflemen allowed other Confederate units to get into position for the upcoming battle (McLaws, 1956). The upper bridges were completed under cover of darkness, and Sumner began deploying his troops within the town during the night of 11th December and the early morning of 12th December.

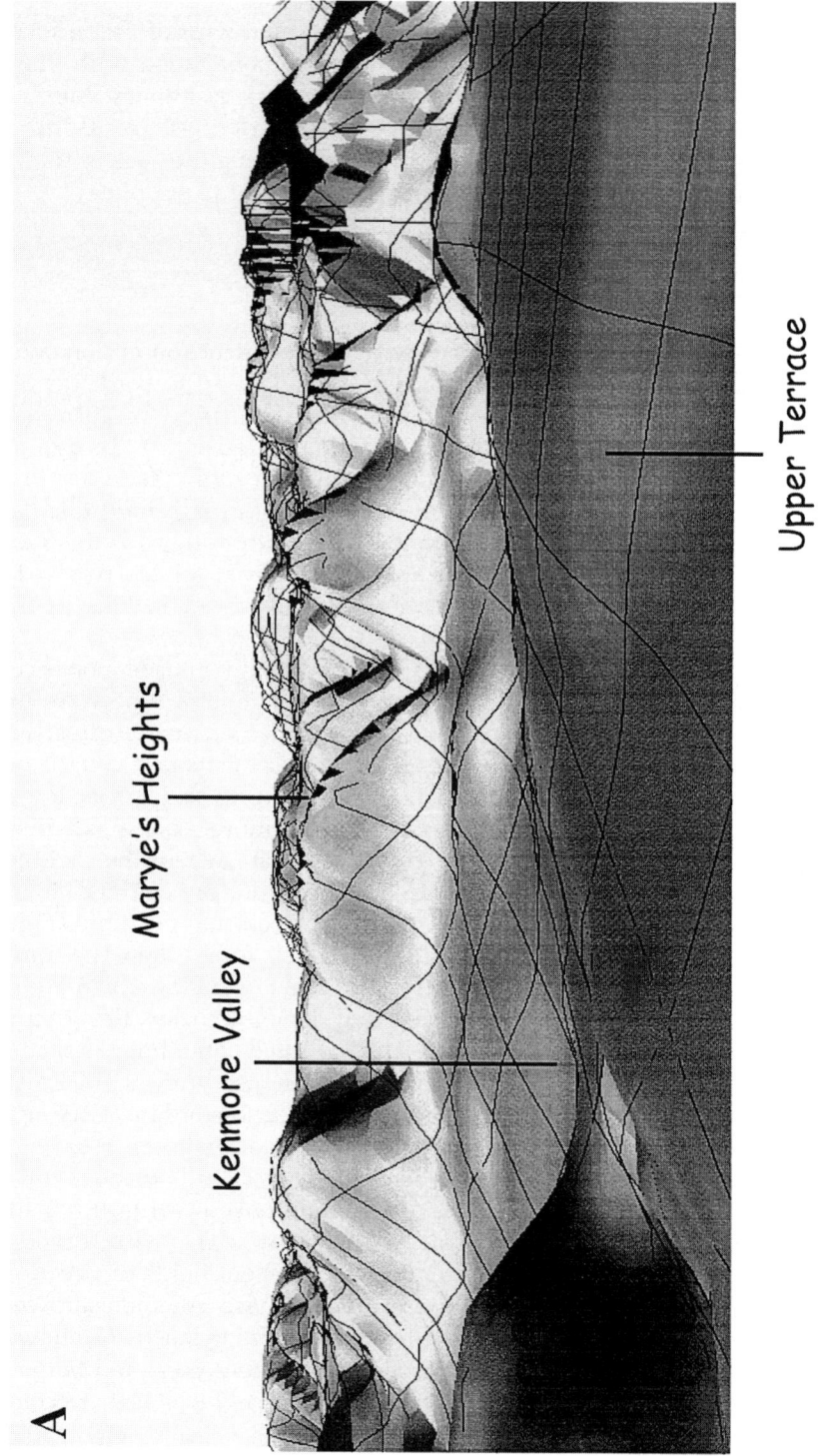

*Figure 10. **A**. 3-D VRML shaded relief model looking southwest from the upper terrace toward Marye's Heights. On the far side of Kenmore Valley is the open slope that leads to Marye's Heights. The grid spacing between each set of parallel lines is 100 m and vertical exaggeration is x8. **B**. View of Marye's Heights from the west side of the upper terrace. Kenmore Valley is in the foreground with the canal ditch and the bluff forming the west side of the valley on the far side. The grassy, open slope, leading to Marye's Heights, together with the stone wall (left) and Marye's Heights, are in the background. Source: National Archives Still Photo Unit, College Park, MD; Matthew Brady studio*

The bluff above the flood plain near the middle bridge is higher than in the town itself and thus easier to defend from ground attack (Figure 17). But there was no significant opposition to bridge building in this area because the terrain beyond the bluff, open fields sloping gently upward to the base of the western ridge, would have been exposed to the full firepower of Union artillery on the opposite bank. The position was initially occupied by Confederate riflemen, but these troops were unable to maintain significant resistance and soon withdrew to safer positions. This bridge was completed in late afternoon on 11th December.

There was also little opposition to bridge building on Burnside's left flank beyond Deep Run where the three lower bridges were constructed (Figure 4A);

again, there was no cover to protect Confederate troops from Union artillery on Stafford Heights. Two of these bridges were completed on schedule by late morning on 11th December, and the third bridge was completed overnight. Franklin's troops did not cross these lower bridges in force until 12th December, although one infantry brigade crossed on 11th December to protect the bridgehead (Whan, 1961). Thus, come nightfall on 12th December, Burnside's army was firmly ensconced upon the west bank, on the flood plain and lower terrace of the Rappahannock, and throughout the town itself.

Lee had moved his remaining two divisions from the south to the main line near Fredericksburg shortly after noon 11th December since bridge building activity had convinced him that Burnside's main attack would be against Marye's Heights and in the area immediately to the south of town (Whan, 1961). Lee, although familiar with the Fredericksburg terrain, made a general reconnaissance with Jackson on 12th December (Nolan, 1995), and the advantageous features of his position confirmed his decision to fight a defensive battle[7]. Throughout that day Confederate forces had fortified their artillery positions and dug rifle pits and connecting trenches, particularly in the south, for protection (Figure 18). These defences, however, remained limited and incomplete because of difficulty digging in the frozen ground. Lee chose a position to observe the upcoming battle on a high hill at the right of Longstreet's line such that he could superintend activities on both flanks.

6. The main battle

The Union attack commenced on the morning of 13th December. Burnside's orders to his 'Grand Division' commanders were that Franklin was to attack first using at least one division to seize, if possible, Prospect Hill, which formed the extreme right flank of Jackson's line. The rest of Franklin's troops, with several of Hooker's divisions in reserve, were then to proceed along the Richmond Road and capture the new road behind the ridge crest that connected the Confederate flanks. Once this had been accomplished and only then (Burnside, *in* Luvaas & Nelson, 1994), Sumner was to attack from the town, also with a division or more, '. . . with a view to seizing . . .' (Luvaas & Nelson, 1994, p. 46) Marye's Heights. Hooker's men were also the reserve for Sumner. Again, as on previous days, heavy fog covered the entire battlefield until mid-morning (McLaws, 1956), thereby delaying the initial attack.

Franklin's Attack
South of Fredericksburg, Franklin's attack against Jackson's corps began when the fog lifted at about 10:00. Confederate Lt. Gen. J.E.B. Stuart's Horse Artillery slowed the initial attack by providing enfillading fire ('Pelham,' Figure 4A), at first with only one gun hidden in a depression surrounded by cedars on the plain near Hamilton's Crossing (O'Reilly, 1993). Union artillery soon responded, however, forcing the battery to retire and temporarily

Figure 11. Marshy area north of Marye's Heights

Figure 12. Approach to Marye's Heights on Hanover Street looking west across Kenmore valley. (Source: National Archives Still Photo Unit, College Park, MD; Matthew Brady studio)

quieting the Confederate barrage. The Horse Artillery action, however, delayed the general Union artillery barrage and forced the Union build-up to be revealed to Jackson (O'Reilly, 1993).

The main Union attack was spearheaded by Maj. Gen. George Meade's division, with Brig. Gen. John Gibbon on his right. The initial advance crossed the Richmond Road and pushed the Confederate line, located behind the railroad embankment, back to the base of the western ridge. The well-concealed Confederate artillery withheld fire during the initial Union artillery barrage, and did not open fire until the advancing Union infantry were well within range, about 720 m from the Confederate guns (O'Reilly, 1993; Jackson, *in* Luvaas & Nelson, 1994). As a result, Franklin's generals did not know precisely where Confederate infantry and artillery units were positioned when the main attack began, and the damage caused by the Union artillery barrage was over-estimated (O'Reilly, 1993).

When Confederate artillery opened fire on the advancing Union infantry, Union artillery was pulled back into position and began firing; so although the location of the Confederate batteries could then be pinpointed, the Union batteries were now being raked by Confederate shelling from three sides (O'Reilly, 1993). Starting about 13:00, Union infantry advanced through stubble fields and across three 1.5 m deep ditches (O'Reilly, 1993). Confederate artillery fire funnelled Union infantry toward the centre of the line in the direction of dense woods that crossed the railroad. Like the artillery, Confederate infantry withheld fire until Union forces were well within musket range (Jackson, *in* Luvaas & Nelson, 1994). Jackson had assumed the woods were impenetrable, and although he had troops on either side and

Figure 13. Confederate dead in the sunken road behind stone wall in 1863. (Source: National Archives Still Photo Unit, College Park, MD; Matthew Brady studio)

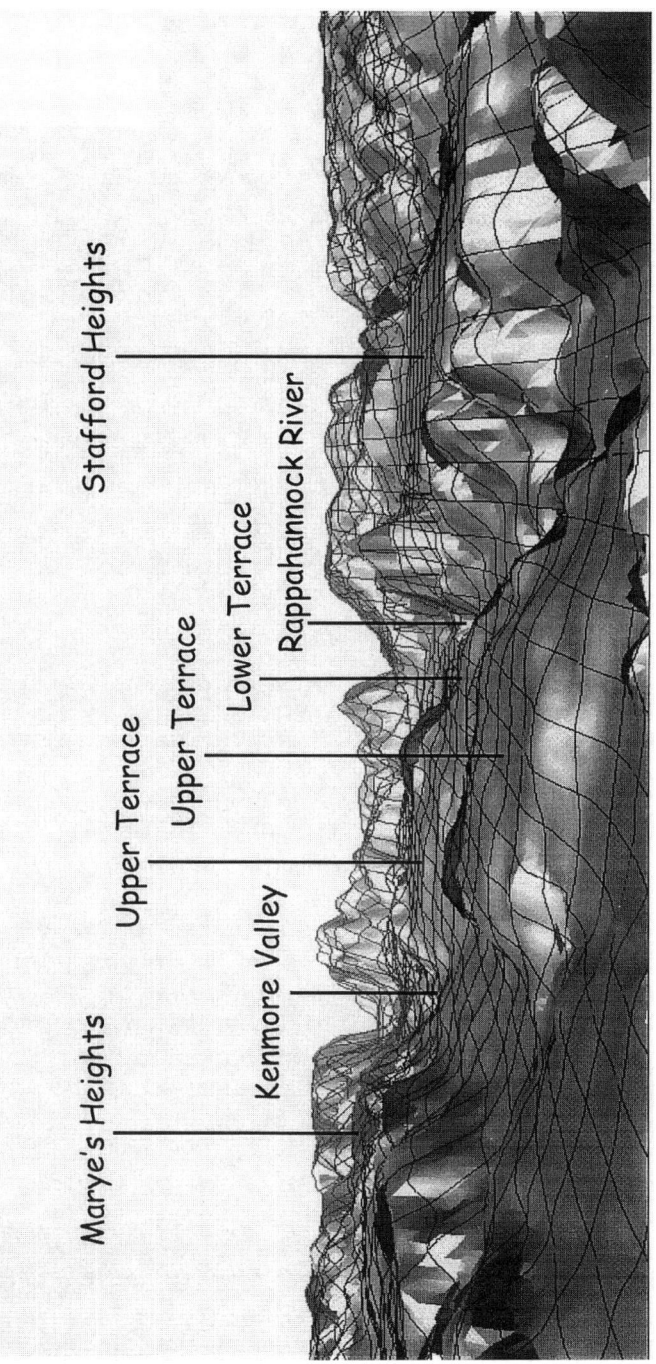

Figure 14. 3-D VRML shaded relief model of the northern sector of the battlefield viewed from the south. The grid spacing between each set of parallel lines is 100 m and vertical exaggeration is x8

Figure 15. Lower pontoon bridges looking west across the Rappahannock in 1863. (Source: National Archives Still Photo Unit, College Park, MD; Timothy H. O'Sullivan, photographer)

on the ridge above, none were positioned in the 550 m wide boggy woodland itself (Whan, 1961; 'The Gap', Figures 4A & 7). Reserves positioned behind this area on the ridge were thought adequate to hold the line if need be (Freeman, 1943). Meade was thus able to penetrate Confederate defences using the dense, swampy woods for protection and temporarily wreaked havoc upon the unsuspecting Confederates. Meade's troops also had the advantage of a rise in the ground (about 6 m high) on the north side of the woods that directed his forces into the unprotected gap (Freeman, 1943). The Union infantry were soon surrounded on three sides by Confederate infantry, however, and were unable to hold their advanced position (Smith, 1956). Meade's penetration went beyond the military road behind Jackson's lines and reached the ridge crest before his men were forced to retreat (O'Reilly, 1993). Gibbon's division made several attempts to push Confederate infantry back from the protection of the railway embankment, and finally succeeded on the third attempt (O'Reilly, 1993). Neither Meade nor Gibbon, however, was able to obtain reinforcements, unlike the front line Confederate units, and as a result, were in the end pushed back (O'Reilly, 1993). The Union infantry were also running low on ammunition.

BATTLE OF FREDERICKSBURG 85

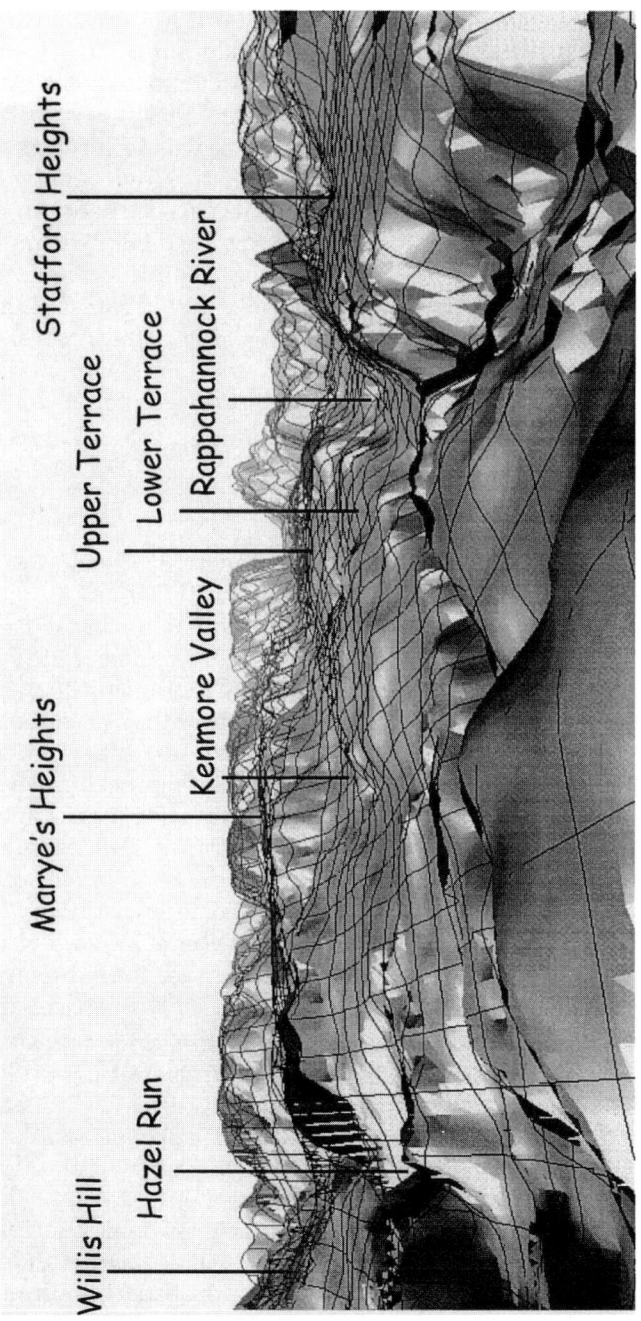

Figure 16. 3-D VRML shaded relief model of the northern sector of the battlefield, viewed from the southeast, looking from the east side of the Rappahannock River. The grid spacing between each set of parallel lines is 100 m and vertical exaggeration is x8

As Meade's and Gibbon's troops streamed back to the main Union line east of the Richmond Road chased by Confederate units, the Confederates re-established their position behind the railroad embankment. At about 14:30, Franklin received orders from Burnside to attack Jackson's position with his entire force (Whan, 1961). Franklin's new attack was to serve as a diversion to the heavy fighting west of town (Whan, 1961). However, by the time the shattered Union infantry had re-grouped, it was too late in the day for a major offensive, so Franklin was unable to provide relief for Sumner.

About twilight, when Jackson realised that Franklin would not renew his attack, he ordered his commanders to advance against Franklin's men. Delayed delivery of orders, the rapid onset of darkness, and the realisation that the Union artillery were as strong as ever, however, forced Jackson to countermand the order shortly after the advance began (O'Reilly, 1993; Jackson, *in* Luvaas & Nelson, 1994).

Sumner's Attack
Regardless of his earlier orders and without knowing the status of Franklin's attack in the south, Burnside ordered Sumner to attack the Confederate position west of Fredericksburg shortly after 10:30 (Whan, 1961). He assumed that Franklin had already attacked heavily and that Lee had weakened his left flank to support his right (Freeman, 1943).[8] The battle for Marye's Heights began with artillery barrages from both Union and Confederate batteries, but neither side's fire could reach the other's position (Stackpole, 1991). Confederate artillery was positioned so that both direct and enfillading fire from the north and south could be concentrated on the Union troops as they advanced (Whan, 1961). Longstreet expected the main attack to occur on his right in the vicinity of Telegraph Hill, which was Lee's headquarters (Luvaas & Nelson, 1994).

The main body of Union infantry marshalled in town about 550 m from the base of Marye's Heights, then descended into Kenmore valley. Crossing this valley and the canal ditch proved difficult because the water was deep, ice-covered, and surrounded by marshy ground; the only practical crossing points were the main road bridges (Couch, 1956; Figures 4A & 12). These routes funnelled the advancing units toward the sunken part of Telegraph Road (Figure 13) and placed considerable restriction on Union troop movements (Hancock & Willcox, *in* Luvaas & Nelson, 1994). Advance along the Richmond Road was not practicable because of enfillading fire from Telegraph and Willis Hills to the south (Whan, 1961; Figure 4), and the land to the north, also covered by enfillading fire, was too wet and boggy for efficient troop manoeuvres (Figure 11). Furthermore, Burnside had decided that the canal in this northern sector (Figure 4A) would prove too difficult an obstacle, so he avoided that area (City of Fredericksburg, 1997). The bluff on the west side of Kenmore valley, about 275 m from the base of Marye's Heights (Figure 10), gave some protection from Confederate artillery fire as the Union infantry advanced, but it provided only limited space for troop deployment (Couch, 1956). Once over the top of this bluff, only a few fences, a cluster of houses and

Figure 17. Flood plain and bluff below terrace at middle pontoon bridge crossing

gardens, and a small rise in the ground about 90 m from the base of the western ridge gave Union troops protection from Confederate fire (Couch, 1956).

As the Union skirmishers approached the base of the hill, less than 100 m from the Confederate lines, they were hit by a wall of musket fire and were forced to seek shelter behind a small rise in the ground (Whan, 1961; Kimball, *in* Luvaas & Nelson, 1994). The musket fire came from the sunken road at the base of Marye's Heights, which could not be seen by the advancing Union infantry (McLaws, 1956; Whan, 1961; Kershaw, *in* Luvaas & Nelson, 1994; Figures 13 & 14). Confederate infantry withheld their fire until the skirmishers were within effective range of their weapons.

Six waves of Union troops (Longstreet, 1956) were sent against the Confederate position, but none succeeded in dislodging Longstreet's men. Brigade after brigade was sent forward, throughout the day, only to be massacred by Confederate infantry hidden along the sunken road. The last assault was made after sundown (McLaws, 1956). Most units got no closer than 100 m from the wall, but several attacks, some of which were bayonet charges (Humphreys, *in* Luvaas & Nelson, 1994; Reardon, 1995), came within 25 m of the stone wall (Hancock, *in* Luvaas & Nelson, 1994).

Figure 18. **A.** *Entrenchments along Jackson's front line, December 1999.* **B.** *Earth works on Marye's Heights, December 1862. (Source: National Archives Still Photo Unit, College Park, MD; Matthew Brady studio)*

7. After the battle

Come nightfall, on 13th December, both Union and Confederate infantry in the southern sector of the battlefield occupied the lines where they had begun the day. Some 70% of Union casualties in the Battle of Fredericksburg occurred on the plain below Prospect Hill (O'Reilly, 1993), and although Franklin's troops occupied ground on the west bank of the Rappahannock, Jackson's defensive position was nevertheless secure.

West of town, Sumner's men were forced to spend that long, cold, December night lying on the muddy ground in front of the stone wall, surrounded by their dead and dying comrades. These men were unable to move for fear of being shot, even to get hardtack from their knapsacks, if they were lucky enough to have brought their knapsacks with them (Stackpole, 1991). The Confederate infantrymen, on the other hand, were ensconced in relative comfort behind their stone wall.

Lee did not pursue the demoralised Union Army for several reasons. First, night had fallen before the battle ended. Perhaps more important, however, he was averse to giving up his splendid defensive position because he fully expected Burnside to renew his attack the next day (Freeman, 1934; Lee, in Luvaas & Nelson, 1994). To this effect he ordered that his entire line be strengthened and fortified (Alexander, in Luvaas & Nelson, 1994). He was also well aware of the commanding presence of the Union artillery on Stafford Heights. When the fog lifted on the morning of 14th December Lee noted no movement and no changes in the deployment of Union troops on either flank. There were minor overnight changes, however, including the erection of additional street barricades, the loop-holing of house walls in Fredericksburg, and the construction of additional rifle pits and fortifications (Freeman, 1934). Heavy skirmishing continued throughout the day on both flanks. Confederate fire at one point was so heavy, particularly on Franklin's front, that Burnside expected a Confederate attack (Whan, 1961). Lee had also tried to entice Burnside into attacking the centre of the Confederate position by pulling units back from the front line in the fields on the plain leading to the wooded ridge, but to no avail (O'Reilly, 1993). Burnside in fact had planned an attack for 14th December, but was convinced by his subordinates that it was impracticable (O'Reilly 1993). So as the day progressed, Burnside pulled most of his troops back from the front line. Below Marye's Heights he maintained only a picket line on the bluff above the west side of Kenmore valley and the canal ditch (Whan, 1961). He also continued to fortify the town throughout the day. Thus both armies now maintained defensive postures.

On 15th December, Burnside requested a truce, which was granted, for further care of the wounded and burial of the dead. That night he moved his troops quietly back across the Rappahannock under cover of a violent storm of wind and rain and removed the pontoon bridges (Freeman, 1943). When Lee first viewed the battlefield on the morning of the 16th, he was surprised to see no Union troops on the western side of the river. He realised then that although the battle was a defensive victory, it had gained the Confederacy nothing. Freeman (1934, p. 473) states, quoting from Lee's letters: "'... [the enemy] suffered heavily as far as the battle went, but it did not go far enough to satisfy me.... the contest will now have to be renewed, but on what field I cannot say." He was deeply depressed that he had not been able to strike a decisive blow, "We had really accomplished nothing; we had not gained a foot of ground, and I knew the enemy could easily replace the men he had lost...'" Lee's conviction that Burnside would attack again the following day kept him from attacking and pursuing the demoralised Union Army, and, in line with Jackson's prediction, the sweetness of a decisive victory was lost.

8. Effective use of terrain

Terrain was a crucial factor in the progress and outcome of the Battle of Fredericksburg. The favourable terrain was under Confederate control, whereas the terrain over which the Union troops had to pass was generally disadvantageous. As the battle progressed on 13th December, the balance for the Confederates shifted from more vulnerable terrain to the south of Fredericksburg to superior terrain to the west of town. The southern terrain was heavily defended, but here the Union forces had some freedom of movement and protection from Confederate infantry and artillery fire. The terrain to the west of town was less heavily defended but Union forces in this area had no room to manoeuvre and little protection from Confederate fire. The main Union advantage in both sectors of the battlefield was the formidable artillery on Stafford Heights.

Lee determined from the outset that the low flood plain and river terraces upon which Fredericksburg is located, that were overlooked by the commanding Union artillery positions on Stafford Heights, were not propitious locations for the Army of Northern Virginia (Lee, in Luvaas & Nelson, 1994). Lee instead opted for a defensive posture on the high ground from which he had clear fields of fire for both infantry and artillery, and then waited for the Army of the Potomac to attack. Furthermore, he recognised the vulnerability of the Union Army in making the river crossing, and effectively employed riflemen to harass the Union bridge-building effort which he knew he could not stop, since the pontoon bridges were beyond the range of his artillery. Finally, he did not wish to shell the town and its inhabitants (Whan, 1961).

Jackson, commanding Lee's right, generally favoured offensive tactics, but his defensive use of terrain, coupled with counter-offensive tactics, were important factors in defeating the Union attack during the Battle of Prospect Hill. He used the dense woodland on the slopes and the railroad embankment to protect his men; his artillery was well placed; and, following the arrival of troops from Port Royal and Skinker's Neck, there were adequate reserves, an unusual situation for a Confederate commander. Unfortunately, frozen ground and a shortage of tools (Whan, 1961) had prevented the completion of

Confederate artillery pits and infantry shelters. On the other hand, Jackson was well aware of the vulnerability of his position with respect to Union attack, and although he began a counter attack against fleeing Union troops late in the afternoon of 13th December, he soon changed his mind and countermanded the order. He realised that Union artillery would decimate his men as they crossed the open, undulating plain below the ridge just as his artillery had annihilated Union troops during their attack on his position earlier that day (Jackson, *in* Luvaas & Nelson, 1994). Terrain considerations thus forced Jackson to maintain his defensive posture on the slopes of Prospect Hill and behind the railroad embankment.

Franklin's forces, on the Union left flank opposite Jackson, used the wooded ravine of Deep Run as protection during their initial advance. The western ridge reaches its lowest elevation in this area, and is fronted by an open plain, more than 3 km in width, that gave Union forces ample room to manoeuvre. The Richmond Road provided a route for deploying troops over the length of the battlefield. The railroad embankment and its ditches provided protection, when needed, although it also acted as an obstacle to forward movement during the initial stages of the battle (Meade, *in* Luvaas & Nelson, 1994). The surface of the plain, although open and sloping gently upward, was undulating, and covered with ditches and hedges, which likewise gave some protection to advancing infantry. Most important, the boggy area of woodland beginning near the railroad embankment allowed Meade's successful penetration of Jackson's line. Finally, the exposed nature of this terrain gave Union artillery clear fields of fire against the defensive Confederate position, although Union artillerymen were unable to locate the Confederate batteries until the latter opened fire because of their position on the densely wooded slopes of the ridge (O'Reilly, 1993).

Unlike Jackson, Longstreet, commanding Lee's right flank, was by his very nature a defensive general (Freeman, 1943). He defended the greater length of Lee's line with a lower troop density than Jackson, but had full confidence in his position. After the third assault on Marye's Heights by Union infantry, Lee expressed concern about the position, and Longstreet replied (Longstreet, 1956, p. 81): 'General, ... if you put every man now on the other side of the Potomac on that field to approach me over the same line, and give me plenty of ammunition, I will kill them all before they reach my line. Look to your right; you are in some danger there, but not on my line.' Longstreet's men were positioned on the wooded slopes of the ridge and in the sunken road, both of which afforded excellent protection. In addition to the stone walls and sunken road, Confederate troops had hastily built earthworks[9], and also used existing obstacles such as stone fences and marshy ground, to the north, to form their front lines.

Sumner, on the other hand, faced numerous obstacles in his advance against Longstreet's position. Troop deployment was awkward because of buildings, and avenues of approach were restricted due to the narrowness of the Fredericksburg streets (Couch, *in* Luvaas & Nelson, 1994). Movement to the front line was also restricted by the canal ditch, which could only be crossed on road bridges, and by marshy ground situated to the north and south of Marye's Heights. This forced advance by brigades, rather than divisions, and funnelled the Union attack toward the strongest Confederate positions. Once across this obstacle, there was also little space for further troop deployments below the bluff on the western side of Kenmore valley. The men then had to cross approximately 275 m of relatively smooth, upward sloping ground, with numerous 'substantial' fences (Hancock, *in* Luvaas & Nelson, 1994), before reaching the Confederate forces situated behind the stone wall at the base of Marye's Heights. Furthermore, good positions for Union batteries were lacking. The attacking Union troops thus had almost no protection, as evidenced by the casualty figures for this sector of the battlefield: Union casualties, 9,000; Confederate casualties, many of which were minor, 1,676 (Freeman, 1934).

The major — and perhaps only — advantage held by Burnside's Army of the Potomac during the Battle of Fredericksburg was the Union artillery with its commanding position on Stafford Heights. The ineffectiveness of Union artillery during construction of the upper pontoon bridges was compensated by strong performance during construction of the middle and lower pontoon bridges. The terrain in these two areas allowed effective use to be made of artillery positioned on high ground and thus facilitated Union dominance in this stage of the battle. Furthermore, the mere presence of Union artillery — superior in strength, calibre, and reliability (O'Reilly, 1993) — was a major influence on Confederate decisions throughout the course of the battle. Jackson delayed his response to the initial Union artillery barrage because he knew that the light Confederate guns were no match for the heavy Union artillery. Jackson thus saved his guns for action against Union infantry (O'Reilly, 1993). This powerful influence is further exemplified by Jackson countermanding the order to pursue retreating Union troops, late on 13th December. In effect, the Stafford Heights artillery positions were Burnside's 'insurance policy,' ensuring that even if defeated in battle, his army would still survive.

9. Conclusions

Lee's Army of Northern Virginia occupied a defensive position on high ground, and Burnside's Army of the Potomac was forced to attack uphill over the low ground. In addition, Union troops had numerous obstacles to negotiate — not least of which was the Rappahannock River. On Burnside's left flank, these included, in addition to the river; the wide, open, smooth plain; two deep ravines; numerous hedges and ditches; the railroad embankment; the Richmond Road; and the curved shape of the ridge that gave Confederate artillery good positions for enfilading fire. On Burnside's right flank the most important obstacles, in addition to the river and the town itself, were Kenmore valley and the canal ditch; limited space for troop deployments; the smooth, open ground below Marye's Heights; numerous fences, houses and gardens; marshy areas both north and south of the main point of attack; the shape of the ridge that allowed Confederate enfilading fire; and most important, the sunken road and its stone walls.

Stackpole contends that the outcome of the Battle of Fredericksburg was linked to Lee's '. . . keen sense of terrain appreciation.' (Stackpole, 1991, p. 172). Lee took a defensive position because '. . . the natural features of the ground were made to order for the purpose . . .' (Stackpole, 1991, p. 271). Other sources and our analysis of the battlefield using modern state-of-the-art-geographical data support this opinion. Lee's judgement, confidence, and effective use of terrain, coupled with Burnside's incompetence, indecision, and limited knowledge of the battlefield upon which he proposed to fight, appear to have been the instrumental factors that controlled the different stages of this conflict and thus contributed in no small part towards the final outcome. It is also clear that analysis of historical battlefield terrains can provide useful lessons for today's commanders in the field of battle.[10]

Acknowledgements

We wish to thank Noel Harrison, Historian, Fredericksburg-Spotsylvania National Military Park, Fredericksburg, Virginia, and Professor Walter E. Pittman, The University of Northern Alabama, for their reviews and helpful suggestions. G.P. Dowling, University of Birmingham, Hilary Foxwell, University of Greenwich, and Chris Lewis, University of Nottingham, helped to prepare the illustrations.

References

City of Fredericksburg. 1997. *Historic Resources along the Rappahannock and Rapidan Rivers*. City of Fredericksburg, Fredericksburg.

Couch, D.N. 1956. Sumner's 'Right Grand Division.' *In:* Johnson, R.U. & Buel, C.C. (Eds) *Battles and Leaders of the Civil War, Vol. III, Retreat from Gettysburg*. Castle Books, New York, 105-120.

Davis, W.C. 1994. *A Concise History of the Civil War*. Eastern National, Civil War Series.

Esposito, Col. V.J. 1959. *The West Point Atlas of American Wars, Vol. I, 1689-1900*. Frederick A. Praeger Publishers, New York.

Freeman, D.S. 1934. *R.E. Lee, A Biography, Vol. II*. Charles Scribner's Sons, New York.

Freeman, D.S. 1943. *Lee's Lieutenants, Vol. II, Cedar Mountain to Chancellorsville*. Charles Scribner's Sons, New York.

Longstreet, J. 1956. The Battle of Fredericksburg. *In:* Johnson, R.U. & Buel, C.C. (Eds), *Battles and Leaders of the Civil War, Vol. III, Retreat from Gettysburg*. Castle Books, New York, 70-85.

Luvaas, J. & Nelson, H.W. (Eds) 1994. *Guide to the Battles of Chancellorsville and Fredericksburg*. University of Kansas Press, Lawrence.

McLaws, L. 1956. The Confederate Left at Fredericksburg. *In:* Johnson, R.U. & Buel, C.C. (Eds), *Battles and Leaders of the Civil War, Vol. III, Retreat from Gettysburg*. Castle Books, New York, 86-94.

Mixon, R.B., Pavlides, L., Powars, D.S., Weems, R.E., Schindler, S.J., & Newall, W.L. 1997. Geology of the Fredericksburg 30'x60' Quadrangle, Northeastern Virginia and Southern Maryland. *In:* Horton, J.W. & Cleaves, E.T. (Eds), *Forum on Geologic Mapping Applications in the Washington-Baltimore Urban Area*. U.S. Geological Survey Circular 1148.

Nolan, A.T. 1995. Confederate leadership at Fredericksburg. *In:* Gallagher, G.W. (Ed.) *The Fredericksburg Campaign: Decision on the Rappahannock*. The University of North Carolina Press, Chapel Hill, 26-47.

O' Reilly, F.A. 1994. *'Stonewall' Jackson at Fredericksburg, the Battle of Prospect Hill, December 13, 1862*, 2nd Edition. H.E. Howard, Inc., Lynchburg.

Reardon, C. 1995. The forlorn hope: Brig. Gen. Andrew A. Humphreys's Pennsylvania Division at Fredericksburg. *In:* Gallagher, G.W. (Ed.), *The Fredericksburg Campaign: Decision on the Rappahannock*. The University of North Carolina Press, Chapel Hill, 80-112.

Smith, W.F. 1956. Franklin's 'Left Grand Division.' *In:* Johnson, R.U. & Buel, C.C. (Eds), *Battles and Leaders of the Civil War, Vol. III, Retreat from Gettysburg*. Castle Books, New York, 128-138.

Stackpole, E.J. 1991. *The Fredericksburg Campaign*, 2nd Edition. Stackpole Books, Mechanicsburg.

Whan, V.E., Jr. 1961. *Fiasco at Fredericksburg*. Olde Soldiers Books, Inc., Gaithersburg.

Notes

1. The Digital Terrain Model (DTM) was built using the Virtual Reality Modelling Language (VRML) and a full description of the processes that were used to generate these images is provided in the Appendix.
2. Southerners named battles after nearby towns, whereas Northerners typically referred to battles in terms of topographic features, usually the names of streams, e.g., Manassas vs. Bull Run, Sharpsburg vs. Antietam Creek. Southern terminology is used throughout this paper.
3. 23,000 casualties reported in a single day!
4. Sharpsburg was a tactical victory for the Union in that they stopped Gen. Lee's advance into Union territory. However, the Union Army failed to disrupt the Southern retreat, so the victory was not decisive.
5. This hill was the headquarters of Confederate General Robert E. Lee during the Battle of Fredericksburg and is now known as Lee's Hill.
6. This was one of the few examples of urban warfare during the Civil War.

7. Burnside, who was unfamiliar with the terrain, made no comparable personal reconnaissance, and, with respect to the terrain his troops would have to cross, relied on information supplied by others and from his reconnaissance balloons (Whan, 1961; Stackpole, 1991; O'Reilly, 1993).
8. This statement is interesting because Burnside had reconnaissance balloons in the air on Stafford Heights near his headquarters at the time which should have been able to determine whether Lee had in fact made these troop movements. The battlefield was no longer obscured by fog and artillery fire had been limited up to this time on his right flank, so smoke should not have obscured the battlefield either.
9. According to Luvaas & Nelson (1994) the utility of 'hasty entrenchments' to protect troops during battle was first proven in the Battle of Fredericksburg. Prior to this battle, entrenchments were used primarily to protect established camps or fortifications and were considered part of the permanent defences.
10. The national battlefield parks were initially set up and administered by the US Department of Defense for the education and training of young officer candidates. Even today, Civil War battlefields, including the Fredericksburg battlefield, are still used for such training.

Appendix: Construction of a Virtual Battlefield

The following text describes the operations and procedures that were used to build our virtual battlefield (Figures 6, 9, 10B, 14 & 16). The main objective was to facilitate a deeper understanding and appreciation with respect to battlefield strategies and the effective use of terrain. It must be stressed that this application of virtual world technologies was in consequence not intended to produce an exact replication of the real world situation; the digital model was instead designed to provide a visual experience that encapsulated and characterised selected features and important aspects of natural terrain features considered pertinent to the Battle of Fredericksburg.

There were twelve main stages in the construction and development process:

1. 1:24,000 7.5 Minute Series digital hypsographic [contour] and hydrographic [river] data were selected and downloaded from the US Geological Survey EROS Data Centre in SDTS (Spatial Data Transfer Standard) format. The battlefield spanned two adjacent map sheets: [1] Fredericksburg Quadrangle and [2] Guinea Quadrangle. Data download site: http://earthexplorer.usgs.gov
2. The downloaded vector data sets were imported into ArcInfo using the public domain software tool 'sdts2cov.aml' that is available from: Software download site: http://mcmcweb.er.usgs.gov/sdts/public_domain.html. The two adjacent map sheets had different projections and

thus required transformation to a common standard [UTM18 map projection; NAD83 datum; GRS1980 spheroid].
3. The 'arcpoint' command was used to derive spot height values from the transformed hypsographic data for each map sheet. Each node on each arc was thus converted into a spot height; the combined set of point data from both map sheets provided 430,927 observations for use in subsequent interpolation operations.
4. The vector data and point data were then exported from ArcInfo and imported into ArcView using the E00 transfer format.
5. The digital model was required to match the battlefield map in Freeman (1943). The original drawing was first scanned and imported as an image into ArcView. This scanned image was then rectified and aligned to our common projection using ground control points, at identifiable nodes, on the river network coverage. The image fitting operation was undertaken using the ImageWarp 2.0 Extension for ArcView: Software download site: http://gis.esri.com/arcscripts/scripts.cfm
6. The enclosing rectangle that determined the full extent of the battlefield model was then established and is defined using the following corner point notation: 282000, 4245000 : 289000, 4235000 which comprises a total area of 70 km^2 [10 km north - south; 7 km east-west]
7. Excel was used to construct two ASCII vector data sets for import into ArcInfo. The first file contained a set of co-ordinates to build a border for the battlefield. The second file contained a set of co-ordinates to build a 100 m x 100 m grid that would criss-cross the bordered area and provide a horizontal surface scale for the virtual world. These vector coverages were generated in ArcInfo and imported into ArcView using the E00 format.
8. The Spatial Analyst Extension in ArcView was used to construct an interpolated surface in the form of a digital elevation model that had a 25 m x 25 m square grid [280 columns; 400 rows; 112,000 cells]. To reduce computational overheads, a sub-set of the spot height point data was selected to build the surface model, which included all point data within the border plus all points within a 0.5 km buffer around it. This data had a reasonable spatial distribution, but the distance units were in metres, and the elevation units were still in feet. There was no ultimate conversion to a common set of measurement units and such action is thus equivalent in final modelling terms to a first vertical exaggeration factor of 3.28. The spline with tension interpolator [tension weight = 0.5; number of points = 12] was used to construct a surface model. This is a general-purpose method that fits a minimum-curvature surface through the input points. It is like bending a sheet of rubber to pass through the points, while minimising the total curvature of the surface. The method fits a mathematical function to a specified number of nearest input points while passing through the sample points. It produces a smooth surface and is best for surfaces that exhibit gentle variation such as elevation, water table heights, or pollution concentrations. It is not appropriate if there are

large changes in the surface within a short horizontal distance because it can overshoot the estimated values. Tension tunes the stiffness of the surface according to the character of the modelled phenomenon. The weight parameter defines the weight of tension and the number of points parameter identifies the number of points per region that is used for local approximation.
9. ArcView 3D Analyst was used to construct a three-dimensional view of the interpolated surface and to further emphasise and accentuate surface variation; for terrain analysis purposes, a second vertical exaggeration factor of 2.5 was applied. This same procedure was then applied to the 100 m x 100 m grid. However, in order to provide a effective scale and for navigation purposes, the grid required an additional offset factor of 25 units to ensure that all points and vectors could be seen above the interpolated surface.
10. The complete view was then exported into VRML 2.0. The different coverages formed different objects in the virtual world that could be viewed and navigated, using a standard web browser, that had the required plug-in, e.g., Cosmo Player: Software download site: http://www.cai.com/cosmo/
11. VrmlPad was then used to adjust and recode the model. The elevation surface was changed from its default set of colours into diffuse [normal colour = R40:G40:B40] and specular [reflective highlights = R80:G80:B80] tones of brightness. Directional lighting effects were also introduced to provide enhanced intensities of shaded relief.
12. The gap between the surface and the grid was then reduced to provide an acceptable compromise for interactive surface wanderings. Fog was also introduced to make distant features appear more realistic [exponential increase; starting distance 15,000 m].

Judy Ehlen
US Army Topographic Engineering Center
7701 Telegraphy Road
Alexandria, VA 22315-3864

Robert J. Abrahart
School of Geography
University of Nottingham
Nottingham, NG7 2RD

Tullahoma: Terrain and Tactics in the American Civil War

Walter Earl Pittman

> **ABSTRACT:** In 1863 the Union commander of the Army of the Cumberland, William S. Rosecrans successfully manoeuvred the weaker Confederate Army of Gen. Braxton B. Bragg out of a strong defensive position in Middle Tennessee with little loss of life. Taking advantage of the unique mountainous topography and Bragg's dependence on a single rail line Rosecrans was able to put his forces astride Bragg's communications. Following after the Confederates and trying to repeat the same trick south of the Tennessee River Rosecrans allowed his corps to become separated in the mountains where they were decisively defeated at the Battle of Chickamauga. Rosecrans created the most effective mapping system developed during the Civil War.

1. Introduction

There is probably no field of battle in the American Civil War where terrain did not play a central role in the outcome. Few military operations, however, were as affected by the constraints of topography as the Tullahoma Campaign of 1863 in Tennessee which set the stage for the climactic battle of Chickamauga. This paper examines its impact.

By the summer of 1863, the Civil War had reached a decisive phase. In the East, General Robert E. Lee's daring invasion of the North ended on the 3rd July at Gettysburg. In the Mississippi Valley, Ulysses S. Grant split the Confederacy in half with the capture of Vicksburg on 4th July, a fatal blow to Southern hopes. However, in the West[1] the second largest army of each belligerent remained passive for much of the year.

2. The Tullahoma battlefield

The Union Army of the Cumberland rested on Murfreesboro (Figure 1) and Nashville, with its long logistics trail stretching back through Tennessee and Kentucky to the Ohio River at Louisville, Kentucky. Both Murfreesboro and Nashville lie in the Nashville Basin, an Ordovician limestone-floored region stretching 450 km (SSW to NNE) from northern Alabama to Indiana (AAPG, 1970). It is about 50 km wide in Middle Tennessee and is generally flat to gently rolling. The land is fertile and prosperous farms were relatively numerous in 1863. The Basin is actually the eroded surface expression of a

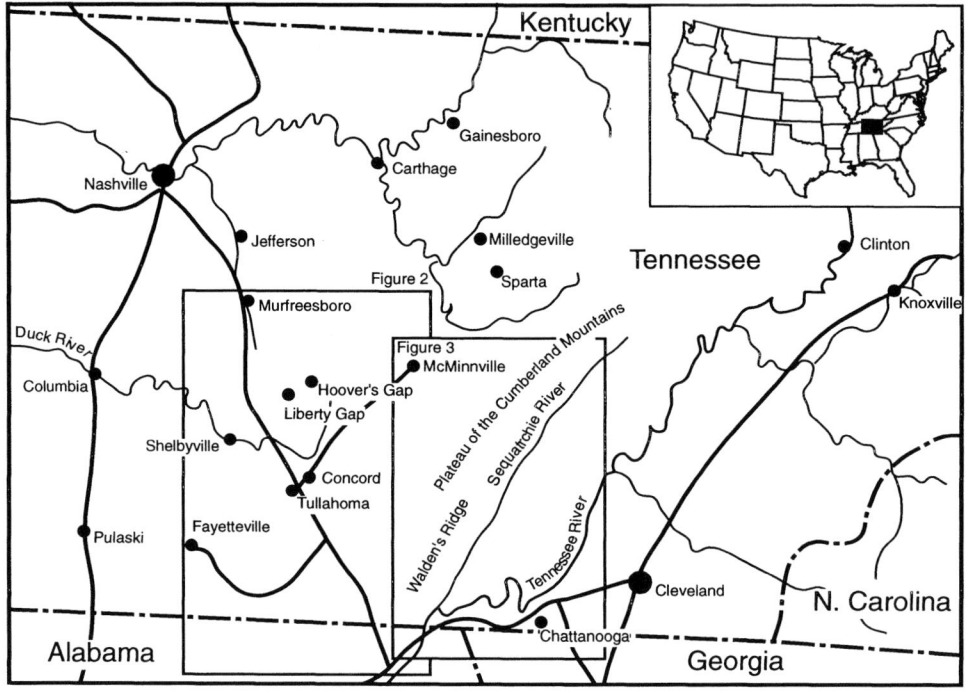

Figure 1. Location map showing the main places mentioned in the text

huge anticlinal structure. The eastern edge of the basin is bounded by the Highland Rim, where more resistant Ordovician to Pennsylvanian (Upper Carboniferous) sedimentary rocks overlay and form cliffs rising 70 to 200 m above the Nashville Basin. Locally, the edge of the rim is eroded and highly dissected. The top level of the rim is covered with a poor thin siliceous soil that is so unproductive that the region is known locally as the 'Barrens'. The soil readily turns to a viscous mud when wet, and it rained seventeen consecutive days during the Tullahoma Campaign (Brown, 1963).

Tullahoma itself stands on the Highland Rim along the line of the main railroad which ran, and still runs, from Nashville to Chattanooga. To the north of Tullahoma, the plateau is cut by the broad basin of the westward flowing Duck River. North of the river lie the towns of Wart race and Shelbyville which respectively lie on the railroad and on the principal highway route south. North of the river, a spur of the rim juts westward. There are four suitable passageways for military forces through this rugged ridge. They are from east to west, Hoover's Gap, Liberty Gap, Bellbuckle Gap

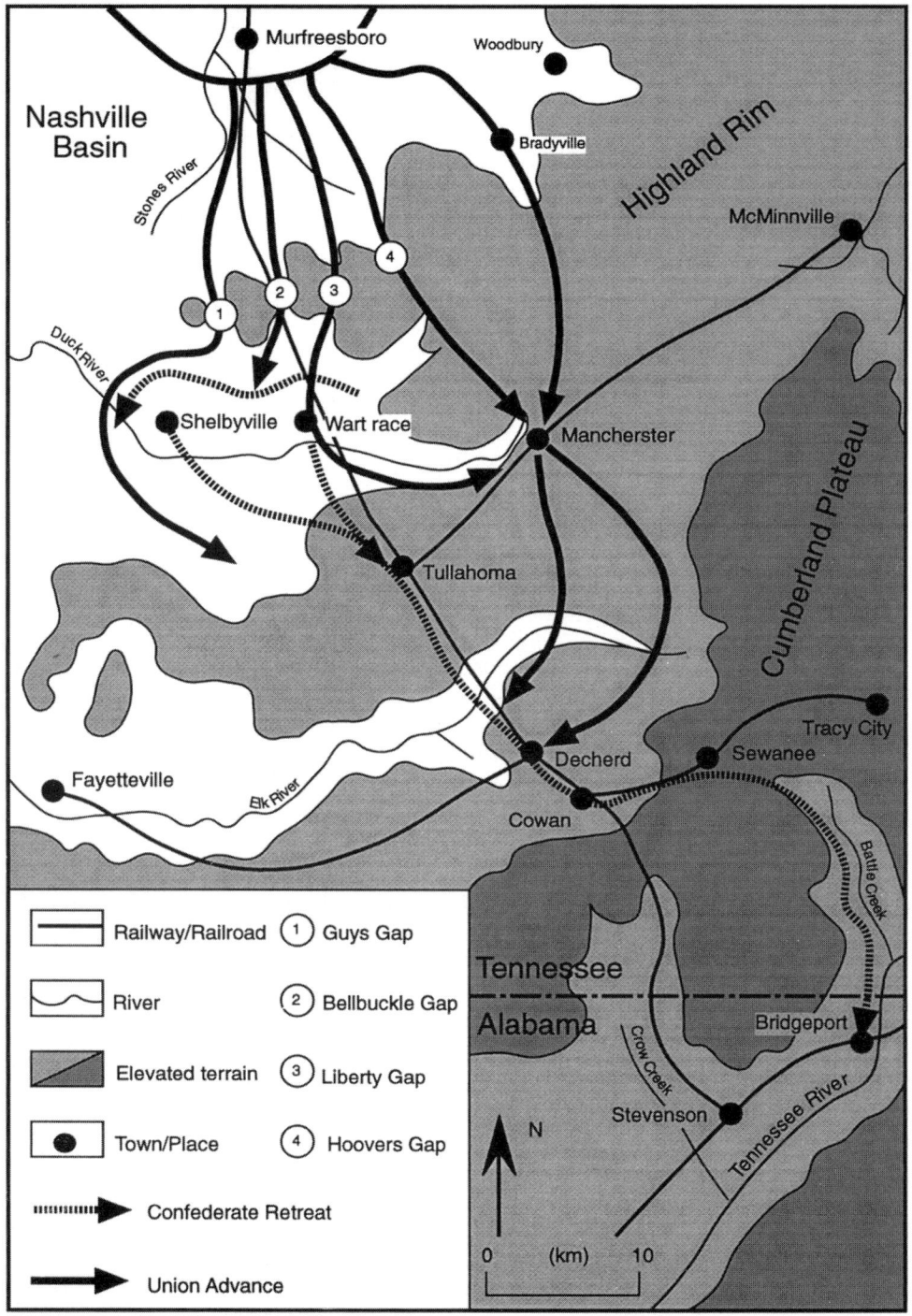

Figure 2. Tullahoma battlefield. (Modified from: Brown 1963)

(which carries the railroad) and Guy's Gap (Figure 2) through which passes the main road from Murfreesboro to Shelbyville (USGS, 1985).

To the east of the Highland Rim rises the Cumberland Plateau with its high, broad, folded ridges stretching for scores of kilometres NNE to SSW. The plateau rises abruptly 300-400 m from the Rim and offers few passageways. The railroad and highway passed from Decherd below Tullahoma (Figure 2) and followed the Boiling Fork Creek to the edge of the 'Mountain' as it was called locally, where the railroad passed though it in a long tunnel as the road climbed steeply to Sewanee (University). It then passed over the broad, flat topped fold mountain and down into the Sequatchie Valley and thence to Chattanooga along the Tennessee River (USGS, 1988).

Like most of the South during the Civil War, the region was thinly populated, heavily forested and roads were few and primitive. Actually, the Tullahoma to Chattanooga region was one of the most desolate in the South, mainly because the poor soil discouraged settlement. Here lurked the Confederate Army of Tennessee, outnumbered, starving, demoralised, and ill-equipped, but still a dangerous foe.

3. General W.S. Rosecrans and the Union Army

The Union army that faced these obstacles was the 84,000 men Army of the Cumberland commanded by General William S. Rosecrans. Born in comfortable circumstances in Kingston, Ohio, in 1819 and entering West Point when he was eighteen, Rosecrans graduated near the top of his class in 1842 and was assigned to the Corps of Engineers, as top graduates customarily were. For the next few years he worked as a military engineer, designing and building military facilities and fortifications and teaching at West Point until ill health caused him to resign his commission in 1854. He turned his attention to coal mining in the Kanawha Valley of western Virginia working as a geologist for an English-American mining company developing new coal mines. He was a self-taught geologist, which was customary in America at the time. Next, Rosecrans founded a company to construct a canal to service the new mines and then became involved in producing illuminating oil from coal (coal oil or kerosene) in Cincinnati, Ohio. He was near fatally burned by an explosion of an experimental apparatus and remained bed-ridden for eighteen months. He recovered and was in the process of rebuilding his failing business when the Civil War broke out. In wartime portraits, scars of his burns can still be seen on his face (Fitch, 1863; Lamers, 1961).

Rosecrans quickly returned to the army when the war broke out. Initially, he commanded the 23rd Ohio Infantry Regiment, which had in its ranks two future US Presidents (Rutherford B. Hayes & William McKinley). Later, Rosecrans' Chief of Staff would be James Garfield, another future President. Quickly promoted to Brigadier General, Roscrans took part in the West Virginia campaign of 1861-1862 (Howe, 1900; Warner, 1964).

Fighting under General George B. MacClellan, Rosecrans distinguished himself in the fighting in western Virginia, which centred in the Kanawha Valley where he had worked for many years and where he was intimately familiar with the topography. Rosecrans largely devised the plan to turn the Confederate defences, and he himself led the Union forces at the Battle of Rich's Mountain, which surprised the Confederate forces and resulted in their expulsion from the area. On MacClellan's departure, Rosecrans was promoted to command the Department of Western Virginia, on 14th February 1862 (O.R, Volume II, pp. 215-219; Cox, 1900).

In May, 1862, Rosecrans took command of the left wing of the Army of Mississippi in its slow advance on Corinth, Mississippi. In June, he was placed in full command of the Army and fought the indecisive battle of Iuka, in June 1862. Promoted to Major General in September, he was given command of the Army of the Cumberland, the second largest of the Union's armies (Warner, 1964; OR., Volume XVI(1), p. 4).

Rosecrans' tenure as an Army commander was short and unhappy. Personally, he was respected. He had a reputation in the Army for exceptional bravery. Rosecrans was a devout convert to Catholicism who kept a priest friend with him at headquarters. His brother, also a priest, later became the Bishop of Columbus (Byrne, 1906). Normally a 'modest, refined, polite and affable' gentleman, Rosecrans, under stress, sometimes became so overexcited that he was reduced to livid, stammering and ineffective incoherence. Most of his men loved him but not many of his officers, (Fitch, 1863; Shanks, 1866; Cox, 1900; Palmer, 1901). These staff officers were slaves to Rosecrans' eccentric personal habits and his irritability. He normally slept late and after breakfast and a lengthy ride through his Army's encampment, he would begin the day's work at headquarters late in the day and persist until it was finished, whatever the hour. Then he liked to gather some of his staff around him for endless theological discussions that often lasted until daylight. Rosecrans would then sleep late, but the staff had to be up early attending to their functions. He also had the bad habit of verbally abusing, in public, officers who had displeased him (Bickham, 1863; Lamers, 1961). But if the officers had mixed feelings about Rosecrans, the authorities in Washington positively detested him. The dislike of Rosecrans was rooted in two causes: his suspected political ambitions, and his strange ineffectiveness as a commanding general.

Actually, Rosecrans had no political ambitions. He did have powerful political connections which was the real reason for his rapid rise in rank. These were Democratic connections, including General George B. MacClellan, who posed a real threat to the apparently ineffective Republican war President, Abraham Lincoln. By 1862, Rosecrans himself was being rumoured as a possible Presidential candidate, and although he was not interested, the suspicious politicians, Abraham Lincoln and his Radical Secretary of War, Edwin Stanton, kept close watch on him (Fitch, 1863; Cox, 1900; Society of the Cumberland, 1902). This was James Garfield's role at Rosecrans' headquarters; spy for the Washington authorities, (OR, Volume XX(3), pp.

230-231). The politicians actually had better reasons to watch Rosecrans. A magnificent organiser who ran his Kentucky-Tennessee military district like a fine watch, Rosecrans proved slow to move, reluctant to fight and ineffective in battle.

After taking command of the Army of the Cumberland in Nashville, in the fall of 1862, Rosecrans remained passively within the city's defences despite a huge numerical and material advantage over the Confederates while he built up more supplies and manpower. Meanwhile, the vastly outnumbered Confederates loitered nearby, their cavalry regularly operating in the city's suburbs. Finally forced into action by his political superiors, Rosecrans moved slowly out of the strong Nashville defences, in the last days of 1862. Just over 28 km away he was smashed by a powerful Confederate attack delivered by General Braxton B. Bragg's Army of Tennessee. Rosecrans was saved from destruction at the Battle of Murfreesboro, or Stone's River, by the magnificent fighting abilities of his men, Confederate weakness, and a lot of luck. He claimed a victory after the starving Rebels withdrew a few days later, but did not try to advance again for six months, although he outnumbered the Confederates by huge and ever growing margins (McDonough, 1980).

4. The Confederate Army and the Union attack

In contrast to Rosecrans' large well-equipped and well-fed army the Confederates were an out-numbered, demoralised, ragged and starving force whose primary food supply came from the region of Tennessee where they were encamped. Nearly half of some units were barefoot and ammunition was so scarce that only three rounds per man could be issued monthly because of the shortage of lead. The cavalry was badly armed, badly mounted, and dispersed over a wide area to control food sources. It was also, despite Rosecrans' claims, seriously outnumbered (O.R., Volume XXIII(1), pp. 648, 927 & 932). Furthermore, the Rebel army was led by a general, Braxton B. Bragg, that few of his officers or men any longer had any confidence in (Johnston, 1878). Most significantly, Bragg had placed his forces in a precarious and unstable strategic position centred on his base at Tullahoma. His forward elements were at Wart race and Shelbyville (Figure 2) in front of which his outpost line guarded the gaps in the rugged hills separating the Duck River basin from the more level and slightly lower land around Murfreesboro. The position was faulty on several accounts: the right flank could be (and was) turned putting Union forces on the Confederate line of communications; it relied upon a rickety one lane railroad that was exposed to attack; there was no line of retreat, mountains and a major river (the Tennessee) were behind the army, between it and safety; and the outpost line was inadequately manned and supported. The possibility then existed that a Union attack on Bragg's right flank might result in the severing of Confederate lines of communication, trapping the army, the South's second largest, north of the Tennessee River (Connelly, 1971). This is exactly what Rosecrans attempted to accomplish.

Yet, for months the Union Army of the Tennessee idled around Murfreesboro to the increasing impatience of the military and civilian authorities in Washington. Rosecrans always had several reasons why his army was not yet ready for battle and they were usually valid to a degree. He lacked men, supplies, wagons, and horses, particularly horses. The weather was too wet or cold, the rivers high, the countryside lacked fodder for horses or his cavalry was on detached service or some other necessary preparation was incomplete. While, there was some truth in all of Rosecrans' complaints about the unreadiness of his forces to do battle, his Confederate opponents were almost infinitely worse off in every category (O.R., Volume XXIII(1), p. 10; Volume XXIII(2), pp. 927-932 & 300-301). While generally indifferent to the physical realities facing its field armies, in Rosecrans case, at least, the Lincoln administration attempted to meet his requirements of manpower, horses and supplies. The entire production of the new Spencer repeating rifles was directed to Rosecrans, for example (Lamers, 1961). Still, Rosecrans hesitated to take the field and to the increasingly strident calls for action coming from Washington he countered by holding a council of war of his generals who, unanimously endorsed his policy of delay. Rosecrans finally began his advance only when peremptorily ordered to by the Secretary of War and even then delayed eight more days (O. R., Volume XXIII(1), pp. 8-10).

The Union advance from Murfreesboro region began on 26th June with a skilfully orchestrated manoeuvre designed to convince Bragg that the main attack was coming across level ground toward Shelbyville through Guy's Gap in the Confederate left, rather than against the Confederate right and Bragg's lines of communications. This movement was led by General David Stanley's cavalry division which swung to the west threatening to turn the Southern left flank. Disguised as a feint, strong Union cavalry forces moved to the east, toward McMinnville. Screened by their movement, the bulk of the Union Army moved against the Confederate right. George Thomas, perhaps the most capable of all Union generals, led his 14th Corps down the Manchester Pike. To the east, around Bradyville, General Thomas Crittenden's 21st Corps massed, poised near Bragg's right flank awaiting developments (Figure 2) (O.R., Volume XXIII(1), pp. 403-409). It began to rain as the Union force started in motion and it would continue to rain, sometimes torrentially, every day of the campaign. Some of the heaviest rains fell between 31st June and 2nd July when Rosecrans' troops were trying to cross the 'Barrens ' (Lamers, 1961).

Thomas' actions proved critical to the success of the campaign. Hoover's Gap was a long defile winding 5 km through a narrow valley in which lay a crude wagon road. The road then wound through the gorge of Matt's Hollow for three more kilometres. Here, the valley was barely wide enough to accommodate the road. In recent years, the valley was widened to accommodate Interstate Highway 24 and today is nowhere less than a kilometre wide. Thomas' forces were spearheaded by Col. John T. Wilder's Indiana Brigade of mounted infantryman (O.R., Volume XXIII(1), pp. 404-405). Dissatisfied with the conservative regular Army officers handling the war, the politically potent Wilder had pulled strings to have his infantry brigade mounted and armed with the new Spencer magazine loaded repeating rifle. This would be the first extensive combat trial of Lincoln's new weapons and its effect would prove overwhelming (Williams, 1935. Sunderland, 1984). Wilder was given an essentially independent command and he used it to the fullest. Finding the Gap only lightly guarded, Wilder brushed the Confederate sentinels out of his way and raced through to seize the entire passage in a sudden *coup de main* (Wilson, 1891). He was now 13 km from his supporting infantry and facing Confederate infantry under General William Bates, who belatedly attempted to close the Gap. The hail of fire from the seven shot Spencers rifles easily turned the attacks back (Wilder, 1908; Connolly, 1987; O.R., Volume XXIII(1), pp. 457-461). Thomas hurried to take advantage of the opportunity, and rushed his infantry into and through the Gap. Then, instead of turning southwest after the retreating Confederates, who were falling back southwest toward their main defences around Wart race and Shelbyville, Thomas pushed his corps to the southeast to join Crittenden at Manchester on 27th June (Figure 2). Here, they were already south of the Confederate defensive line and closer to Bragg's base of supplies at Tullahoma than Bragg was himself.

With Stanley's horsemen sweeping around their left and Hoover's Gap open on their right, the Confederates fell back on their defences at Shelbyville and Wart race intending to give battle there. This allowed Alexander McD. McCook's 20th Corps to push through Liberty Gap and follow the retreating Rebels to Wart race (O.R., Volume XXIII(1), p. 412 & pp. 586-587). To the west, General Gordon Granger's Reserve Corps pushed through Bellbuckle Gap and then turned southwest toward Shelbyville. They found Bragg's army evacuating Shelbyville and Wart race to concentrate at Tullahoma, Granger followed the retreating army but McCook's Corps turned east after crossing the Duck River, toward Manchester to join the rest of Rosecrans' army. These forces were already moving south toward Bragg's main line of communication, the rickety single track railway stretching to Chattanooga (Kniffen, 1884-1888).

5. The Confederate retreat

The rapidity of Rosecrans' movement surprised the Confederates. Although he had been warned by William Hardee, one of his corps commanders, on 5th June that Hoover's Gap was too far forward to defend, Bragg had taken no action. Bragg had learned almost immediately (27th June) from his scouts that his right flank had been turned and had begun to concentrate his scattered forces around Tullahoma (Figure 2). His troops arrived just ahead of the Yankees. They hurriedly began to dig in at Tullahoma. When he learned of Rosecrans' continued movements south toward his line of communication, Bragg fell back across the flooded Elk River on 31st June and then across the Tennessee River by 3rd July (Figure 2; O.R., Volume XXIII(1), pp. 584-585, & pp. 891-894).

The Confederate retreat had been covered by the cavalry of General Joseph Wheeler who was forced to fight without his best units: John Hunt Morgan's brigade, off on its famous raid into Ohio and Nathan Bedford Forrest's division which was dispersed to the west. Mistakenly believing that Forrest was trapped north of the Duck River, Wheeler with 500 men, recrossed the river to hold a bridgehead open. They were trapped there by Stanley's entire division and those who escaped (including Wheeler) did so only by galloping their horses off a 14 m cliff into the flooded river to swim across, under heavy fire. Forrest with his usual resourcefulness had already crossed further downstream and was guarding Bragg's flank (Wyeth, 1914; O.R. XXIII(1), pp. 528-532).

Rosecrans tried again to swing around Bragg toward the rail line that was the Confederate lifeline, at Decherd. Thomas pushed his cavalry toward the line, followed closely by infantry (O.R. Volume XXIII(1), pp. 432-433). Although slowed by the heavy rains and mud, the Union forces reached the rail line before Bragg's army was clear of their position, then made no serious effort to cut it. Bragg was allowed to retreat unmolested across the Tennessee River with his supply train intact (Connelly, 1987). The Confederate corps commanders were furious with Bragg for his failure to turn and fight at the narrow defile below Sewanee (University; Figure 2) that led up onto the Cumberland Plateau (O.R., Volume XXIII(1), pp. 632-625). Both commanding generals seemed content to end the campaign without a decisive engagement.

Nevertheless, Rosecrans' accomplishments were impressive. At a cost of only 560 casualties (85 killed) he had driven the Confederates from Middle Tennessee, an important food producing region. But his political masters were not satisfied and immediately resumed hounding him to move forward again. Vainly, he pointed out the impossibility of amassing the supplies needed to cross the Tennessee River, in a barren country at the end of a ramshackle, 460 km railroad exposed to enemy cavalry attacks for most of its length. Rosecrans even appealed directly to President Lincoln in an extraordinary letter (O.R., Volume XXIII(1), pp. 427-428). It was to no avail and soon a radical spy, the Assistant Secretary of War, Charles A. Dana was sent to watch over him as a sort of Republican commissar. There already was one Radical spy in Rosecrans' army, his chief of staff and future American President, James Garfield, who soon betrayed his commander to the Washington politicians (Van Horne, 1988).

The Union forces reached the Tennessee River by 27th July (Figure 2) and hastily repaired the railroad which the retreating Confederate failed to destroy properly. Supplies were forwarded from Nashville and accumulated as quickly as possible. The army was ready to move again by 12th August. (O.R. Volume XXX(1), pp. 47-65).

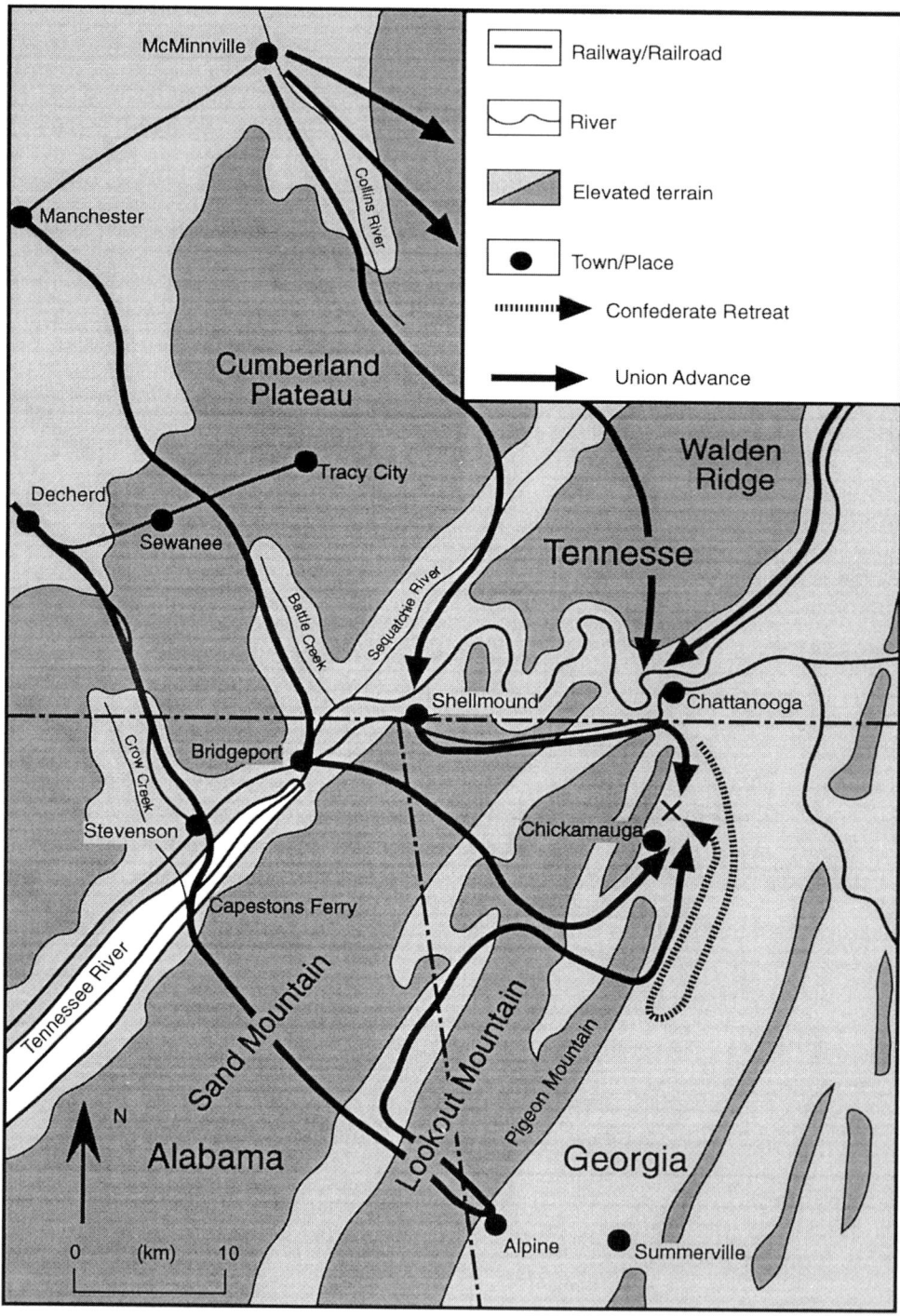

Figure 3. Chickamauga battlefield. (Modified from: Brown 1964)

6. The Tennessee River and Chickamauga Creek

The Tennessee River flows sharply southwest from Chattanooga below huge, long (100-150 km), ridges which rise to an altitude of 700 m (Figures 1 & 3), namely Sand Mountain and Lookout Mountain. The drainage between the ridges, Lookout Valley and Chattanooga Valley flows to the north where the streams join the Tennessee River at Chattanooga. The mountains are steep-sided, flat-topped ridges up to 10 km in width that trend SW-NE and have very few usable passageways through or onto them. They are actually large, breached anticlines plunging to the southwest with the valleys comprising the eroded sections. Lookout Mountain terminates in an escarpment looming 720 m over Chattanooga from the south, around which the Tennessee River loops before heading southwest for nearly 130 km. The beds of primarily Cambrian and Ordovician sedimentary rocks dip steeply to the southeast and the ridges are capped by cliff-forming conglomerates. These are no true passes through either of these mountains, only a few, narrow, widely separated, 'gaps' where the approaches are slightly less steep than along the rest of the cliff (Brown, 1964).

By moving directly across the Tennessee River at Bridgeport and Stevenson, Rosecrans placed himself in a position some 50 km south of Chattanooga where he threatened Bragg's only line of communication, the railroad running southeast from Chattanooga to Atlanta. Because of the rugged mountains, it was a move Bragg did not expect and the river was inadequately picketed by Confederate cavalry (Cozzens, 1992).

Secretly moving supplies and bridging materials, Rosecrans achieved complete surprise, crossing unmolested and almost undetected, beginning 29th August. He had fixed Bragg's attention to the north with a feint toward Chattanooga from the north across the only approach to the city over reasonably level ground. Rosecrans hurried his army across the river but failed to push quickly toward Bragg's exposed line of communication. Bragg, soon recognising his peril, abandoned Chattanooga and concentrated his army at Lafayette, Georgia. where it protected his railroad line of communication. Without any effective cavalry scouting, Rosecrans recklessly pressed on after what he believed to be a hastily retreating enemy. However, Bragg was not retreating, he was watching and waiting for Rosecrans, so he could strike as Rosecrans debouched from the mountains. And he was being reinforced with James Longstreet's Corps from General Robert E. Lee's Army of Northern Virginia. By the day of battle the Confederates had nearly as many men as Rosecrans, an extremely rare occurrence during the Civil War (Tucker, 1961).

Without cavalry reconnaissance and without even sketchy maps of the mountainous area, Rosecrans recklessly pressed ahead. His corps were soon separated at the only three gaps passable for troops over a 67 km distance along the west face of the 135 km long Lookout Mountain massif (Figure 3). One of these corps, the southern most and most exposed, was itself scattered with its divisions up to 13 km apart over mountains and out of mutual supporting distance. The main Confederate forces, massed around Lafayette were closer to any of the Union corps than any other Union corps was. Why the Confederates failed to take advantage of this potentially catastrophic alignment remains an open historical question rooted in tensions within the Southern Army of Tennessee. Vacillation by Bragg and failure of his generals to obey their orders temporarily saved Rosecrans (Connelly, 1971). Rosecrans was allowed to concentrate his forces along Chickamauga Creek where they were nevertheless decisively defeated because of an egregious tactical error on Rosecrans' part on 19th September in one of the largest and bloodiest battles of the Civil War. The remnant fled to Chattanooga where they were trapped. In the aftermath of battle, Rosecrans was fired from command of the Army (Baumgartner & Strayer, 1996).

7. Rosecrans and the terrain of the Tullahoma

For a few weeks Rosecrans was left in limbo, but on 22nd January 1864, he was made Commander of the Department of Missouri, a less important sector (O.R., Volume XXXII(2), p. 182). He was not successful there either, unable to control either the Confederates or his critical superiors. He was finally able to stop and drive back into Arkansas the weak Rebel invading force of General Sterling Price in 1864, but the damage had already been done. Lincoln, Stanton, General Henry Halleck, and Grant had become exasperated at Rosecrans' prolonged inactivity while Price had roamed over much of central and northern Missouri (O.R., Volume XLX(1), p. 35, Volume SL, p. 372, Volume XLI(4), p. 126 & p. 153, Volume XLI(3), pp. 307-317). He was fired again and this time was not reassigned. He finally resigned his commission in 1867 (O.R., Volume XLI(4), p. 32; Warner, 1964). Rosecrans enjoyed a successful post-war career. He moved to California where he established himself as a civil and mining engineer. He was Ambassador to Mexico in 1868-1869 and was involved in the development of Mexican railroads for the next decade. In 1880, he was elected to Congress and remained until 1885 rising to the Chairmanship of the Military Affairs Committee. In 1885, he was appointed Register of the US Treasury, dying at his ranch in California in 1898, apparently still haunted by the ghosts of Chickamauga (Chambers, 1961).

The historical view of William Rosecrans has generally been that while he was an effective military bureaucrat, in the field, he was slow to act, careless and excitable. The defeat at Chickamauga was caused by the reckless scattering of his forces through the mountains out of mutual supporting distances combined with a major tactical error on the battlefield. In fact, most historians condemn Bragg for failing to take advantage of repeated opportunities to destroy Rosecrans' army piecemeal during the Tullahoma-Chickamauga Campaigns as he flung his isolated corps across the mountains.

There has always been a minority view, most common among veterans of the Army of the Cumberland, that Rosecrans' campaign, in which he unhinged two major Confederate defensive positions (Tullahoma &

Chattanooga) by threatening Confederate lines of communication and without a major battle are among the most brilliant in the Civil War. One writer, a geologist (with the US Geological Survey) not an historian, Andrew Brown, not only claims that these campaigns are brilliant but that Rosecrans, who is a near military genius, was able to carry them out because of his geological training and his knowledge of the geology of the region (Brown, 1963, 1964, 1965).

Exactly what geological and cartographic information was available to Rosecrans in 1862-1863 is unclear. Even less clear is what, if any, he actually used. Brown assumes that Rosecrans knew of, had access to, and used existent geological maps of Alabama, Georgia and Tennessee. The only proof Brown offers is an assertion that a large scale map of the region made by the Army of the Cumberland's chief Topographical Engineer, Captain Nathaniel Michler, before the campaign incorporated data from the Safford map of 1855. The map itself has been long lost and only later versions exist. Brown provides no corroborating details (Brown, 1965).

The Safford map originally was published in 1855 to accompany the first report of the State Geological Survey of Tennessee by the State Geologist, James M. Safford of Cumberland University. It was a black and white map, at a scale of 1:7,874. Safford, who worked four years part-time surveying Tennessee geology before becoming the State Geologist, had a large volume of information available for his report. As early as 1828, Dr. Gerald Troost of the University of Nashville, who became the first State Geologist in 1831, had begun a systematic long-term survey of the geology of Tennessee. In this he was assisted by some outstanding geologists including Arnold Guyot, David Dale Owens and Safford himself. Troost published eight geological reports from 1831-1851. The fifth included a state geological map and an E-W vertical section. The eighth centred on railroad routes from Nashville to Chattanooga, the precise location of the Tullahoma Campaign. Also, the State Engineer, A.M. Lee, surveyed the same region (and others) for possible railroad and turnpike routes (Glenn, 1912).

8. Discussion

What, if any, of this information was used by Rosecrans cannot be determined from existing sources. Contrary to Brown's assertions, there is no need for him to have used any of this specialised geological information to do what he did. He could, and probably did, arrive at the same plans of operation from simple time and distance calculations using an outline road map. His chief of staff, the duplicitous James A. Garfield did and no one ever accused Garfield of knowing anything about geology (Garfield, 1964). Furthermore, it appears from the Official Records, that the map of the operational area was not completed until the Tullahoma Campaign was over (O.R., Volume XXIII(2), p. 611). It is, of course, possible that Rosecrans had access to and used the information before the map was completed, but there is no evidence for this.

Rosecrans did make major efforts to create accurate maps of his operational area. In this, he was far ahead of any other general of the era. While still in command of 14th Corps in northern Mississippi and western Tennessee, Rosecrans' staff developed an excellent set of maps later used by US Grant in his unsuccessful early campaigns against Vicksburg in 1862 and 1863. When Rosecrans became the commander of the Army of the Cumberland, he created a Topographical Department, first under Captain Nathaniel Michler and later William E. Merrill. Mapping was made a high priority and the cartographers were given every resource possible. Using pre-war state geological survey maps as their base, they supplemented these observations with those from cavalry reconnaissance, civilian informers, and other sources of information about areas still under Confederate control. The intelligence was quickly reduced to usable maps. The Army carried with it 'a printing press, two lithographic presses, one photographic establishment, arrangement for map-mounting and a full corps of draftsmen and assistants'. One of Rosecrans' officers, Captain W.C. Margadent, invented a blueprint process for the quick (and expensive) reproduction of maps in the field (O.R., Volume XXVI(1) p. 656, Volume XX(2), p. 115, Volume XXX(3), p. 112; Brown, 1965).

The results were an impressive set of maps of eastern Tennessee and northern Alabama and Georgia. Some of these still exist, as they were incorporated into the Atlas of the Official Records. Rosecrans' map making enterprise also helps to explain two other historical questions: the slowness of his Army in movement and the curious ineffectiveness of his cavalry. Rosecrans' Army carried a huge, lavishly equipped headquarters which must have slowed his movements. He required more baggage wagons, horses and mules to move his forces than any other army in the Civil War. To feed the many animals he had to widely disperse his forces in a barren, mountainous region which set the stage for disaster. And, although his rate of movement was surprisingly fast for such a ponderous force, about 13 km a day, he could not keep up with the retreating Confederates (Hagerman, 1992).

Similarly, the ineffectiveness of Rosecrans' cavalry can, at least partly, be explained by its preoccupation with the cartographic efforts to the neglect of its primary functions. Confederate cavalry normally dominated the battlefields in the first three years of the war although the Federal cavalry were more numerous. 'Our armies have been ridden round time and again our trains captured, buildings burned, communications cut, and we never succeed in destroying or capturing the force which does the damage, and never will except by fortunate accident' complained the Quartermaster General of the Army, Montgomery C. Meigs, who was being pressured relentlessly by Rosecrans for more and more horses (O.R., Volume XXII(2), p. 302). Not once during Rosecrans' advance from Murfreesboro to Chickamauga, a period of six months, did the numerically superior Union cavalry penetrate the Confederate cavalry screen; nor did it block a road or cut a railroad during the Tullahoma Campaign. Although Union cavalry unites twice reached the Nashville to Chattanooga railroad in strength, they failed to make any serious effort to cut the Confederate main line of supply (Lamers, 1961). At Chickamauga, the Union cavalry failed to detect the nearby presence of Bragg's concentrated army until too late. Rosecrans blamed his cavalry commander David Stanley for its failure, but Rosecrans was ultimately responsible and it is likely that his preoccupation with guarding his supply lines and assigning his horsemen cartographic duties tied Stanley's forces too close to the Army (O.R., Volume XXIII(1), p. 52). On the other side, Bragg did not even have a map of the Chickamauga area when the battle opened, but his cavalry faithfully and accurately carried out their scouting and screening functions (Martin, 1883).

There are few campaigns during the American Civil War in which topography played as important a role as the Tullahoma Campaign. It is not a well known campaign; there were no large bloody battles during the campaign and the end result of a brilliantly successful manoeuvre was a disaster for Union arms. It yielded no immediately decisive results. Yet compared to other campaigns, such as W.T. Sherman's complicated advance on Atlanta against J.E. Johnson's Confederate defenders in 1864, the Tullahoma Campaign involved a relatively stronger Confederate Army which occupied stronger defensive positions. It was also accomplished more quickly with a lower loss rate. Of course, the Atlanta Campaign proved to be the death blow to the Confederacy while the Tullahoma Campaign led to a major Union defeat at Chickamauga. While it was indecisive in the long run, in this one case, at least, William G. Rosecrans, manoeuvred his Union Army of the Cumberland with great skill and imagination.

References

AAPG. 1970. *Mid-Atlantic Region. Geological Highway Map.* American Association of Petroleum Geologists, Tulsa, Oklahoma.

Baumgartner, R. A. & Strayer, L. M. 1996. *Echoes of Battle: The Struggle for Chattanooga.* Blue Acorn Press, Huntington, West Virginia.

Bickham, W.D. 1863. *Rosecran's Campaign with the Fourteenth Army Corps.* Moore, Wilstach, Keys and Co., Cincinnati.

Brown, A. 1963. Geology and the Tullahoma Campaign of 1863. *Geotimes* 8, 20-25, 53.

Brown, A. 1964. The Chickamauga Campaign of 1863 and Geology. *Geotimes* 9, 17-21.

Brown, A. 1965. A geologist - general in the Civil War. *Geotimes* 10, 8-16.

Byrne, L. G. 1906. Bishop Rosecrans. *The 'Old Northwest' Genealogical Quarterly* IX, 311-313.

Connelly, T. L. 1971. *Autumn of Glory: The Army of Tennessee, 1862-1865.* I. Louisiana State University Press, Baton Rouge.

Connolly, J. A. 1987. *Three Years In The Army of The Cumberland.* Indiana University Press, Bloomington.

Cox, J. B. 1900. *Military Reminiscences of The Civil War*. Charles Scribner's Sons, New York.

Cozzens, P. 1992. *This Terrible Sound; The Battle of Chickamauga*. University of Illinois Press, Champaign - Urbana.

Esposito, Brig. Gen. V. J. (Ed.) 1959. *The West Point Atlas of American Wars*. Frederick A. Praeger, Publishers, New York.

Fitch, J. 1863. *Annals of the Army of the Cumberland*. J.B. Lippincott & Co., Philadelphia.

Garfield, J. A. 1964. *The Wild Life of the Army*. Michigan State University Press, Lansing.

Glenn, L.C. 1912. The Growth of Our Knowledge of Tennessee Geology. *The Resources of Tennessee* 2, 167-219.

Hagerman, E. 1992. *The American Civil War and the Origins of Modern Warfare*. University of Indiana Press, Bloomington.

Howe, H. 1900. *Historical Collections of Ohio*. I. C.J. Krehbiel & Co., Cincinnati.

Johnston, W.P. 1878. *The Life of Gen. Albert Sidney Johnston*. D. Appleton & Co., NY.

Kniffen, G.C. 1884-1888. Maneuvering Bragg Out of Tennessee. In: Anon. (Ed.), *Battles and Leaders of the Civil War*. The Century Co., New York, III, pt II, 635-639.

Lamers, W. S. 1961. *Edge of Glory: A Biography of General William S. Rosecrans, U.S.A*. Harcourt, Brace and World, New York.

McDonough, J. L. 1980. *Stone's River: Bloody Winter in Tennessee*. University of Tennessee Press, Knoxville.

Martin, W.T. 1883. A Defense of General Bragg's conduct at Chickamauga. *Southern Historical Society Papers* XI, 201-206.

O.R. 1880-1901.*The War of the Rebellion: A Compilation of the Official Records of the Union and Confederate Armies*. 128 vols, U.S.Government Printing Office, Washington D.C.

Palmer, J.M. 1901. *Personal Recollections of John M. Palmer*. The Robert Clarke Co., Cincinnati.

Shanks, W.F.G. 1866. *Personal Recollections of Distinguished Generals*. Harper & Brothers. New York.

Society of the Cumberland, 1902. *Burial of General Rosecrans, Arlington Cemetery, May 17, 1902*. The Robert Clarke Co., Cincinnati.

Sunderland, G.W. 1984. *Wilder's Lightening Brigade - And Its Spencer Repeaters*. The Book Works, Washington.

Tucker, G. 1961. *Chickamauga; Bloody Battle in the West*. Bobbs - Merrill Co., New York.

United States Geological Survey 1985. *Murfreesboro, Tennessee*. 30 x 60 Minute Quadrangle Map. 1:100,000-scale metric topographic map.

United States Geological Survey 1988. *Chattanooga, Tennessee*. 30 x 60 Minute Quadrangle Map. 1: 100,000-scale metric topographic map.

Van Horne, T. B. 1988. *History of the Army of the Cumberland*. Broadfoot Publishing Co., Wilmington.

Warner, E.J. 1964. *Generals in Blue: Lives of the Union Commanders.* Louisiana State University Press, Baton Rouge.

Wilder, J.T. 1908. *The Battle of Hoover's Gap. Paper of John T. Wilder. Read Before the Ohio Commandery of the Loyal Legion, November 4, 1908.* In Sketches of War History 1861-1865. Papers Prepared for the Commandery of the State of Ohio, Military Order of the Loyal Legion of the United States 1903-1908. Montfort and Co. Cincinnati.

Williams, S.C. 1935. General John T. Wilder. *Indiana Magazine of History* 31, 174-176.

Wilson, Capt. G.S. 1891. Wilder's Brigade of Mounted Infantry in the Tullahoma -Chickamauga Campaign. *In:* Anon. (Ed.), *War Talks in Kansas, A Series of Papers Read Before the Kansas Commandry of MOLLUS.* Broadfoot Publishing Co., Wilmington.

Wyeth, J.A. 1914. *With Sabre and Scalpel: the Autobiography of a Solider and Surgeon.* Harper and Brothers, New York.

Notes

1. During the Civil War the region between the Appalachian Mountains and the Mississippi River, comprising the states of Kentucky, Tennessee, Mississippi, Alabama and Georgia was called the West.

Walter E. Pittman
Department of History and Social Science
University of West Alabama
Livingston, AL 35470

The Mountain is their Monument: An archaeological Approach to the Landscapes of the Anglo-Zulu War of 1879

Tony Pollard

> **ABSTRACT:** The Anglo-Zulu War occupies a special position in British military history; not least because a single day, 22nd January 1879, saw one of the worst defeats ever inflicted on an imperial army by an indigenous force, and one of the most celebrated examples of a small force overcoming overwhelming odds. But the battles of Isandlwana and Rorke's Drift were only two events in a war which although lasting only six months saw a number of notable engagements within contrasting types of terrain, including the battles of Nyezane, Hlobane, Khambula, Ntombe and Gingindlovu, culminating with the final defeat of the Zulu army by a British square at Ulundi on the Hahlabathini Plain in the shadow of the Zulu king's homestead. There was also a siege, with Zulu forces keeping more than 1700 British troops penned inside their hastily constructed fort at Eshowe for two and a half months, until relieved by a re-invasion force. This paper provides an introduction to the Anglo-Zulu War Archaeology Project, a joint enterprise between the Department of Archaeology at Glasgow University and Heritage KwaZulu-Natal. Fieldwork began with the survey of the British fort at Eshowe in September 1999. The project will adopt a wide landscape based approach in order to place sites of archaeological interest (battlefields, forts, camps, Zulu homesteads etc.) within their wider context. The influence of terrain on the progress of the war is discussed, as is the differing perceptions of landscape held by both the Zulu and the British, the origins of the latter being traceable back to early traveller's writings such as Gardiner's *Journey to the Zoolu Country* of 1836.

1. Introduction

This paper presents a preliminary statement on the Anglo-Zulu War Archaeological Project in KwaZulu-Natal, South Africa. The project is a long term collaboration between the Department of Archaeology, University of Glasgow, and Amafa aKwaZulu-Natali (Heritage KwaZulu-Natal). It is the first large scale archaeological project geared toward the investigation of sites related to the war of 1879.

The sites to be investigated, which include forts, battlefields and Zulu settlements, are situated in the northern part of KwaZulu-Natal known as

Zululand (although annexed by Natal in 1897 and much later incorporated within KwaZulu-Natal the place name is still in common usage). The project's first field season took place in September 1999 at the British fort at Eshowe (Figure 1) and the results of this work are summarised elsewhere (Pollard, 2001). The present contribution concentrates on the planned investigation at Isandlwana, and stresses the importance of a landscape approach to the archaeological study of battlefields. In future years the battlefield at Isandlwana will become the main focus of investigation. It was here, on 22nd January 1879, that the British army suffered its worst ever defeat at the hands of a technologically inferior indigenous force.

2. The Anglo-Zulu War: a brief outline

The Anglo-Zulu War[1] of 1879 was one of a number of late 19th century colonial wars prosecuted throughout the British Empire in the name of Queen Victoria. The Zulu Kingdom was founded in the early 19th century by Shaka, and its southern neighbour, Natal, was annexed from the Boers by the British in 1843 and declared a separate British colony in 1856. Relationships between the British and the Zulu people had usually been fairly amicable but the same could not be said for those between the Boers and the Zulu. The two groups had been at war with one another intermittently since the Boers first began to make incursions into western Zululand in the late 1830s as they searched for a new homeland at the end of their 'Great Trek' (it should be noted that at that time the Zulu kingdom extended much further south than the Thukela River which had become its southern border by the time of the Anglo-Zulu War).

However, by the late 1870s the British colonial government in southern Africa was eager to pursue a policy of confederation, which on a very simplistic level meant that independent tribal kingdoms were brought under its direct control. Although the brainchild of Lord Carnarvon, the Colonial Secretary in England, confederation of the Zulu Kingdom, the greatest indigenous power in Southern Africa, became something of a personal quest to the British High Commissioner in southern Africa, Sir Henry Bartle Frere. Confederation required the dismantling of the Zulu military system, in which every male was obliged to do military service under the king. The potential threat posed to Natal by the Zulu army was something that Frere was keen to emphasise in gaining support for his enterprise.

In December 1879, the Zulu king, Cetshwayo kaMpande, was presented with an impossible ultimatum by the British colonial authorities and only a month in which to comply with it. To do so he had to disband his regiments, an act that would destabilise not only his own position as king but also undermine the entire fabric of Zulu society as it had existed since Shaka's time. Having failed to elicit a positive response to the ultimatum Frere ordered Lord Chelmsford, the commander of British forces in southern Africa, to invade Zululand in January 1879.

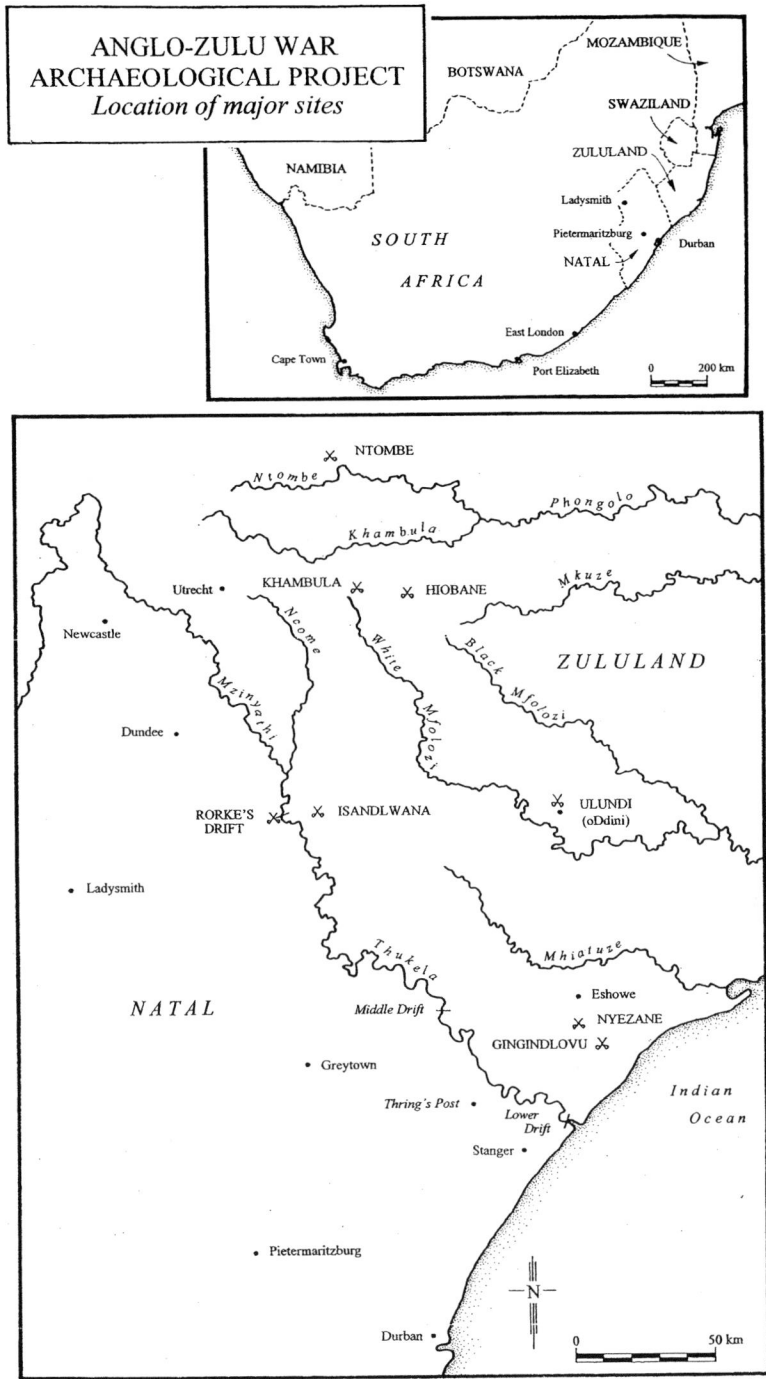

Figure 1. Zululand and major sites mentioned in text

The invasion, which began on 11th January, involved three columns, with a further two remaining in Natal to defend against a possible Zulu counter-invasion. Chelmsford crossed the Buffalo (Mzinyathi) River into Zululand from Rorke's Drift (Figure 1) at the head of the central column of almost 5,000 men and 300 wagons. Following a successful assault on the homestead of a local chief called Sihayo[2] on 12th January the column moved on to make camp on the east facing slopes of the distinctively shaped mountain known to the Zulu as Isandlwana.

The column, which included the 1st and 2nd Battalions of the 24th Regiment, various native and colonial contingents and artillery, began to arrive at the mountain on 20th January. Chelmsford, despite advice from Boers with plenty of Zulu fighting experience, including Paul Kruger, did not 'laager' the camp by circling the wagons or throwing up a defensive ditch. This failure has been much debated by historians (e.g. Laband & Thompson 1979; Laband & Mathews 1992; Knight 1995a), but it would appear that the stony nature of the ground, the effort of moving the wagons into position and the fact that he did not intend to be there for more than a day or so were enough to put the General off the idea. All historians are of the opinion that when considering this issue we should not forget one more factor, a serious underestimation of the military capability of the enemy.

Chelmsford's ultimate objective was the Zulu royal homestead at oNdini (Ulundi), over 64 km to the east, but he was keen to engage the Zulus in battle wherever the opportunity may arise. The British army had previously defeated the Xhosa under Sandile in the Cape and Chelmsford believed the same tactics of engaging in the open would also succeed with the Zulus. A patrol led by Major Dartnell of the Natal Mounted Police detected what appeared to be a large Zulu force moving through the hills to the southeast of the camp. Believing Dartnell to have found the main Zulu impi (army) Chelmsford marched out of the camp at Isandlwana with more than half of his force early on the morning of 22nd January. Leaving the 1st Battalion 24th Regiment, one company of the 2nd Battalion, 24th Regiment; the Natal Native Contingent, various colonial units and two guns of the Royal Artillery in the camp, under the command of Lieutenant-Colonel Pulleine, Chelmsford marched out across the open plain, which apart from a conical-shaped koppie (hill) ran uninterrupted to the east for around 24 km.

Dartnell's assumption about the position of the Zulu impi was fatally wrong. In fact, around 20,000 warriors were now concealed in a valley just 8 km to the north east of the camp, having arrived there from Ulundi on the morning of the 21st. Before departing on his wild goose chase, which may have been a deliberate Zulu ruse (Laband & Mathews, 1992), Chelmsford sent orders that Lieutenant-Colonel Durnford R.E. and his mounted troops of the Natal Native Horse were to come up from Rorke's Drift. Durnford arrived at the camp at around 10.15 AM to be told that Zulus had been seen moving over the Nqutu Plateau, to the north of the camp beyond the Nyoni Heights. On his arrival Durnford was the senior officer, but chose to leave the camp to Pulleine and with his mounted men advanced to the east of the mountain for several kilometres, while also sending a patrol back up onto the plateau. At around 11.30 AM the patrol, under the command of Lieutenant Raw, stumbled across the main Zulu impi, which after being fired upon advanced towards the camp.

Back in the camp the men of the 24th Regiment formed in firing lines extending up to 1,000 m in front of the tents, most of them facing the slope down from the Nyoni Heights to the north. By the time they crested the horizon the Zulus had formed into their classic 'horns of the bull' attack formation, with the left horn streaming around the Conical Koppie in front of the camp (Figure 2). The main body of the Zulu force, the chest, came down off the Heights onto the main firing line to the left (north) of the camp. As this deployment took place, the right horn passed the camp. Their movement was observed only by two rifle companies sent up onto the ridge to provide forward protection. In silence the Zulus of the right horn moved down a valley out of sight of the mountain before turning south and straddling the road to Rorke's Drift, thus closing the obvious escape route for the soldiers in the camp. The first troops to fall foul of the Zulu onslaught were those manning a rocket battery, which under the command of Major Russell had accompanied Durnford from Rorke's Drift. With only enough time to let off a couple of rockets the battery was over run by the left horn as it fanned out beyond the Conical Koppie.

By now Durnford and his mounted men were engaged with the left horn as its regiments rounded the Koppie and turned west toward the camp. For a while his troopers held the Zulus back from the cover of a donga about 1.5 km in front of the camp. But even with support from a company of the 2nd Battalion, 24th Regiment, under the command of Lieutenant Pope, and well-aimed fire from one of the field guns, the Zulu advance could not be checked for long and Durnford and his men retreated into the camp. Many of the foot soldiers of the supporting 2nd Battalion, 24th Regiment, did not make it that far, and the now unstoppable Zulu advance barely paused as it swallowed them up.

Despite withering volley fire the Zulus eventually managed to press home their charge on the main firing line. Zulu success at breaking through the British left has traditionally been put down to ammunition shortages, the firing lines being deployed well in advance of the ammunition wagons, and quartermasters apparently refusing to issue ammunition to any one but men from their own companies (Morris, 1965). The official inquiry into the episode was also quick to criticise the Natal Native Contingent for not standing firm in the face of the Zulu advance. However, the rapid deployment of the left horn and the retreat of Durnford's men had by this time compromised the camp to the south and must surely have prompted the order for an orderly retreat to a more viable position (Knight, pers comm.). The red-jacketed soldiers began to fall back, probably pausing to fire as they hurriedly returned to the camp. The natural line of retreat was back to the Saddle, a low spur also referred to as the Nek, which joins the south end of the mountain with a rocky knoll (usually referred to as Stony or Black's Koppie). From there, if things went very badly, the next move would be on to the road to Rorke's Drift. But by now the Zulus had gained the upper hand and it was on the Saddle or close by that the defence broke down and many of the famous last stands, including those of Durnford and Captain Younghusband (24th Regiment) and their men took place.

With all hope gone the survivors, finding the road to Rorke's Drift blocked, made their way through the only gap left in the Zulu encirclement, across the broken ground to the south-west of the Saddle. Among them was Lieutenant Melvill of the 1st Battalion, 24th Regiment, who had been assigned the task of taking the regimental (Queen's) colours to safety. He was killed in the attempt, and both he and Lieutenant Coghill, who had come to his assistance, were later awarded the first posthumous Victoria Crosses for their bravery. A lucky few made it to the Buffalo River and back into Natal, at a place forever after known as Fugitive's Drift; just one example of how the events of the 22nd January 1879 have become permanently attached to the landscape. Melvill and Coghill made it across the river but were killed on the Natal side, and the colours swept downstream.

By two o'clock in the afternoon the battle, which in its latter stages had been fought under a sky darkened by a partial eclipse of the sun, was all but over and the camp in possession of the Zulus. Of the 1,707 men and 67 officers left in the camp by Chelmsford a total of 1,357 men lay dead. The number of Zulu dead is uncertain, with estimates varying between one and three thousand. But it did not end there.

Late in the afternoon, the Zulu reserve of around 4,000 men, under the command of the king's half brother, Prince Dabulamanzi kaMpande, crossed the Buffalo into Natal. The battle of Rorke's Drift was about to begin. As fighting raged at Rorke's Drift Chelmsford returned to the camp under the mountain, his men snatching what sleep they could among the dead. In the face of such disaster the General had little choice but to retreat across the Buffalo and lick his wounds.

The right flank column, under the command of Colonel Charles Pearson of the Buffs (3rd Regiment), met with more success, at least initially. Pearson and around 3,000 men crossed the Lower Drift on the Thukela River into eastern Zululand. On the same day as the defeat at Isandlwana, the column, strung out with its supply wagons in convoy, was attacked by a strong Zulu force that charged down from the heights of Nyezane.

After a brief but fierce engagement, in which the British army tested the recently acquired American Gatling gun in battle for the first time, the Zulus were forced back and Pearson's advance continued. His objective was a Norwegian mission station, known as KwaMondi, at Eshowe where his orders were to establish a storage depot that would supply operations in eastern Zululand. It was very apparent to Pearson that heavy concentrations of Zulus were at large in the area, a factor that placed his exposed force under great threat. To minimise risk the Colonel, immediately upon his arrival at KwaMondi (Eshowe), set his troops the task of throwing up a fort around the mission buildings. Most of his supply wagons and non-crucial personnel were then sent back to Natal and he, along with over 1,700 men, began to sit out a siege which was to last almost three months.

The besieged garrison had to wait until 3rd April for relief. Before arriving at Eshowe the relief column, under the command of Lord Chelmsford, had to fight off a Zulu impi some 12,000 strong at Gingindlovu, 24 km to the southeast. With the lesson of Isandlwana hard learned Chelmsford ordered his men to dig in and create a fortified laager. Against such defences and massed rifle fire the Zulus had little chance and were cut down in great numbers before being routed by a merciless cavalry pursuit.

Chelmsford's success at Gingindlovu had been preceded, on 29th March, by a British victory at Khambula in northern Zululand. Here, in a well-defended position, just over 2,000 men of the left flank column, under the command of Colonel Wood, threw back an attack by around 15,000 Zulus. Khambula is generally agreed to have been the most decisive engagement of the war (e.g. Smail 1969; Laband & Thompson 1989), marking the turning point in favour of the invading army and sounding the death knell for the Zulu Kingdom. Never again would the pride of the Zulu army throw itself so relentlessly against the massed firepower of its British invaders.

The final blow came on 3rd July, close to King Cetshwayo's royal residence at oNdini (Ulundi). The British, having been reinforced by troops shipped out from England and India, advanced in a square, flanked by artillery and Gatling guns. In the face of incredibly heavy fire issuing from all sides of the square the tired Zulu army suffered its inevitable final defeat. The Battle of Ulundi marked the end of the Anglo-Zulu war and Cetshwayo himself was captured soon after.

3. The battlefield as landscape

Military historians tend to discuss battlefields in terms of terrain, rightly so as this is how military tacticians view the ground over which they do battle. To them the battlefield is composed of a topography of hills, ridges, rivers, and forests all of which may influence the disposition and movement of troops, and in doing so play a vital role in determining victory or defeat.

Moving beyond the concept of terrain, the landscape is a cultural construct, inhabited and given meaning by people and it is suggested here that archaeological approaches to battlefields must acknowledge this fact. This approach is rooted within the school of cultural geography (e.g. Duncan & Ley, 1993) but more specifically that of phenomenological archaeology (e.g. Tilley, 1994). Both approaches regard the landscape as not just a stage on which human existence is played out but as a frame of reference that is conceptualised, experienced and remade by those who inhabit it and pass through it. As Tilley writes: 'The landscape is an anonymous sculptural form always ready, already fashioned by human agency, never completed and constantly being added to, and the relationship between people and it is a constant dialectic and process of structuration: the landscape is both medium for and outcome of action and previous histories of action. Landscapes are experienced in practice, in life activities' (Tilley, 1994, p. 23). It just so happens that with battlefields these life actions, at least for a short while, are geared toward the creation of death.

It is important that archaeologists consider the entire cultural history of a battlefield and not simply the few hours or days during which the battle was fought. We must be aware that the idea of a battlefield as discussed by historians or archaeologists is itself an artificial creation, an artefact. The battle is not being fought as we visit the site, nor is it the only human activity that has ever taken place there. In a way, the very act of us visiting a place, following signed routes and reading guide books, is as much a part of giving a place meaning as a 'battlefield' as any of the fighting which took place there.

There can be no doubt that battlefields are special places, emotionally charged and a focus for memory. For this reason an important part of the Anglo-Zulu Archaeological Project will be to collect oral evidence from people living on and around these sites; what do they mean to them, what do they know about the history, did any of their ancestors take part in the war?

4. Isandlwana battlefield

A basic aim of the Anglo-Zulu War Archaeological Project will be a detailed survey and investigation of the Isandlwana battlefield. The site, or at least 800 hectares of it, was proclaimed a protected Battlefield Reserve in 1989 and has become a popular visitor attraction, with a small museum established in the St. Vincent's Mission in the local township, to the north of the mountain. Despite the presence of homesteads and settlements in the vicinity, which include a recently built hotel (specifically geared to attracting tourists to the battlefield) and a number of schools, the battlefield still appears pretty much as it must have done on that fateful day. As battlefields go, Isandlwana must surely be one of the most eerie and atmospheric, not least because of the strange, sphinx-shaped mountain that looms over the plain on which the events of 22nd January 1879 unfolded.

Some idea of the way in which topographic features are integrated into a concept of landscape can be gleaned from the different ways in which the British and the Zulu viewed the mountain. To the British it was very much the image of the sphinx that struck an immediate chord. This was not surprising as the 24th Regiment's insignia, featuring on helmet and collar badges, was a sphinx — a reference to its service in an earlier Egyptian campaign. Some soldiers may have regarded the similarity as a good omen, while others may have held the opposite opinion. The Zulu interpretation of the mountain is a little less straight forward, as one might expect in the case of people who had been aware of the mountain long enough to give it a name, although the British apparently referred to it as the 'Little Sphinx' (Gon, 1979). The name Isandlwana has been translated in several ways; including

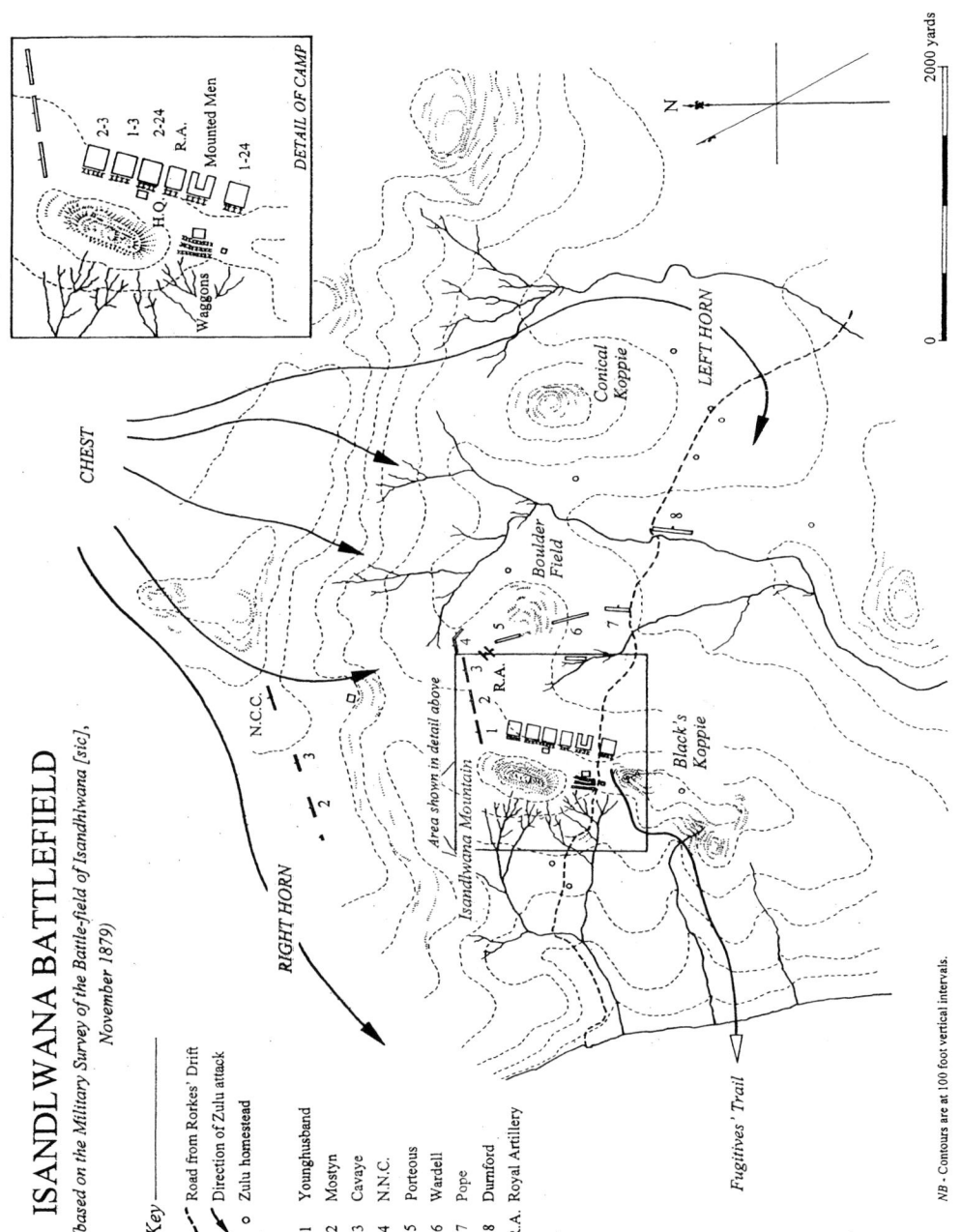

Figure 2. Isandlwana – redrawn from an original 1879 British military map

'like a hand', but the most often recited is 'like a little house'. This is actually a complex, twice removed reference, with part of a cow's stomach sometimes referred to in the Zulu language as 'like a little house' and it is to this anatomical resemblance rather than the architectural to which the name refers (Knight & Castle, 1993). The Zulu name thus embodies references to day-to-day life, with cattle being a mainstay of the Zulu economy and an important symbol of social status. To the casual observer however the mountain may not appear to resemble any of these things. Joshua Luke, the young son of my co-field director, Len van Schalkwyk, calls it 'Shoe Mountain', and there can be no denying that it does look like a giant boot or shoe, perhaps even more so than a sphinx.

Beyond the shadow of the hill, the first thing to strike the visitor is the sheer scale of the place. The plain over which the British deployed and the slopes down which the Zulu charged cover an area of at least 5 km by 5 km, with only a portion of this contained within the Reserve. The second is probably the modern settlements, with Zulu houses clustered around the edges of the Reserve — a number of settlements were relocated to the other side of the fence with the founding of the Reserve. This relocation was negotiated and agreed with local people and the Mangwebuthanani Tribal Authority among others (Konigkramer, 1999). It would be tempting to assume that the scene was somewhat different in 1879, with not much in the way of settlement in the area, but this was not the case. As the military survey of 1879 shows there were a number of Zulu homesteads distributed across the battlefield (Figure 2).

The remains of some of these homesteads or kraals (imizi) can still be seen today, as circular platforms of cobbled stones and stamped clay, all that is left of the traditional beehive huts that once punctuated the landscape. Most of these dwellings were abandoned when the British invasion force crossed over from Natal, and those people who chose to stay in their homes were subject to harassment by British patrols seeking intelligence. Other homesteads were put to the torch. Circular earth banks inside the Reserve probably relate to post-war settlements, as they are not marked on the military map of 1879 and are of a construction style not adopted by the Zulu until the late 19th, early 20th century. Rectangular wall foundations definitely relate to later structures. Some of these may represent structures abandoned at the founding of the Reserve in 1989. An important aim of the project will be to excavate a sample of these earlier and later settlements. It is hoped that this work will shed some light on the impact of the British invasion and later break-up of the Zulu Kingdom on Zulu social life.

Historians have repeatedly pointed out that the Zulus displayed great discipline and dexterity in deploying their huge force so effectively after being discovered before they were ready to attack (e.g. Coupland, 1948; Selby, 1971). Although it is recognised that the Zulu high command were kept well informed of enemy activity by a network of spies and efficient scouts (Coupland, 1948; Knight 1995a), what has not really been considered is the role of local knowledge (although see Knight, 1995a) and understanding of the landscape in their efficient deployment and decisive attack. Warriors from the locality took part in the battle and their experience of the terrain must surely have provided important intelligence for the Zulu high command. It should not be forgotten that Sihayo himself, an important local chief, was with the king at the time of the British attack on his homestead and one of his sons, Mehlokazulu was a junior commander of the iNgobamakhosi Regiment during the battle. Some idea of the extent of this knowledge comes across in rare Zulu eyewitness accounts collected by Bertrand Mitford (in 1882) and others after the war.

A warrior called Uguku from the umCijo Regiment describes how 'The first shell took effect in the ranks of my regiment, just above the kraal of Baza.' He then goes on to describe how, 'The Undonke (a Zulu regiment) ran out towards the kraal of Nyzenzane (a local induna or headman) on the road to Isandlwana and Rorke's Drift. The engagement now became very hot...The infantry then opened fire on us, and their fire was so hot that those of us who were not in the donga retired back over the hill. It was then that the uNokenke and uNodewengu regiments ran out towards Nyenzane's kraal' (Emery 1977, p. 86-87). Further, an unknown warrior, of the uNokenke regiment recounted that he was close to his own homestead near the koppie (probably the Conical Koppie) where Russell and his rocket battery came to grief (Emery, 1977).

Such accounts suggest that known landmarks, in this case named Zulu homesteads, were used by the impi as a means of co-ordinating and focusing the attack, with individual regiments converging on these sites. Whether this indicates a pre-determined plan of attack in place at this time is less certain but the fact that the impi deployed so effectively and so quickly certainly points to this. Laband (1998) has suggested that some of the regiments held back long enough to receive a last minute briefing from their commanders, Ntshingwayo and Mavumengwana. However, none of these more disciplined regiments, some of which at least were held back in reserve, appear to be those mentioned by name in the quotations above, which were among the first to bolt forward. But the fact that the younger, rasher regiments advanced through places which individuals could name, along with the ability of those which held back to make good any inadequacies in the attack, does suggest that an overall plan had been discussed and agreed previously, and at least verbally sketched out at the regimental command level.

People had been living around the mountain for many years before 22nd January 1879, and we should not allow the events of the battle, which lasted just a few hours, to entirely overshadow the long-term occupation of this landscape. A preliminary walk over the battlefield in early 1999 resulted in the identification of major scatters of prehistoric stone tools, on the plain between the mountain and the Conical Koppie. These tools included Middle Stone Age blades, tens of thousands of years old, and also hand axes, which could be anything up to a million years old. To hold these things in your hand while looking at the mountain certainly helps to put the battle into perspective. Some British officers were themselves aware of the antiquity of human activity on the ground over which they fought in Zululand, and took small collections of stone tools back home with them (Mitchell, 1998). Further evidence of the long-term use of the site comes from the summit of the mountain itself. Here, scatters of pottery sherds and iron slag point to the use of the mountain top for ritual purposes during the early Zulu period (early 19th century) but these discoveries will be discussed more fully in a forthcoming paper.

5. The battlefield as monument: the creation of an artefact

The history of the site after the battle, which includes the digging of graves and the construction of cairns and monuments can be regarded as the time during which the battlefield becomes an artefact. Certain places, such as the Saddle, have become a special focus for attention at Isandlwana. It is here that many of the formal monuments are located. On 22nd January the wagons were parked on the Saddle, astride the road that ran back to Rorke's Drift (this role is mirrored today in the location of the visitor car park in the same place).

The road brought the column to the foot of the mountain and it was this route that the fugitives from the camp hoped would carry them to safety. However, as they crossed over the Saddle the retreating troops found the road to Rorke's Drift blocked by the right horn of the Zulu impi as it carried out its encircling movement. It is here that disciplined defence finally broke down and retreat turned to rout. The Saddle, where many of the bodies of the dead were encountered after the battle, became an obvious focus for monumentalisation in the years that followed the war. In March 1914 a granite obelisk was dedicated to the dead of the 24th, close to the site of a mass grave of soldiers from the 24th Regiment. The solid stone platform on which the obelisk stands may include stones from this earlier cairn within its construction, with the lower stones of the rougher cairn remaining in place at the foot of the plinth on its eastern side (Knight, pers. comm.). Other whitewashed stone cairns can be seen scattered across the Saddle, and also punctuating the Fugitive's Trail and the main battlefield itself.

Burials of the British and colonial dead were carried out on several occasions, the process being entirely prohibited for some time after the battle as the field lay in enemy territory. The first return to the site of the camp did not take place until March 14th when a patrol led by Major Black of the 2nd Battalion, 24th Regiment attempted to retrieve the missing regimental colours, during which time they were fired on by small parties of Zulus. When Black had returned to Isandlwana with Chelmsford after the battle, on the evening of 22nd January, he was ordered to secure the koppie which surmounts the southern end of the Saddle — previously known as the Stony Koppie or Mahlabamkhosi — the hill was thereafter renamed Black's Koppie. Black returned to the battlefield again on 15th May, by which time he had been

Figure 3. Contemporary artist's impression of the British return to the Isandlwana battlefield in May 1879

promoted to Lieutenant-Colonel, and it was during this foray that the body of Major Smith, commander of the Royal Artillery battery, was discovered and buried on the Fugitive's Trail.

On 21st May, a much stronger force of British troops, under General Marshall, commander of the newly formed Cavalry Brigade consisting of the 17th Lancers and 1st (King's) Dragoon Guards, freshly arrived in Zululand, came back to the mountain. This time they came to bury the dead and while they were at it, retrieve the serviceable wagons for use in the looming re-invasion of Zululand. This expedition was recorded in sketches by the war artist Melton Prior and was later worked up into the dramatic image (Figure 3) published in the *Illustrated London News*. Today, this illustration stands as an emotive and informative record of how the site must have appeared in the months that followed the battle. The view is taken from Black's Koppie across the Saddle and up to the foot of the mountain. As well as the wagons there is the other debris of war, including the skeletons of men and animals, broken ammunition boxes, barrels and scattered books. Perhaps most poignant of all is an open suitcase in the foreground, probably the property of a dead British officer, which by now contains nothing but a brush. In the middle distance a group of soldiers can be seen building a burial cairn from stones.

This visit was also photographed at the time but it is difficult to make out details picked out in the drawing, which appears to have been compiled from several sketches and shows the debris and people crowded within a much-foreshortened view.

Black returned to continue burial work over several days in June. However, the bodies of the 24th Regiment were not buried until September, at the request of the 2nd Battalion, who perhaps understandably wanted to bury its own soldiers. The last of the bones were put to rest, after being collected in sacks, in 1882 during an effort to tidy up the battlefield prior to a visit by Queen Victoria. The visit never took place but this exercise seems to have begun the tradition of whitewashing the cairns. Various maps and plans of the sites of graves exist from this time. There appears to be little doubt that a number of cairns have become lost to erosion since then, while it is also possible that some cairns are not real — merely bogus cairns created for some unknown reason or the product of over zealous whitewashing parties.

The Zulu, having possession of the field from the afternoon of 22nd January, had much more time in which to dispose of their own dead. The war correspondent Norris Newman noted that the majority of Zulu dead had been removed from the field by the time of the first British return (Knight & Castle, 1993). Many Zulu were hurriedly buried close by in shallow scoops scraped into the sides of dongas, while others may have been carried home. Mehlokazulu kaSihayo, already noted as one of chief Sihayo's sons and a junior commander of the iNgobamakhosi Regiment during the battle, told Bertrand Mitford in 1882 that, 'the dead Zulus were buried in the grain silos in two kraals; some in dongas, and elsewhere' (quoted in: Knight & Castle 1993, p. 35). At the time of the battle the grain (mealies) had not yet been harvested but was ripe in the fields and stored supplies were coming to an end, thus leaving the grain storage pits open and empty and therefore a possible receptacle for the dead. For the Zulu people, the use of grain silos — food stores are powerful symbols of life — as places for burial can only have served as a premonition of what this war would ultimately cost them.

As has been the case with a number of British and colonial cairns, Zulu burials have been exposed as seasonal rains erode the banks of dongas on the northern side of the battlefield. Over the years a number of these bones have been collected and interred within an ossuary next to St. Vincent's Church, and the site is now generally referred to as the 'grave of the unknown warrior'. The dongas (dried streambeds) can be seen cutting across various parts of the battlefield and seasonal water erosion is a serious problem. Accordingly Amafa aKwaZulu-Natali — the provincial organisation responsible for cultural heritage management — has implemented a programme of erosion control, constructing gabion barriers in the most affected areas. This act in itself adds a further dimension to the act of battlefield creation and can be viewed as an attempt to 'fossilise' the ground over which the battle was fought.

It took a 120 years for the Zulu dead to be commemorated on the battlefield. Their monument was unveiled by the present Zulu king, Goodwill

Zwelethini, as part of the 120th anniversary commemoration of the battle on 22nd January 1999. The impressive bronze sculpture is based on a Zulu necklace presented to warriors for valour in battle known as an *iziqu*. It is located near the entrance to the Reserve, which is close to the modern Zulu settlement beyond the northern boundary.

6. Archaeological potential of Isandlwana battlefield

Work of American archaeologists at the site of the battle of Little Bighorn (Scott et. al., 1989), which was fought just six years before Isandlwana and carries with it some of the same historical and cultural legacy, has demonstrated how effectively archaeological analysis can provide a fresh insight into the course of battle. Using archaeological techniques we can question not only historical accounts but also the popular images and stereotypes created as battles enter into the wider consciousness. The American project has partly debunked the popular image of Custer dying with his men around him in a noble last stand, and has repainted the battle as a series of small, desperate fights as men were scattered across the field in a final struggle for survival. However, such reassessment depends entirely on the presence of battle-related artefacts on the battlefield.

Artefact scatters were noted on the 1879 military survey map of the field at Isandlwana, with one annotation reading 'cartridge cases lying behind the boulders', with this concentration presumably influencing the mapmaker's positioning of the firing line. Likewise, the reference to the 'left bank of spruit strewn with cartridge cases' has been taken to mark Durnford's stand in the donga (these annotations have, for the sake of clarity, been left off Figure 2 which is redrawn from the 1879 map). Not surprisingly, these high concentrations of surface finds are not to be seen on the field today (it was normal practice for the army to pick up their own expended brass cartridge cases so they could be recycled, but such an exercise would obviously require survivors). As is the case with many battlefields the process of artefact attrition begins before the gun smoke has cleared, with weapons and personal effects looted from the dead and supplies liberated by the victor.

Looting plays a vital role in determining what remains on the field to become part of the archaeological record related to any battle, and given that the act is often an integral part of any conflict we should not be surprised at this. At Isandlwana looting by the Zulus was not merely an act of property accumulation or supply procurement. To the Zulus warfare and killing carried with it deeply rooted elements of taboo which required ritual cleansing, and this process included wearing the clothes of a defeated foe, something which could result in its own acts of deposition. As a Zulu combatant called Mpashana said: 'When a man kills another in battle, he takes from him all or part of his wearing apparel and discards his own. If he has stabbed more than one, he takes something from each of the dead' (Laband, 1985, p. 18). On a more simplistic level the British were avid souvenir and trophy hunters (looting by another name) and regimental mess walls were soon covered with shields and weapons collected from battlefields in Zululand.

More problematic is the continued looting of battlefield artefacts and this is a major concern at Isandlwana and other Anglo-Zulu War sites. Donald Morris in his popular book *The Washing of the Spears* (Morris, 1965) describes small Zulu boys selling bullets at Isandlwana to visitors as long ago as the early 1960s. This practice still continues today but it is the increased availability of metal detectors and the fashion for the collection of battlefield artefacts in the UK and USA that has caused the greatest threat to the archaeological potential of these sites. These factors, along with high levels of poverty in KwaZulu-Natal, are potentially a recipe for disaster as far as these important cultural sites are concerned. The interpretation of a battle through archaeology obviously relies upon the location of artefacts being known. Unfortunately, once removed from their original context poorly or unrecorded artefacts become worthless as data and become nothing more than objects, however desirable they may be. It is now illegal to remove objects from these sites but the law is difficult to enforce and looting continues.

Despite the forgoing, it is hoped that enough debris related to the battle remains in-situ to allow a meaningful archaeological interpretation of the battle. Some of the research issues which this programme of investigation hopes to address include resolution of the debate over the actual position of the British firing line and insight into the ammunition issue — does the archaeological evidence bear out the idea that ammunition ran low or gave out all together? Survey and artefact collection, which will include prehistoric and other archaeological remains, will be integrated within a wider programme that will include the examination of Zulu settlements and the recording of stories and memories of local people.

7. Conclusions

Much of this paper has been by necessity speculative, but I am hopeful that the first season of work at Isandlwana will make an important contribution to our understanding of the Anglo-Zulu War as a historic event and social process. It should not be forgotten that the majority of historical works on the Anglo-Zulu War, of which there are perhaps hundreds, are written from a British perspective. There is obviously some truth to the old axiom that 'history is written by the victors'. I believe that this project is one way in which the Zulu side of the story may be brought to the fore, dealing as it does with a landscape and a place which long before and ever since the battle has been populated by and given meaning by the Zulu people.

Although the arguments summarised here may have been somewhat simplistic I have attempted to distinguish between terrain and landscape and to emphasise the importance of the latter in the archaeological consideration of the battlefield at Isandlwana.

Acknowledgements

I would like to thank my colleagues and friends who worked at Eshowe during the first season of fieldwork in September 1999: John Arthur, Iain Banks and Jane Clarke. Neil Oliver has been involved throughout the project from its inception, latterly as a co-director at Isandlwana. Ian Knight has given much food for thought and I look forward to working with him as a co-director on the project in the years to come. The project would not have been possible without the co-operation and assistance of Jenny Hawke, curator of the Zululand Historical Museum, Eshowe, along with the staff of the museum. Warm thanks to Graham and Meg Chennels and their family and staff at the George Hotel, Eshowe for their hospitality. The Anglo-Zulu War Archaeological Project is carried out with the full co-operation and collaboration of Amafa aKwaZulu-Natali, and the assistance and co-operation of Director, Barry Marshall and Chairman, Arthur Kronigkramer is gratefully acknowledged. Len van Schalkwyk, formerly Assistant Director of Amafa is joint field director at Isandlwana and has helped make our time in South Africa such an amazing experience. The British Academy provided funding for the work outlined in this paper. Equipment and administrative support was provided by GUARD and the Department of Archaeology, University of Glasgow, most notably Jen Cochrane, and Mel Richmond. Finally, thanks to Beth for finding me there.

References

Cope, R. 1999. *Ploughshare of War: The Origins of the Anglo-Zulu War of 1879*. University of Natal Press, Pietermaritzburg.

Coupland, R. 1948. *Zulu Battle Piece: Isandlwana*. Tom Donovan, London (1991 reprint).

Dominy, G. 1993. 'Frere's War'?: a reconsideration of the geopolitics of the Anglo-Zulu War of 1879. *Natal Museum Journal of Humanities* 5, 189-206.

Duminy, A. & Guest, D. (Eds) 1989. *Natal and Zululand: from earliest times to 1910*. University of Natal Press, Shuter & Shooter, Pietermaritzberg.

Duncan, J. and Ley, D. (Eds) 1993. *Place/Culture/Representation*. Routledge, London.

Emery, F. 1977. *The Red Soldier*. Hodder and Stoughton, London.

Gon, P. 1979. *The Road to Isandlwana*. Jonathan Ball, Johannesburg.

Guy, J. 1979. *The Destruction of the Zulu Kingdom: the Civil War in Zululand, 1879-1884*. University of Natal, Natal.

Knight, I. 1995a. *The Sun Turned Black: Isandlwana and Rorke's Drift, 1879*. Willaim Waterman Publications, Rivonia.

Knight, I. 1995b. *The Anatomy of the Zulu Army*. Greenhill Books, London.

Knight, I. & Castle, I. 1993. *The Zulu War Then and Now*. Battle of Britain Prints International, London.

Kronigkramer, A.J. 1999. A messsage from Amafa aKwaZulu-Natali. Preface to Knight, I., *Dead Was Everything: A retrospective of the Anglo-Zulu War*, Amiara, Pietermaritzburg.

Laband, J. 1985. *Fight Us in the Open*. Shuter and Shooter, Pietermaritzburg.

Laband, J. 1997. *The Rise and Fall of the Zulu Nation*. Arms & Armour Press (1998 reprint), London.

Laband, J. & Mathews, J. 1992. *Isandlwana*. Centaur Publications, Pietermaritzburg.

Laband, J. & Thompson, P. 1989. The Reduction of Zululand, 1878-1904. *In:* Duminy, A. & Guest, B (Eds), *Natal and Zululand: from Earliest Times to 1910*. University of Natal Press, Shuter & Shooter, Pietermaritzburg, 193-233.

Mitchell, P. 1998. Archaeological collections from the Anglo-Zulu War in the collections of the British Museum. *Southern African Field Archaeology 7*, 12-19.

Morris, D. R. 1965. *The Washing of the Spears*. Jonathan Cape, London.

Mountain, A. 1999. *The Rise and Fall of the Zulu Empire*. kwaNtaba publications, Constantia.

Pollard, T. 2001. Place Ekowe in a state of defence: the archaeological investigation of the British Fort at KwaMondi, Eshowe, Zululand. *In:* Freeman, P.W.M. & Pollard, T. (Eds), *Fields of Conflict: Progress & Prospect in Battlefield Archaeology*, BAR Institute Series 958, Oxford, 229-237.

Scott, D.D., Fox, R.A, Connor, M.A. & Harmon, D. 1989. *Archaeological Perspectives on the battle of the Little Bighorn*. University of Oklahoma Press, Norman.

Smail, J.L. 1969. *With Shield and Assegai*. Howard Timmins, Cape Town.

Tilley, C. 1994. *A Phenomenology of Landscape: Places, Paths and Monuments*. Berg, Oxford.

Notes

1. The rather cumbersome term 'Anglo-Zulu War' has generally been adopted by historians in favour of the more popular 'Zulu War' simply to distinguish it from the many 19th century conflicts that involved the Zulu people. Knight (1995b) lists no fewer than 14 campaigns between Shaka's ascension in the 1820s and the Anglo-Zulu War in 1879. These included several conflicts between the Zulu and the Boers, and various Zulu civil wars, including the struggle that brought Cetshwayo to the throne in 1872. Nor was the Anglo-Zulu War the end of it, with the last civil war of the early 1880s finally bringing about the destruction of the old Zulu Kingdom and the Bambatha rebellion of 1906 taking conflict well into the 20th century.

2. The political motives behind Confederation and the Anglo-Zulu War were complex and cannot be discussed in any depth in this paper. They included the need for land and raw materials including coal, gold and diamonds; access to a large workforce (for the extraction of raw materials) and a need to control the strongly independent Boers. For an in depth consideration of these and other political issues surrounding the Anglo-Zulu War Jeff Guy's The Destruction of the Zulu Kingdom (1979), Richard Cope's *Ploughshare of War* (1999) and Graham Dominy's *'Frere's War'?: a reconsideration of the geopolitics of the Anglo-Zulu War of 1879* (1993) are thoroughly recommended.
3. It was the incursion by Sihayo's sons into Natal in 1878 in pursuit of two of their father's supposedly adulterous wives, and the latters' subsequent murder, which provided one of the excuses for the British invasion in 1879. One of the stipulations of the British ultimatum was that the king hand over these men for trial. At the time of the attack by the British Sihayo himself and many of his men were at Ulundi preparing for war. Word of Chelmsford's attack on Sihayo's homestead reached Cetshwayo, and judging this to be the most aggressive of the three invading columns he despatched his main impi to engage them at Isandlwana (no doubt having received much valuable intelligence from Sihayo and his followers). The relative ease with which the Zulu were defeated during this skirmish may also have encouraged a sense of complacency among the British high command which was to lead to the lack of adequate defences at the camp at Isandlwana (Knight pers. comm.).

Tony Pollard
Department of Archaeology
University of Glasgow
Gregory Building
Lillybank Gardens
Glasgow, G12 8QQ

Maps and Decisions: Buller South and North of the Tugela, 1899-1900

Martin Marix Evans

ABSTRACT: In London, and certainly after arrival in South Africa, Sir Redvers Buller could and should have seen the Intelligence Department, War Office (IDWO) map No. 1449, Sketch Map of Country round Ladysmith, published in 1899. It is misleading about the drifts (fords) across the Tugela River, and the heights and positions of hills near Colenso, Spioen Kop and Pieters where the British fought from December 1899 to February 1900. Thomas Pakenham writes of Field Intelligence Department maps that were made for Buller at the time, but these were also faulty. Thus, against Botha on the Tugela, Buller was fatally ill-informed. Natal north of Ladysmith was well mapped. The Military Sketch of the Biggarsberg, IDWO No. 1223, was published in colour, 16 sheets, 1 inch to the mile, in 1897. Along the roads and railways the detail is excellent, further afield less satisfactory. When Buller faced the task of ousting the Boers from Liang's Nek in June 1900 he had all the data required to plan a flanking movement through Botha's Pass, supported by 5 inch guns of the Royal Garrison Artillery on Van Wyk Hill, and, acting with similar dash at Alleman's Nek, forced the Boers to withdraw from Natal. This paper reviews Buller's poor reputation in the light of these new findings.

1. Introduction

The reputation of Sir Redvers Buller, one of the leading British generals of the Boer War of 1899-1902, was severely damaged by the perception of his generalship; and in particular the adverse reporting of the correspondent to *The Times* newspaper, L.S. Amery. His reports of the battles of Colenso, Spioenkop and Vaalkrans are often cited, but subsequent engagements in which Buller enjoyed success are usually neglected. The factors that differ between these early failures and the later victories include Buller's learning how to manage the support of infantry with artillery in the new conditions, and the modification of infantry tactics to counter the effectiveness of the newly introduced high-velocity rifle, subjects that lie outside the compass of this paper. Substantially neglected, however, is the influence of the differing information available to him about the nature of the terrain on the ground and consequently the aim of this paper is to try and rectify this neglect.

2. Maps and the Boer War

The principal interests of the British prior to the outbreak of war in 1899 were to make good the woeful insufficiency of maps of the Boer Republics and to create improved maps of the borderlands of their colony of Natal. Captain S.C.N. Grant, R.E., assisted by another engineer, Captain H.R. Gale and by Captain W.S. Melville, Leicestershire Regiment, was sent in 1896 to carry out a large scale topographic survey of the Biggarsberg mountains in Northern Natal, assumed to be the area in which, should war break out, fighting would take place. The 21 sheets of 1 inch-to-the-mile (1:63,360) maps were published in colour the following year as Intelligence Department, War Office (IDWO) Number 1223, *Military Sketch of the Biggarsberg and of the Communications in Natal*. The detail shown along roads and railways is excellent, but the work was seriously incomplete in omitting coverage of the terrain south and east of Ladysmith. Away from the lines of communication the work becomes vague. The officers who prepared these maps were caught in the siege of Ladysmith and Buller had no access to them.

Of the terrain south of Ladysmith maps did exist, but they were principally educational maps or cadastral maps, compilations of farm surveys of doubtful accuracy and intended to establish boundaries of properties. Topographical information was lacking. *Russel's School Map*, five miles to the inch, was issued as a general map of the colony. There was an IDWO map, Number 1449, published in 1899 (Figure 1) and apparently a final version of the earlier IDWO 1308, which did not, however, show drifts (fords) over the Tugela River. The map is entitled *Sketch Map of Country Around Ladysmith* and the edition held at the Bodleian Library in Oxford is the revised printing of January 1900. This map should have been known to Buller and his immediate subordinates, though I can produce no direct evidence that this was so. On arrival on this front, Buller set up a Field Intelligence Department including Captain A. Kenny Herbert and Captain H.C. Simpson which compiled a map of the area along the Tugela which had to be taken to advance to Ladysmith. This would fill in the missing southern coverage of Grant's IDWO 1223. It is mentioned by Thomas Pakenham in *The Boer War* (Pakenham, 1979 pp. 208, 216, 227 & 283) and was, apparently, a blueprint in the collection of Lieutenant-Colonel A. Sandbach. The Public Record Office has a blueprint map relating to Colenso (Sheet K3, PRO CO 700 Natal 26 [2]) which appears to match Pakenham's description but which is dated 3rd January 1900. If this is the map on which Pakenham bases his observations it would not have been available in time for the Battle of Colenso. It shows three drifts west of Colenso, as appear on IDWO 1449, and another bridle drift to the west of those and east of Robinson's Wagon Drift, which is numbered 3a on the blueprint.

Major G. S. Elliot, R.E., Captain Elton and the range-takers of two batteries of the Royal Field Artillery carried out the best survey they could manage, given the presence of the enemy, before the Battle of Colenso. Elliot set out the ranges and bases on the plane table and checked with the rays in

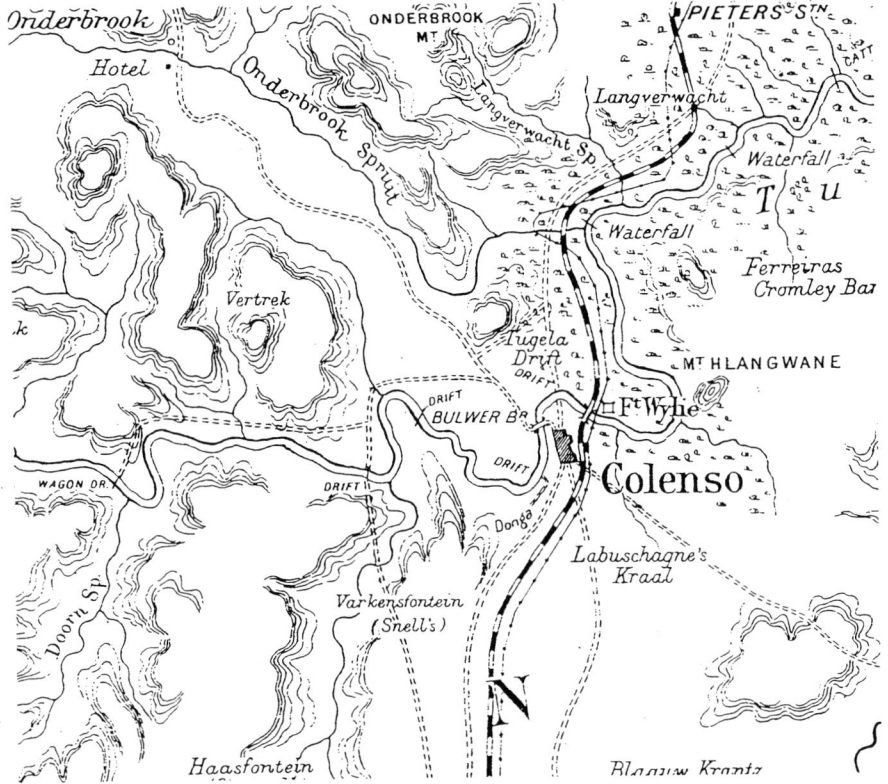

Figure 1. A detail from IDWO No. 1449 of 1899, revised 1900. (Bodleian Library, Oxford, 621-11t 1 [7])

the usual way to produce a very creditable sketch map (Elton, 1901/2) showing all the drifts marked on the subsequent blueprint. There is a little more detail than IDWO 1449, but not enough to be significant.

3. Topography of the Tugela

It was of course the case that Victorian soldiers rarely had detailed maps of the theatre of operations and had to depend on scouting and observation to gather information on the terrain. From the ridge now occupied by the Clouston Cemetery, which contains the remains of the fallen of this battle and which at the time was Buller's observation point, the land falls away gently to the meandering line of the Tugela. On the west it runs close under the hill by Vertrek farm, swings away and back again forming 'The Loop', a meander pointing towards the hill, before running south east and north east in another loop west of Colenso, from which the road north east through the hills to Ladysmith passes over it by Bulwer Bridge. The railway passes east of Colenso village, over the river and north, to the west of Fort Wylie which the Boers held, to wind on through the confusing array of hills to Ladysmith. To the east of the village and south of the river the hill of Hlangwane (Nhlangwini) offers an enfilading position against forces attacking the river, but this is hard to see from Buller's position as it is difficult to distinguish from the hills beyond. It appears that local people, such as the farmer, Pringle, who rescued an ammunition wagon after the battle, were not consulted.

In the event, on 15th December 1899, Major-general Hart took his Irish Brigade forward to cross the river on the left flank. It is supposed that he had been ordered to cross the bridle drift (ford) south of Vertrek on the track shown on IDWO 1449, but was misled by an African guide who claimed, wrongly, that a drift was to be found in the Loop. Hart's willingness to be guided as he was, could very well have been encouraged by Elliot's map — which clearly shows a drift close to the legend describing Bulwer Bridge to the east of Colenso on the revision of IDWO 1449 (Figure 1). The modern map does not concern itself with such primitive crossings as drifts, but a farm is shown, Tugela Drift, to which a road leads. This road, projected to the river, would take one to the alleged drift position on IDWO 1449.

On the eastern flank Colonel Long took two Royal Artillery Field batteries well forward to a position approximately above the first 'o' in Colenso on IDWO 1449, well ahead of the infantry (Figure 1). The Boer's Mauser fire from the river bank and from Nhlangwini cut his men to shreds. The twin crises led Buller to order a complete withdrawal before their losses became intolerable. It was a sensible decision, but the reverse at Colenso, taken together with the other defeats of 'Black Week', Magersfontein and Stormberg, cost Buller his position as Commander-in-Chief.

In the next month attempts were made to breach the Boer lines at Spioenkop and at Vaalkrans. The former was another fight lost through lack of appreciation of the terrain, though at the end of the day the Boers also thought they had lost the battle. Vaalkrans was well mapped by Field Intelligence on 1st February 1900 (PRO ADM 1/7491) but the area covered by the map did not live up to its title: Sketch of the Enemy's Position. As Major F. W. von Wichmann's map (Bodleian Library E54:11 [40]) of the dispositions of the troops on the day of the battle, 5th February, shows, there was a 155 mm Creusot gun to the south east which was to make the British position on Vaalkranz untenable. Further, the relative importance of the various hills is not clear from the Field Intelligence map. British artillery support from Zwart Kop was significantly better than in the previous encounters and the map might thus have contributed to the avoidance of another Spioenkop. The eventual breach of the Tugela line was made as a result of conducting a sequence of operations by infantry with close artillery support from heavy guns within visual range, an early example of the lifting barrage.

After the relief of Ladysmith, and now under the distant, and largely indifferent, overall command of Lord Roberts, Buller advanced into northern Natal. The Boers at first offered resistance on the Biggarsbergs, but were quickly and elegantly rolled up by the British. They then fell back to the scene of their triumph in the First Boer War, Langs Nek, to await the inevitable British assault. Buller made a demonstration of force before this pass through to the Transvaal, and then arranged a truce to discuss possible peace terms. Meanwhile part of his army was quietly deploying westwards. A 5 inch gun of 16th Battery, Southern Division, Royal Garrison Artillery, was positioned on Van Wyk hill and, when the truce expired, opened fire on Botha's Pass through the Drakensberg Mountains to the west. Buller himself accompanied his men over the mountains in a fast flanking movement that left the Boers at Langs Nek helpless and obliged to retreat, abandoning their defence of their country's border. It was an operation that would have done credit to a crack Second World War commander.

4. Conclusion

The key difference between the battle of 1881, which the Boers won, and the encounter in 1900 was the availability of IDWO 1223, Grant's 1 inch map. Not only did it permit the planning of the operation, but also the laying-down of accurate artillery fire at a distance. The terrain was favourable to defence, but knowledge of it from precise mapping, I suggest, tipped the balance towards the British.

Acknowledgement

I am grateful to Colonel M.A. Nolan for his comments after the first presentation of this paper, his account of War Office mapping activity and his suggestions of additional sources of information. Any errors or omissions remain my responsibility. Maps were consulted in the archives of the Public Record Office (PRO), Kew, and the Bodleian Library, Oxford.

References

Elton, F.A. 1901/2. The Boer Position on the Tugela, 1899. *Proceedings of the Royal Artillery Institute* XXVIII, 397-398.
Jewitt, A.C. 1992. *Maps for Empire: the First 2000 Numbered War Office Maps.* The British Library, London
Marix Evans, M. 1999. *The Boer War: South Africa 1899-1902.* Osprey Publishing, Oxford.
Marix Evans, M. 2000. *Encyclopedia of the Boer War.* ABC-CLIO, Oxford.
Pakenham, T. 1979. *The Boer War.* Weidenfeld & Nicolson, London [Reprinted by Abacus, 1992].
Powell, G. 1994. *Buller: A Scapegoat?* Leo Cooper, London.

Martin Marix Evans
3 Murswell Lane
Silverstone, Towcester
Northants, NN12 8UT

The Thirty-Years War, 1914-1945: Mapping the Battlefields of the Past for the Construction of the European Future

Edmon Castell & Sònia Roura

ABSTRACT: The Great or First World War inaugurated an era of slaughter on the European Continent. This thirty year period between 1914 and 1945 can arguably be considered the bloodiest in the history of humankind. Entire libraries deal with the military, economic and diplomatic aspects of this period of Total War. There are many reminders left in the landscape: fortifications, trenches, cemeteries, all of these are mute witnesses of the descent into barbarism. The most recent projects make battlefields accessible and comprehensible through conceptual and technical means, and show that these historical spaces are a highly effective resource for comprehending European history during the first half of the 20th century.

1. Descent into hell: the thirty-years war

Over the last 200 years, the European continent has been subject to numerous armed confrontations: wars to conquer territory, wars of resistance, civil wars, wars to gain liberty. Nonetheless, on 28th June, 1914, not even the young Serb, Garilo Princip, could imagine the magnitude of the consequences that the assassination of an Austrian Archduke would trigger in Europe: the first great slaughter in the industrialised world in history (Le Goff, 1996, p. 76).

The shooting at Sarajevo, 'like a gunshot at a concert', caused a chain reaction. Austria declared war on Serbia, and this factor initiated the unstoppable spiral of war. Russia took the side of its ally Serbia. Germany, allied with Austria, declared war on Russia and its ally, France. The neutral borders of Belgium and Luxembourg were occupied by the German army, and Great Britain, as guarantor to these countries, intervened in the conflict, which was no longer exclusively limited to the European scenario.

The war, on a European and world scale, became a total, all-out war. In successive phases, the war was transformed into a strategy for annihilation of the adversary, into a struggle to wear out the enemy countries and their economic resources more than a confrontation between armies on 'the fields of death'. As the war drew on, it involved more and more the complete mobilisation of industrial, technological and other resources. This same concept of a war of total mobilisation eliminated the distinction between combatants and non-combatants. The application of scientific knowledge to the means of mass destruction during the first half of the 20th century radically changed the character of war and threatened the very survival of humankind.

With the Great War, the world began to sink into barbarism. It was the beginning of the bloodiest era in history to this day. Some historians consider the cycle of events beginning in 1914 and ending in 1945 as a sort of European 'civil war', Pavone (1999, p. 259) calling it 'The Thirty-Years War'.

It was a period in which liberal European regimes had great difficulty in insuring political stability because, after the end of the Great War, those societies were characterised by deep economic crisis — increasing after 1929 — and strong social movements. Liberalism was questioned on all flanks. In Europe the conservative classes lived under the fear of revolution. And if finally, in countries such as Italy and Germany, fascist regimes came into power, it was because dominant sectors confided in this form of government to get out of the dilemma they felt they were in (Carme Molinera & Ysàs, 1999).

By the same token, the dynamics of political confrontation and social tension occurring in Spain during the 1930's must be viewed in the wider context of European radicalisation during the first half of the 20th century. In this sense, the establishment of the Franco Regime in 1939 as a result of the use of military insurrection and of the victory of the conservative and reactionary armed forces in the Spanish Civil War (1936-1939), fits fully into the history of Europe at that time.

Individual episodes, such as the initial Republican offensive and resistance at the Battle of the Ebro River (1938) display this European dimension in various ways. It can be seen in their strategic objectives, given that the Spanish Republican leaders sought to internationalise the conflict in view of the foreseen outbreak of the Second World War. It is also evident in the participation of foreign forces in military operations, with Italian and German aircraft jointly bombing the Republican combatants, whose forces in turn consisted of international volunteers. The final outcome of the war also showed this European dimension, as the peace treaty was signed in the city of Munich by the governments of Great Britain, France, Italy and Germany (September 1938), and it stipulated that a large part of Czechoslovakia should be handed over to Germany. The Peace Treaty of Munich dealt the definitive blow to the Republican offensive initiated on the Ebro three months earlier and led the way to a Fascism which was not to be stopped, in neither Central Europe nor the Mediterranean region.

2. Remember their example and sacrifice[1]

In some aspects, this historical terrain is very familiar. Entire libraries deal with the military, economic and diplomatic events during this period of Total War. Nonetheless, the memories of this entire period are not only to be found in history books or in the minds of those who survived. At the very scenes of events, at the battlefields, these memories are part of the landscape. The past is visibly written in the landscapes of the Somme River, the Ebro River,

Normandy or Auschwitz. The memory of the Great War is expressed in a military and national sense through the monuments erected to the dead soldiers. The memorials erected during the decade of the 1920's in the memory of the Great War slowly became immense centres of mourning for those fallen for the homeland. The majority of the monuments erected in the 'fields of honour' are of an epic nature, showing the unconditional commitment of the soldiers, and an identification with certain values and attitudes considered exemplary, such as heroism, etc. The causes of the confrontations are never explained, nor the interests of each side in the conflict. Even today, in Verdun, the memorial of Fleury-devant-Douaumont describes 'the bloody struggle of the adversaries', in the name of different ideals, but with similar sacrifice. Hence, this memorial testifies to what happened here: 'the everyday heroism, the physical and moral suffering, a willpower generating the highest of virtues'.

In a similar fashion, the scenes of the Battle of the Ebro became an immense space of mourning, in which the victors, Franco's forces, rendered homage to the sacrifice of the fallen 'in the lands Spain to construct a legendary and traditional empire' and reminded the population of the origin of their power, their total victory at the Battle of the Ebro and in the war. The victors of the Spanish Civil War covered some symbolic spaces of their counter offensives with funeral monuments. In addition to these strategic spaces, the victors occupied many areas of towns with more crosses and ostentatious monuments which commemorated 'the combatants who attained glory'. These relics remind the population of the 'contract' that the victors signed with the blood of their honoured dead, and appear with an aura of immortality which invites to passivity (Castell & Falcò, 1999).

Furthermore, in the peace of cemeteries, the funerary monuments — or the geometry of millions of white crosses — hardly serve to evoke the individuality of the combatants who fought on the battlefronts. They were created to evoke a great slaughter. Nonetheless, these material remains become unfamiliar with the passage of time. Seventy years later, 'who remembers the millions of soldiers who suffered or died in the mud at Flanders between 1914 and 1918?' (Leguineche, 1999).

The iconography of the past is linked to positivist traditional historiography according to which 'history is confused with war, and this in reality has been a manner of perceiving history in general which has prevailed over many other' (Villar, 1991, p. 9). It is a concept still present in numerous adaptations of historical spaces, such as the permanent exhibit planned for the new Museum of the Army in Toledo's Alcazar palace, which attempts to narrate the 'real history of Spain, and the military point of view is one of the most important. We hope we will be capable of making real history comprehensible, with its good sides and its bad sides, without emphasising periods which have not been the happiest of our past'[2].

3. Who would want to preserve this?

Battlefields, when considered as historical spaces associated with wartime events, have only been studied within the field of industrial archaeology, insofar as they provide objective information 'on the adaptation of the various forces involved in a conflict to the terrain, the different specialised systems of military engineering, the types of fortifications and adaptation of shelters, casemates and trenches or the location of artillery, provisional airstrips and a long list of other military constructions and remains' (Rovira, 1988). Generally, this research stresses the technological advance that all armed confrontations entail.

If adequately signposted and adapted for visitors, battlefields become a useful didactic resource for gaining historical knowledge, for their content as well as for the procedures and values which they can communicate. Above all, cultural landscapes are a material source of knowledge on societies and social processes located in the past. Instead of confining students' contact with history to the pages of a book, analysis and interpretation directly in historical spaces carries an emotive charge which frees us of encyclopaedic knowledge. A simple *in situ* reconnaissance allows one to better understand history, because the topography contained in the mind is suddenly transformed into something real at the scene of the action. In the same manner, at the extermination camps at Birkenau, 'millions of shaken lives come to our mind because of the immediacy of the place' (Becker, 1998, p. 245).

In the second place, learning about history does not consist exclusively of accumulating historical concepts. Studying that which is tangible provides us with more opportunities to find patterns for comprehending the past. Historians' work procedures are the very foundations of that which we understand by history as a discipline. The procedures constitute the basis which allows students to form hypotheses, find documents, contrast sources, make judgements, in sum, to be capable of critically analysing social reality.

In the same fashion as the genre of the military comic, the possibilities for didactic use are not limited solely to 'the capacity to graphically reconstruct the circumstances of the conflict, but above all, lie in the orientation and concepts which these stories transmit' (Ordònez, 1999, p. 64). Battlefields, like comics, reflect diverse points of view which can be analysed and evaluated in the classroom. In these interpretative historical spaces, students can understand the causes and motives explaining the different behaviour of people and social groups in their historical context.

The most recent museological innovations are oriented towards this approach. The *Historial de la Grand Guerre* in Peronne, France (inaugurated in 1992), explains the First World War, its origins and consequences through the parallel histories of Germany, France and the United Kingdom, offering a global cultural vision of the war that goes from the experience of soldiers to that of civilians in the rearguard of the countries involved. In this manner, the *Historial* in Peronne, initially designed as a museum specialising the Battle of the Somme (1916), has become an important centre which, equipped with abundant pedagogical resources, facilitates the keys to understanding the whole period of the Great War.

By the same token, battlefields can become spaces for interpreting European history during the first half of the 20th century. The period between 1914 and 1945 represents a time when all aspects that had characterised 19th century, liberal capitalist, society disintegrated. It was an era of European — and worldwide — wars which were followed by social revolutions and the fall of former empires. It was a period in which the world economy was on the verge of collapsing, while democratic institutions were being defeated everywhere (Hosbawm, 1995). In Europe, the history of the second half of the 20th century was written about the defeat of Fascism. Its identity was created on the basis of this victory.

Although museological adaptations having to do with battlefields on the European continent continue to be of very unequal quality — or simply non-existent — the process of European integration brings forth the need to reconsider this entire period, which brutalised several generations. We are at the end of what may be considered as the cruellest century in European history; such a past does not simply disappear. It is present in all of our countries...'[3]. Indeed, if we want to understand and, finally, overcome the memory of the disastrous history of Europe during the first half of the 20th century, we must revisit the thirty-year period which activated the forces that distanced us from a democratic, unified and supportive Europe.

References

Becker, A. 1998. Musées ouverts, traces des guerres dans les paysages. *In:* Joly, M.-H. (Ed.) *Des Musé es d'Histoire pour l'Avenir.* Noêsis, Paris.

Carme Molinero, P.Y. 1999. La instauració d'un nou ordere politic, socioeconòmic I cultural; el primer franquisme. *Jornades Sobre la fin de la Guerra Civil,* in press.

Castell E & Falcó, L. 1999. Interpretar la batalla del Ebro. Los campos de batalla como recurso didàctico. *Iber* 19, 85-97.

Hobsbawm, E. 1995. *La época de la guerra total. Historia del Siglo XX.* Critica, Barcelona.

Ordóñez, A.A. 1999. Historietes en el camp de batalla. *El Corrreu de la UNESCO* 247, 64.

Le Goff, J. 1999. *Europa explicada als Joves.* Anagrama-Empúries, Barcelona.

Leguineche, M. 1999. El ultimátum era a un mundo. *El Pais,* Serial pullout No 5.

Pavone, C. 1999. La guerra en la historia. Apuntes para una investigación sobre la guerra total en el siglo XX. *In:* Lorenzo, A. V. (Ed.), *La guerra en la Historia.* Ediciones Universidad de Salamanca, Salamanca.

Rovira, J. 1988. Un nou aspecte de l'arqueologia contemporània catalania. L'arqueologia de la Guerra Civil. *I Jornades sobre Arqueologia Industrial de Catalunya.*

Villar, P. 1991. *L'Historiador I les Guerres*. Eumo Editorial, Barcelona.

Notes

1. Inscription on the monument erected by supporters of Franco at Punta Targa, by the Infantry Regiment of Our Lady of Montserrat, in memory of the Battle of the Ebro.
2. Statements by Eduardo Serra, Spanish Minister of Defence, reported in *El Pais*, Wednesday 3rd February, 1999.
3. Statements by Nobel Peace Prize winner, Fritz Stern, reported in *El Pais*, Monday 18th October, 1999.

Edmon Castell & Sònia Roura
Departament de Didactica en les Ciencias Socials
Universidad de Barcelona
Edificio de Llevant
despatx 121
08035 Barcelona

Terrain and the Gallipoli Campaign, 1915

Peter Doyle & Matthew R. Bennett

> **ABSTRACT**: The Gallipoli Campaign of 1915 was one of the most strategically significant theatres during the Great War of 1914-1918. A land system analysis of the Gallipoli Peninsula was carried out, and five land systems, based on aspects of geology, geomorphology, hydrogeology and vegetation, were identified. The landings of 25th April 1915 were made at Cape Helles and Anzac Cove, with objectives to capture the high ground. The land system analysis demonstrates that these landing places were disadvantaged by terrain, with steep, deeply incised slopes, narrow beaches and inadequate water supplies. A later landing at Suvla Bay in August 1915 had more terrain advantages, with wide landing beaches and locally available water supplies, but the tactical advantages of a lightly held terrain were not exploited.

1. Introduction

The Gallipoli campaign of 1915 ranks as one of the most controversial of the Great War. Often seen as a 'side-show' to the main events of the Western Front, it was a costly gamble intended to knock Turkey out of the war and to command the main supply line to Russia. The land-based Gallipoli campaign evolved from a plan sponsored by the First Lord of the Admiralty, Winston Churchill, to 'force' a way through the narrow Dardanelles using ships alone, knocking out the fortresses commanding the straits, and ultimately threatening Constantinople (Istanbul). A combination of factors led to the failure of this plan, most particularly the ineffectiveness of naval artillery fire against land targets, the presence of mobile batteries on both the European and Asian shores and ineffective minesweeping by the allies (Figure 1). This ultimately caused an escalation of the conflict, and a commitment to deploy troops in an invasion of the Gallipoli Peninsula.

The Gallipoli campaign is controversial because it is usually considered as having been hastily conceived, and poorly executed. Perhaps above all, the allied general staff had a poor understanding the nature of the terrain that they were expected to fight on. During the planning stages, and in the initial landings, they had inadequate and poorly-surveyed maps, and little aerial reconnaissance. It is argued that there was little or no direct knowledge of the terrain, and little time to assemble expert advice, particularly with respect to the provision of resources and the nature of the ground conditions. This contrasts with the extensive understanding and use of terrain by the Turkish troops, which led ultimately to their victory in the campaign, and the withdrawal of allied troops.

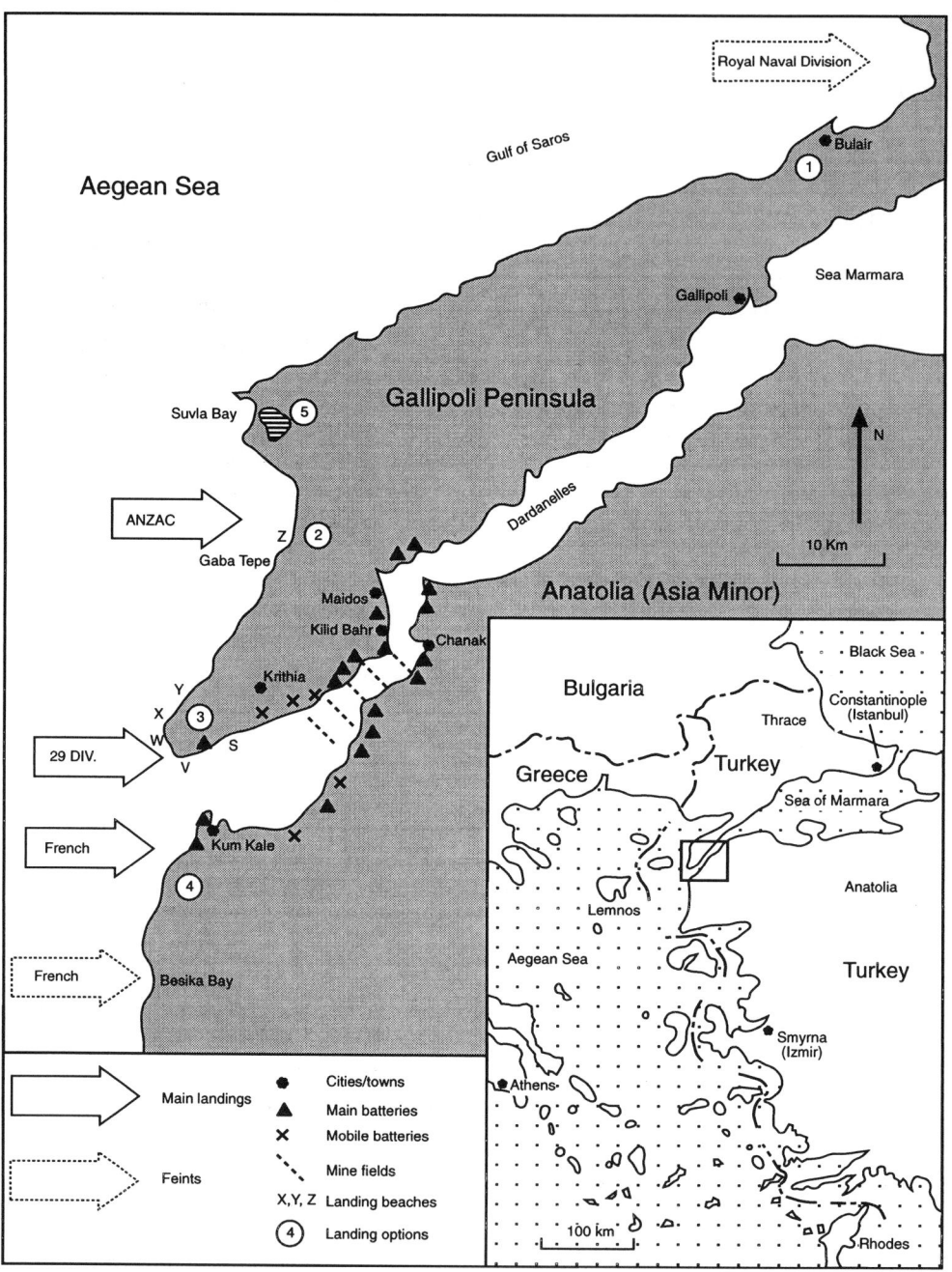

Figure 1. Map of the Gallipoli Peninsula and the Dardanelles, showing landing options of the 1915 Campaign, and the main landings and feints made on 25th April 1915

The purpose of this paper is to outline the gross terrain characteristics of the Gallipoli Peninsula, and consider its impact on the Gallipoli Campaign of 1915. It represents a summary of our earlier paper on the subject (Doyle & Bennett, 1999), updated to examine in a little more detail current views on the controversial initial reconnaissance carried out prior to the landings in April 1915. As in our previous paper, land system mapping is used in order to characterise the typical terrain units involved in terms of their geology, geomorphology, surface 'going' characteristics, vegetation and hydrogeology — all of which influenced the tactical use of ground. An overview of the terrain characteristics of the Gallipoli Peninsula is given below, followed by the details of land system classification. These elements are then discussed in relation to the strategic and tactical aspects of the campaign, before determining their contribution to the overall allied defeat.

2. Topography of the Gallipoli Peninsula

Climate and vegetation

The climate of the region is typically Mediterranean, with mild winters, the mean January air temperature being normally between 7-9° C, and hot summers, with average air temperatures exceeding 25° C in July and August. There is a marked summer drought, although annual precipitation is normally between 600-700 mm at sea level, rising to in excess of 1,000 mm in mountain regions (United Nations, 1982). The vegetation reflects the Mediterranean climate, developed by human clearance from the original mixed woodland, and comprises low dense herbaceous and aromatic shrubs of *garrigue* type. There are few wooded areas, and trees of evergreen oak and pine are usually isolated and scattered (Naval Intelligence Division, 1942).

Geology

The Gallipoli Peninsula forms part of the Alpine Pontide range. The Pontides have a strong east-west structural grain, and comprise ancient crystalline massifs developed in Anatolia, and folded Mesozoic-Cenozic sediments in Thrace and basement margins of Anatolia. The most dominant feature is the North Anatolian Fault zone, separating the European and Anatolian plates, which runs under the Sea of Marmara and crosses the Peninsula to the Gulf of Saros, forming the northern, rifted and strongly rectilinear margin of the peninsula and separating it from the rest of Thrace (Sengor & Yilmaz, 1981; Elmas & Meriç, 1998). This fault zone has predominantly strike-slip movement, and is complex, as other branches of it form the Dardanelles and the Sea of Marmara. This fault zone is still active today (Crampin & Evans, 1986). During the Neogene movement of the fault developed a transtensional basin which produced the Sea of Marmara, with a maximum depth of 1000 m, and led to the deposition of the thick Neogene sediments on either side of the Dardanelles (Crampin & Evans, 1986; Perincek, 1991; Elmas & Meriç, 1998). The Gallipoli Peninsula is therefore mostly composed of Palaeogene and

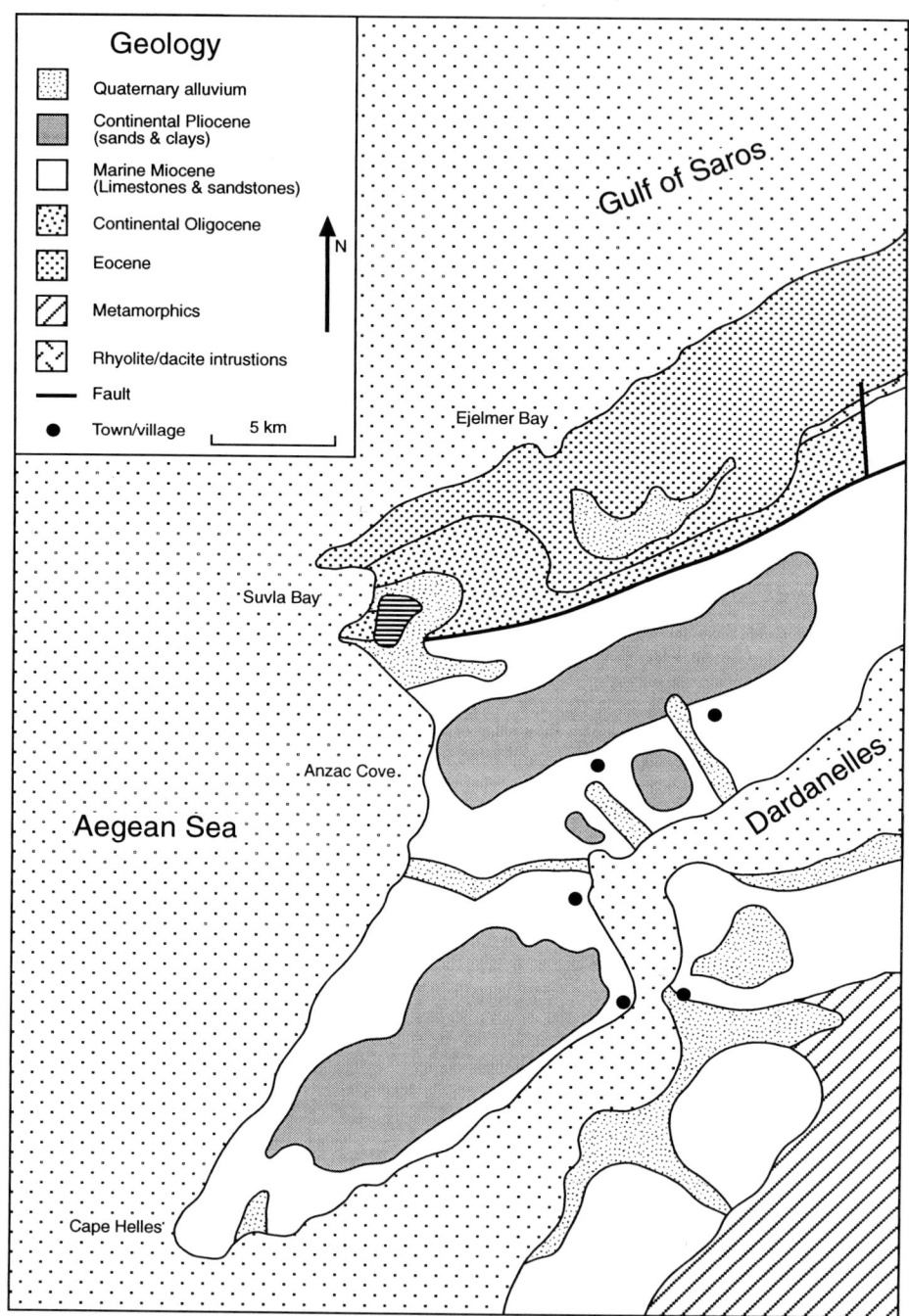

Figure 2. Geological map of the Gallipoli Peninsula and the Dardanelles

Neogene sediments in a simple, relatively undisturbed relationship (Erguvanli, 1957; Sengor & Yilmaz, 1981; Ternez et al., 1987; Elmas & Meriç, 1998; Figure 2).

The oldest basement exposed in the Peninsula is an outlier of metamorphic basement north east of Bolayir (Bulair), at the point where the Peninsula is at its narrowest. The Dardanelles are bounded on both sides by a Palaeogene to Neogene sedimentary basin with relatively undisturbed limestones and sandstones (Figure 2). Further to the east the mainland of Anatolia exposes basement of Palaeozoic-Mesozoic high grade metamorphic rocks, and marine Miocene lies in nonconformity with these close to the present day port of Çanakkale (Figure 2). The oldest Tertiary beds are those of Middle and Upper Eocene sediments which form the northern coast of the Peninsula, paralleling the Gulf of Saros. Some limited acid intrusive rocks are found associated with these. These Eocene beds are succeeded by continental Oligocene deposits, and marine sediments of Miocene age. Continental Pliocene caps most of the upland areas south of Suvla Bay, and Quaternary alluvium and related sediments are found in valleys (Figure 2).

Relief

The relief of the southern part of the Gallipoli Peninsula is relatively subdued, the dominant topographic elements being a series of ridges in the north and two northeast-southwest trending plateaux in the south (Figure 3). The northern ridges are formed from folded Palaeogene sandstones and limestones, and further north, Cretaceous rocks. This contrasts with the southern plateaux which are formed from Pliocene sediments overlying bedded Miocene limestones. The margins of these plateaux are heavily dissected, forming a complex network of sharp-crested interfluves. In most cases the slopes are heavily vegetated. The exception is the northern margin of the Sari Bair plateau which is marked by a fault line scarp. Beneath the steep upper face of the scarp the slopes are heavily gullied and are barren of vegetation, forming classic 'badland' topography. In the southeastern part of the peninsula, the slopes of the Kilid Bahr Massif are strongly gullied, in some cases forming deep ravines. These ravines exploit the structural grain of the Peninsula, to give a parallel-alignment to the drainage of the southern peninsula (Figure 3).

Hydrology

The majority of rivers within the southern Gallipoli Peninsula are seasonal, and most valleys are dry for much of the year. Exceptions occur in the northern part of the study area, on the margins of the Suvla Plain, where there are some perennial streams. All the major lithological units have potential as aquifers. Few detailed hydrological studies have been carried out in the Gallipoli Peninsula itself, as it is relatively unpopulated, and most studies have concentrated upon the Ergene Basin to the north, strategically important for the supply of water to Istanbul (Karatekin, 1953; Pamir, 1953; United Nations, 1982). However, it is clear from studies of Neogene and

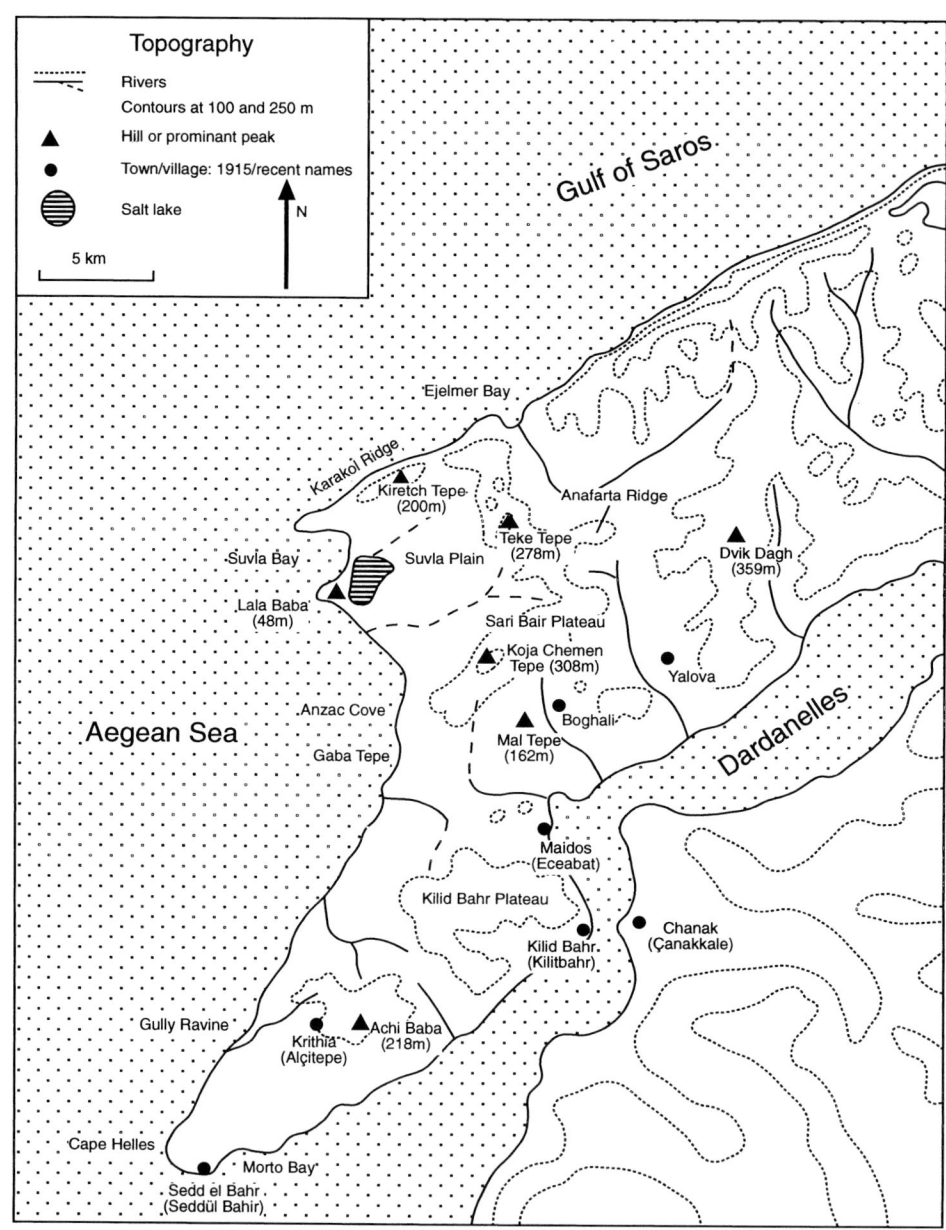

Figure 3. Topographic map of the Gallipoli Peninsula and the Dardanelles, showing the main areas of habitation, upland areas, and rivers in 1915

younger sediments on the southern margin of the Dardanelles that the main aquifer potential lies with the Miocene limestones and the Quaternary alluvial deposits (United Nations, 1982; Figure 4).

3. Terrain classification

The terrain of the Gallipoli Peninsula can be classified into a series of land systems based upon geomorphology, geology, hydrogeology and vegetation characteristics. The six land systems recognised by Doyle & Bennett (1999) are described briefly below, and are outlined in Figure 5.

Land system I
This land system consists of northeast-southwest trending plateaux which are dissected by deeply incised valleys. Land system I is developed on Pliocene continental sediments which consist of interbedded fluvial sands and lacustrine clays. This provides a relatively firm and dry 'going' surface, although disturbed finer grained sediments are prone to wind transport during the summer months. In most cases, land system I is characterised by relatively dense low growing *garrigue* shrubs typical of the Mediterranean coastal areas. The hydrological characteristics of this system are locally variable, dependant on the disposition of locally porous and impervious strata. Ground water is carried at depth, usually within limestone aquifers. Surface ground water is scarce, although it may occur in the form of perched water tables within the Pliocene sediments. This land system is typical of the Sari Bair and Kilid Bahr plateaux. A steep, northwest-facing fault scarp which trends southwest-northeast defines the northern boundary of the Sari Bair Plateau. The lower slopes of this fault line scarp are heavily dissected by numerous dry valleys with sharp interfluves. Valley sides are cut by gulls and rills. Typical slope angles range from 20-40° on the valley sides. This land system merges downslope into land system II. The plateau top south of the fault line scarp is dissected by northeast-southwest trending valleys. High points form along ridges, typified by the peaks of Chunuk Bair and Koja Chemen Tepe, with elevations of 261 and 308 m respectively (Figure 3). To the southwest of the plateau area there is an undulating dip slope (5-10°), deeply incised by dendritic channel systems. Much broader plateau surfaces typify the Kilid Bahr Plateau. This plateau has a mean elevation of 150 m and is dissected by steep sided river valleys. The level of plateau dissection is much less than that of the Sari Bair Plateau, although the valleys are generally much deeper. The Kilid Bahr Plateau also has an asymmetrical profile with steeper northwest facing scarp slopes and a gentler southeastwards dip slope dissected by incised northeast-southwest trending valleys. These valleys form a strongly rectilinear drainage network which exploits the underlying structural grain (Figure 4).

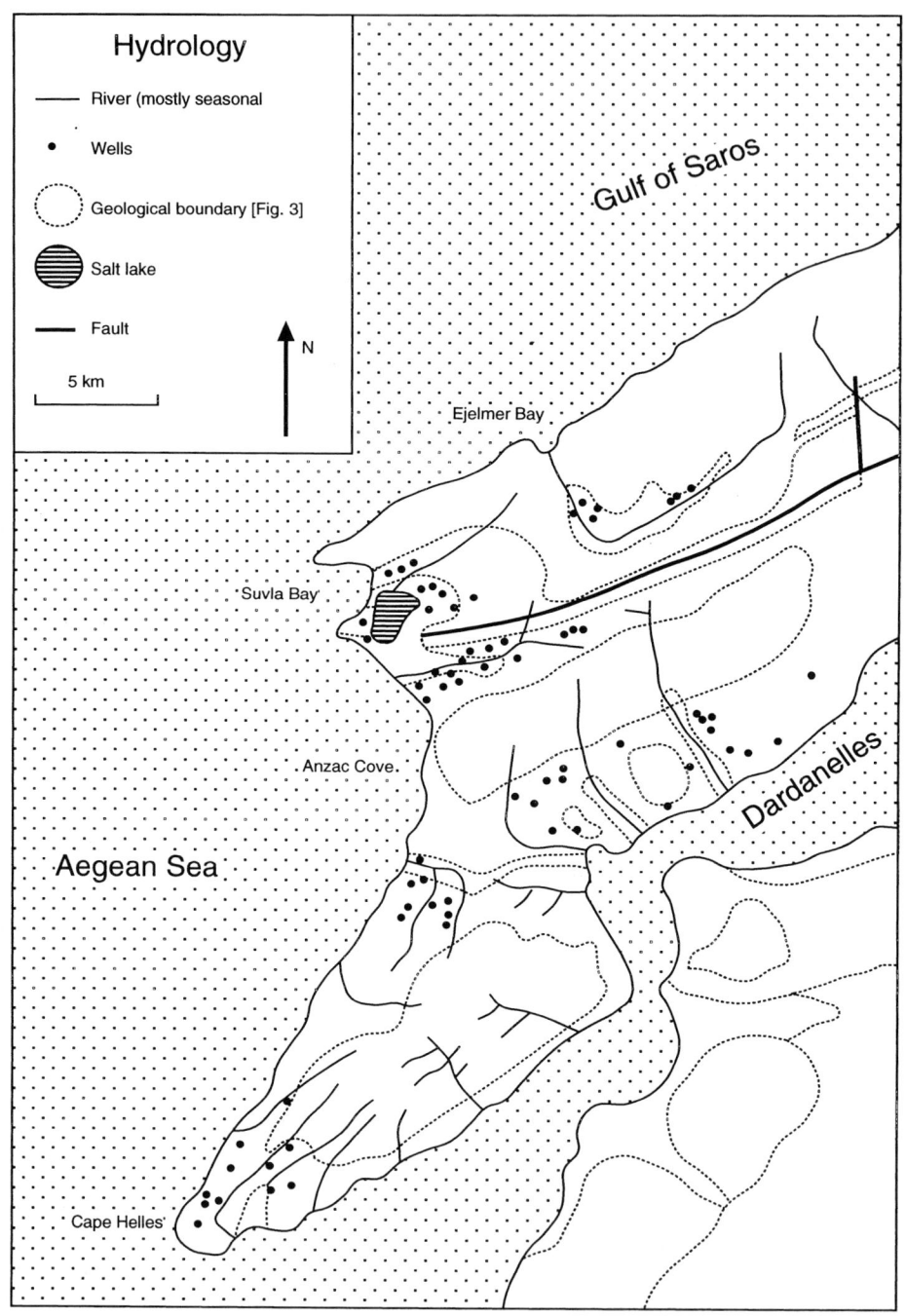

Figure 4. Hydrological map of the Gallipoli Peninsula showing the main rivers and the position of wells in 1915. For geological boundaries see Figure 2

Land system II
This land system is found at the base of land system I, and is transitional between the low lying slopes of land systems III and IV, and the steeper slopes of land system I. These slopes have relatively gentle slope angles and are usually dissected to a varying degree by a number of prominent dry valleys. Interfluves are broad and undulating. This land system is mostly developed on sub-horizontal bioclastic limestone, sandstones and marl units of Miocene age (Figure 2). It provides satisfactory 'going' surfaces which are mostly dry and firm, especially were there is insubstantial soil development, and particularly in the coastal areas. Most slopes are covered with dense, low *garrigue* scrub. Alluvial fans may occur with this system at the mouths of prominent valleys. Intersection of this land system with coastal areas produces slopes which are sub-vertical and largely unvegetated sea cliffs (*c.* 55-60°). Hydrological characteristics are variable, but limestone units have the greatest potential as aquifers (Figure 4).

Land system III
This system is characterised by low lying, flat or gently undulating valley floors filled with alluvium (Figures 2 & 5). They may be traversed by poorly defined valley/channel systems. Relatively few have flowing rivers. The majority of these valleys are cultivated and have been cleared of dense vegetation. The valley floors provide variable 'going' surfaces, although in dry valleys it may be relatively firm, although may be prone to lifting by wind where exposed. Hydrological characteristics are dependant on the underlying geology and the thickness of the alluvium. The alluvium itself may form a potential aquifer (Figure 4), although close to coastal areas, groundwater is seasonally contaminated by sea water incursion.

Land system IV
This system consists of a low lying coastal plain, typified by the Suvla Plain, with gentle slopes and areas of inland drainage (Figures 3 & 5). In the Suvla Plain, water flow is highly seasonal and is impounded by a system of coastal sand dunes, and an area of elevated terrain at Lala Baba, to form a salt lake. Low lying areas close to the lake may retain water and are usually marshy. The alluvial plain associated with the lake is mostly cultivated, with few trees, although uncultivated areas may be covered by dense *garrigue* scrub (Prior, 1985). Where this coastal plain is traversed by perennially flowing rivers, the river corridor/valley is densely vegetated. The Salt Lake is known to desiccate during high summer (August). The coastal plain 'going' surfaces are variable, as they are usually soft and water saturated close to the lake, although seasonal desiccation produces a surface traversable by foot in August. In the inland areas, the going surfaces are passable in most cases.

The Salt Lake is too saline to provide potable water, due primarily to seasonal evaporation. Groundwater is present in the alluvium, although close to the sea it may be contaminated by saline incursions.

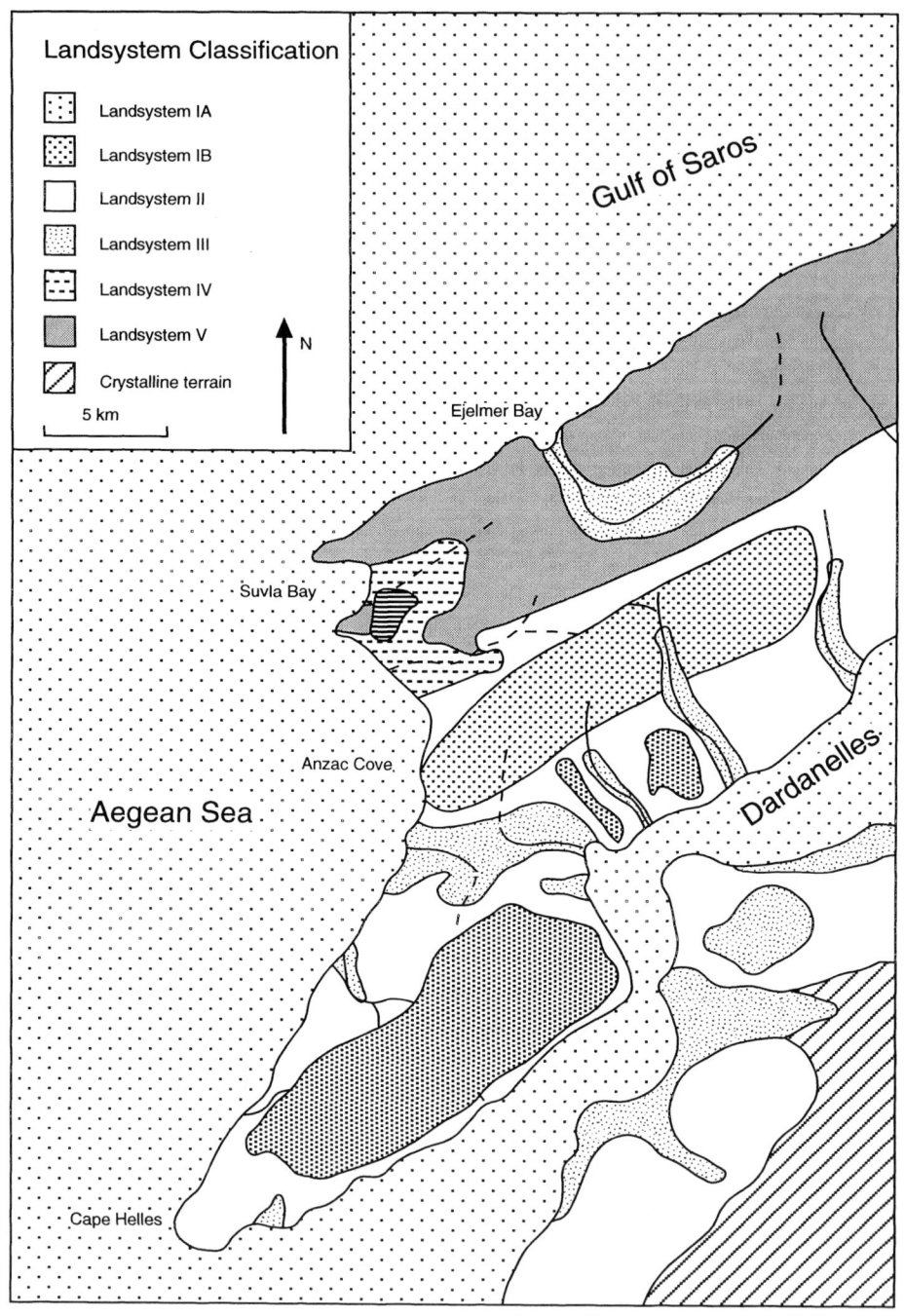

Figure 5. Land system classification map of the Gallipoli Peninsula and the Dardanelles

Land system V
Land system V comprises linear ridges trending approximately east-west and northeast-southwest. Elevations are typically up to 200 m. There are two principal linear hills within this system: the Anafarta Ridge to the southeast; and the Karakol Ridge to the northwest, running parallel to the northeast coast, between Suvla Point and Ejelmer Bay (Figure 5). Both ridges are associated with the strike of soutwestwards dipping Oligocene and Eocene marine sediments and faults associated with the North Anatolian Fault system, and are bounded by the rifted coast of the Gulf of Saros in the north (Figure 2). The orientation of the ridges and the development of the drainage pattern indicates the strong east-west oriented structural control of the region. Both major ridges have an asymmetrical cross section, with gentler slopes to the southeast, and steeper slopes to the northwest. Northwest slopes are dissected by shallow, parallel aligned valleys, most marked in the main Karakol Ridge. The gentler dip slopes are only dissected to a minor degree by tributary stream networks feeding low-lying areas. The two ridges are linked by a series of flat topped hills, with elevations of up to 250 m, which trend northwest-southeast, which also show some limited dissection (Figure 3). In general, the slopes are broad and open, although usually covered by dense *garrigue* scrub vegetation. Going surfaces are mostly firm and dry.

4. Military implications of the land system analysis

The land system classification highlights a number of terrain aspects which are of importance in planning a military operation from scratch within the Gallipoli Peninsula. These are: vantage points; ground conditions; traversability and 'going' surfaces; beach width; and hydrogeological characteristics.

Vantage/refuge points
The highest areas providing the greatest vantage points are provided by land systems I and V. Land system I comprises asymmetrical plateaux with steep north west facing slopes. Particularly important is the fault line scarp of the Sari Bair Plateau (land system I) which provides a direct line of sight from the subsidiary peaks of Koja Chemen Tepe and Chunuk Bair to the Suvla Plain below. Little refuge is available for troops at the base of this slope, especially in the numerous deeply incised valleys and broad surfaces provided by land systems II, III and IV. The south east facing slopes of land system I have a gentler gradient and vantage is gained from the plateaux surfaces such as Achi Baba with a good field of observation. Ekins (2000) has demonstrated that the narrows of the Dardanelles may be adequately viewed from neither the summit of the Sari Bair Plateau nor Achi Baba. As capture of these points was of immense significance to the attackers, intended to provide for artillery observation, clearly this demonstrates that the capture of this high ground would have been ultimately futile.

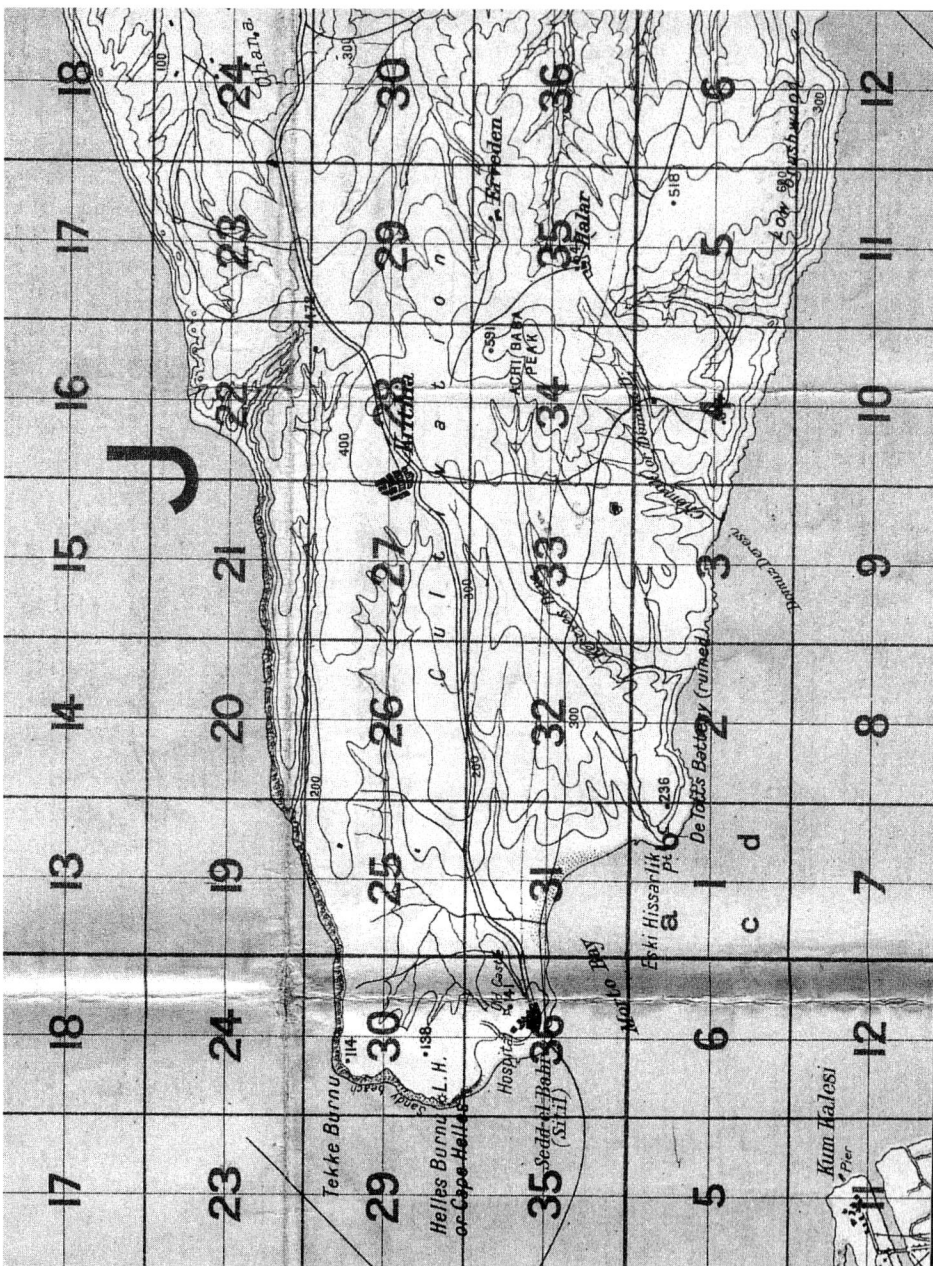

Figure 6. Section of the 1:63,686 scale GSGS map of 1908, showing the Helles Sector of the Gallipoli Peninsula. Lacking in detail, its wide contour interval had the effect of downplaying highs at Cape Helles, and in subduing the prominent linear valley known as Gully Ravine (Aspinall-Oglander, 1929)

The parallel valley systems provides good opportunity for refuge. The ridges of land system V (Karakol and Anafarta ridges) also provide significant vantage potential, particularly with respect to observation of the Suvla Plain and valley bottoms, although the east-west orientation of the ridges is a complication and requires careful positioning for defensive positions.

Ground conditions for defensive works
Ground conditions for defensive works and the nature of 'going' surfaces which controls trafficability are variable. Land system I provides suitable ground conditions for trench, dugout and tunnel construction as it is composed of dry Pliocene sands and clays with the water table at depth which are capable of maintaining the integrity of the steeply cut faces. Problems arise through the density of vegetation cover and consequent root system penetration. Land systems II and V are composed of harder materials which are capable of supporting defensive systems, but which require heavier tools and more intensive excavation works. Deep systems would encounter seasonal difficulties with water table fluctuations. Land systems III and IV are composed of alluvial sediments which would require a greater degree of revetment, and which would be seasonally wet.

Traversability and 'going' surfaces
In all cases, with the possible exception of land systems III and IV, slopes are difficult to traverse. In particular, the dissected slopes of land systems I and II are especially problematical. In land system I badland and fault line scarp topography prevents easy communication and traversability, relevant to the north western slope of the Sari Bair Plateau. In both plateaux of this land system the depth of valley incision on the south eastern slopes creates a broken terrain in which it is easy for attacking troops to lose contact and for communications to breakdown. Finally, the steeply incised, narrow and flat bottomed valleys and steep coastal cliffs of land system II at Cape Helles are a significant barrier and hazard to mass troop movements and landings. Given the overall aridity of the landscape suitably firm 'going' surfaces are provided by all land systems with the exception of seasonally wet land systems III and IV. Most suitable for heavy traffic is land system II, given its limestone substrata.

Beach width
The most suitable landing beaches are associated with land systems III and IV. These provide broad expanses of beach with a relatively gentle gradient which are not directly associated with coastal cliffs and other height/slope disadvantage. Suvla Bay (land system IV) is typical of this, but other, less expansive examples are seen either side of the Gaba Tepe promontory, and in Morto Bay (land system III). Beach development with the coastal expression of land systems I and II is mostly unsatisfactory, either forming narrow beaches with steep sea cliffs, or, in the case of land system II, narrow beaches

associated with steep, incised and easily enfiladed valleys, typical of the beaches at Cape Helles.

Hydrogeological characteristics
All of the land systems are composed of geological materials which have potential to be aquifers. However, the most important materials are the limestone strata of land system II and the alluvial sediments of land systems III and IV. Beach sediments associated with all land systems have also some potential as aquifers, although seasonal salt water incursion is a problem. The poorest aquifer is that associated with land system I, the water table being at depth within the limestone units, and surface groundwaters present only where there is the possibility of perched water tables.

Summary
The land system analysis highlights two major points which need to be considered in planning an invasion, but which were not necessarily to hand in the planning of the 1915 campaign. Firstly, with exception of the high ground surrounding the Suvla Plain (Karakol and Anafarta ridges), the remainder formed a distinct land system (I) which although suitable for the construction of defensive positions, has limited groundwaters for the supply of troops. Clearly then capture of the high ground should be executed rapidly to prevent stagnation into static warfare. Secondly, there are few satisfactory landing beaches associated with land system II, given the steepness of the sea cliffs; those of land systems III and IV are wider and more suitable to mass deployment of troops.

5. Mapping and reconnaissance in 1915

One of the most controversial aspects of the Gallipoli Campaign is the quality of staff work, particularly in relation to the preparations for the initial landings. These aspects are picked over in detail by historians (e.g. Aspinall-Oglander, 1929; Rhodes James, 1965; Lee, 2000). In his recent biography of General Sir Ian Hamilton, Lee (2000) has painted a picture of a soldier with an appreciation of terrain gained over years of active service, but who was hampered by the lack of adequate or extensive plans or orders in taking over his command. Staff work in developing the campaign has similarly been criticised, and it is popularly reputed that topographical information was limited to the equivalent of commercially available Baedecker type guides, bought in Alexandria (Moorhouse, 1956). Despite this, it is clear that Hamilton had at his disposal some aerial reconnaissance reports and photographs of defences constructed by the Turks in the run up to the landing (Hamilton, 1920; Aspinall-Oglander, 1929); that he had carried out ship-borne observations of the coast; and that he had maps of the Peninsula, in two sheets produced by the Geographical Section, General Staff (GSGS 2285 covered the Helles to Suvla Bay areas) in 1908, at a scale of 1 inch to 1 mile (1:63,686), with contours at an approximate 100 feet vertical interval (Geographical Section, General Staff, 1908; Figure 6); and that these were backed up by a contemporary report on the state of the defences of the Peninsula, although it has been suggested that this last item was not available to Hamilton.

More than anything else it is the adequacy of the 1908 maps that has concentrated attention on the adequacy of mapping in general in the campaign (Dowson, 1920; Aspinall-Oglander, 1929; Mumford, 2001). The scale of the 1:63,686 map meant that details of topography were limited, and this was to prove difficult for the assaulting troops, particularly in the Helles sector (Aspinall-Oglander, 1929; Figure 6). Other features, such as the prominent gully referred to as 'Gully Ravine' was also subdued on the map (Figure 6). It cannot be doubted therefore that the best maps available to Hamilton's forces were inadequate. However, the 1908 sheets were to be superseded by captured Turkish maps, and by a concerted effort to produce adequate maps by aerial reconnaissance early in the campaign after the landings had been made (Dowson, 1920; Mumford, 2001). These later maps were of high quality and successive editions of 1:20,000 scale trench maps provided a valuable resource for the battles which took place at Gallipoli throughout the remainder of 1915 (M. Nolan, pers comm.).

Water supply was a major issue throughout the campaign, and adequacy of supply was to be decisive (Aspinall-Oglander, 1932). Adequate reconnaissance of sources of ground water supplies was clearly a priority given the warmth of the climate, and it is clear that even a cursory examination of the Peninsula would demonstrate the presence of dry watercourses. The 1908 maps did note the presence of some wells, however (Geographical Section, General Staff, 1908). Clearly the allied armies were forced to plan for the provision of an adequate supply of potable water, which was to be derived from three sources: (1) surface water; (2) ground water; and (3) imported supplies.

The search for potable water supplies initiated some of the earliest reports on the geology of the Gallipoli Peninsula, commissioned directly from the Geological Survey of Great Britain (Strahan, 1919; Flett, 1937). Three geologists of the Survey (Pocock, Cunningham & Whitehead) were apparently sent to the Peninsula and were instructed to report on the water supply of the peninsula, and this survey, never published, was submitted to the War Office (Strahan, 1919; Flett, 1937). No record of this document survives today in the Geological Survey or Public Record Office. However, Arthur Beeby-Thomson, a consultant geologist, was attached to the Mediterranean Expeditionary Force (M.E.F.) from 1915-1918, and had responsibility for the provision of on the spot reconnaissance for potable water. His official reports were also apparently never published, but he was later to discuss aspects of exploration for ground waters in the Mediterranean lands based on his experiences (Beeby-Thomson, 1924), and this provides an invaluable insight into the exploration for water.

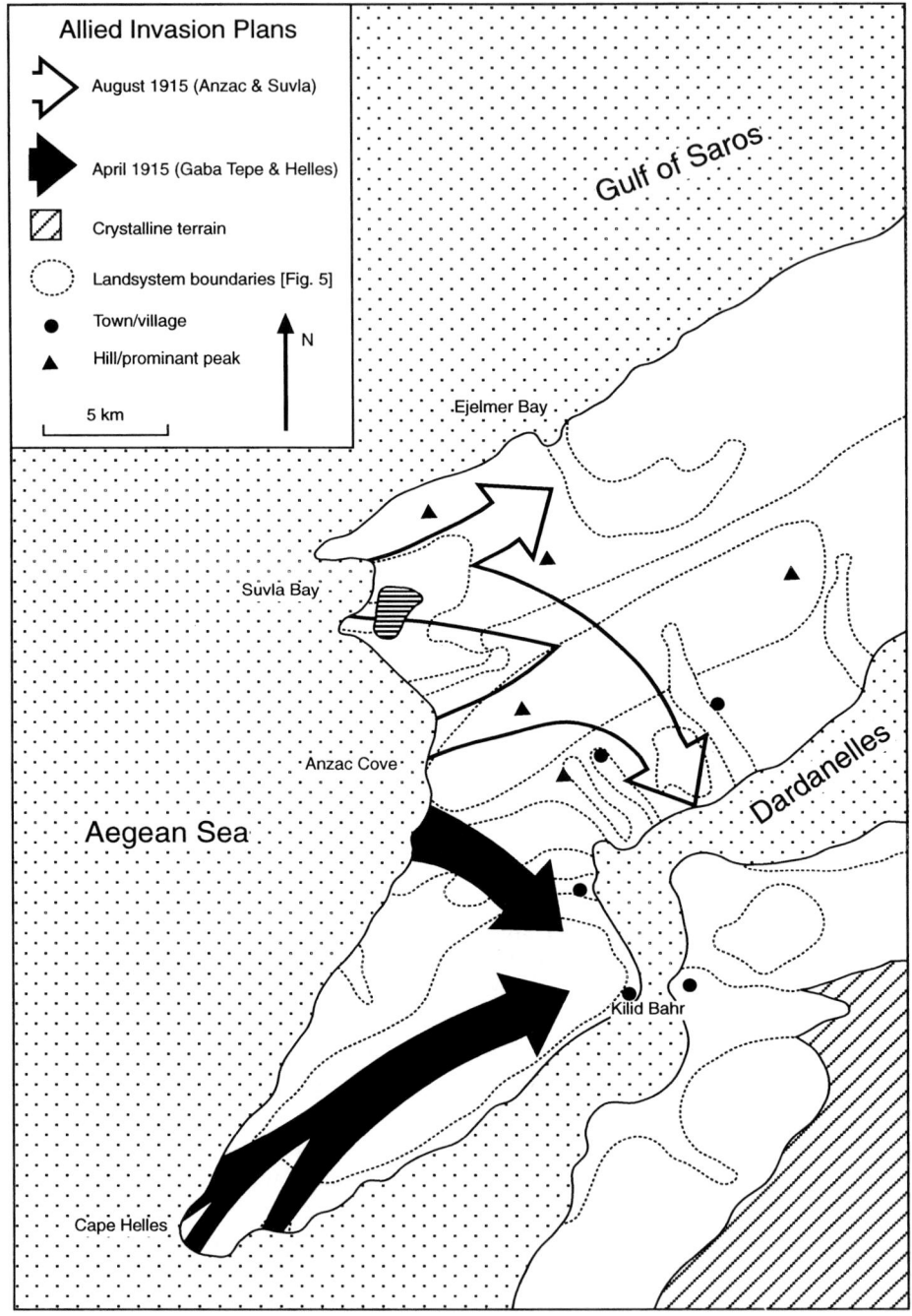

Figure 7. Map of the Allied invasion plans for the Gallipoli Peninsula, in April 1915 and in August 1915. For land system boundaries see Figure 5

6. Discussion

Choice of landing beaches
In a paper read to the Royal Geographical Society on April 26th 1915, D.G. Hogarth outlined the geography of the war with Turkey (Hogarth, 1915). This paper, read a day after the landing had been made and undoubtedly in ignorance of the land-based assault, made the following conclusion: 'All the western end of the Gallipoli Peninsula is of broken hilly character, which combines with lack of water and consequent lack of population and roads to render it an unfavourable area for military operations. No general, if he had the choice, would land a considerable force upon it at any spot below the narrows' (Hogarth, 1915, p. 461). Clearly, this was a view which upheld the need for a landing at Bulair. Yet despite the view of an expert — later to be attached to the Admiralty — and given that the strategic aims of the campaign were to simply support the naval operation through the capture of the high ground, and removal of the threat of shore batteries, a landing in the southwestern part of the peninsula was necessary, with a consequently limited range of options available to the Allied commander in chief.

However, as Masefield (1916) was to argue in a near contemporary account, if the peninsula was to be captured then General Sir Ian Hamilton had no real choice but to land where he did. The capture of the high ground was of clear importance, and Hamilton's plans, to capture the Kilid Bahr Plateau using a pincer movement from Gaba Tepe and Helles and exploiting the low slopes of land system II (Figure 7) appears sound. However, the beaches at Helles, developed at the mouth of the narrow ravine-like valleys of land system II, are clearly more suited to the defender, and the clearer understanding of terrain demonstrated by the Turkish troops led to the decimation of the British 29th Division during the 'Battle of the Beaches'. Clearly then, and with the benefit of hindsight, it would seem that a more objective use of terrain at Cape Helles would have been: (1) the steeper but undefended slopes of the northwest facing coast; and (2) the wider expanse of beach and alluvial sediment typical of land system III at Morto Bay. The first option is only really an option where the cliffline was undefended, as given the slope and narrowness of the beach defenders would have ample opportunity to pick off the attackers. Morto Bay was presumably not extensively used because of the difficulty of approaching the beaches without being fired upon by the coastal batteries. The wider expanse of beach and the much greater extent of open country would have enabled a deeper and wider deployment of troops, while still rising up the slopes of the Kilid Bahr Plateau towards Achi Baba.

The choice of Gaba Tepe makes sense from the perspective of terrain, using as it does the low slopes of land system II, but the use of the Anzac beaches does not. It must therefore be recognised of course that this was beyond the control of the commander once the landing was in place. Here, at the only place where land system I reaches the sea, the landing was made (Figure 7). As it happens, the landing here was a surprise, but as with Cape Helles in the south, the advantage was not pressed home, and after the initial surprise the tactical advantage soon passed to the defenders, controlling as they did the high ground on the tops of land system I.

The real mystery is why neither commander fully appreciated the value of Suvla Bay as a landing site earlier in the campaign (Figure 7). This land system (IV) has the benefit of both relatively wide beaches for the landing of men and material, and the Suvla Plain — flat but with sufficient cover for rapid deployment, and with a ground surface sufficiently dry for movement on foot or with animal transport, although damp and muddy in places. There is, however, debate over whether Suvla was difficult to traverse due to the density of scrub, something supported by Rhodes James (1965), and more recently, Prior (1985). Suvla was some distance from the main objectives — the Turkish guns on the Dardanelles shore — and that the allies had little knowledge of either the terrain or the landing beaches themselves. For example, the Admiralty had little information upon the nature of shoaling of the coastal area, and the military commanders had minimal understanding of the nature of important water resources (Steel & Hart, 1994). It was also considered too open, with the original attacking force too small to hold it while attacking the heights (Hamilton, 1920). When the British finally landed in August 1915, the control of the high ground to the east (land system V), which ultimately could be used to outflank the Turkish positions on the Sari Bair Plateau overlooking Maidos and the Dardanelles, was in their grasp. Poor generalship led to the loss of this advantage, to the subsequent costly and unsuccessful battles for the heights of the Anafarta Ridge and the Sari Bair Plateau, and ultimately to the admission of defeat by the allies.

Water supply

Once ashore, and having become resigned to a stalemate at Anzac and Helles at least, the local use of terrain was reasonably handled. The most important problem lay in the provision of water supplies from groundwaters. Surface water supplies were extremely variable and for the most part, insufficient. There are few flowing rivers on the peninsula, the majority of them being seasonal. Ground water supplies were more promising. The majority of ground waters were derived from: (1) limestone/sandstone aquifer at depth; (2) beach and alluvial sediments; (3) sandstone aquifers in the north, and; (4) from perched water tables in the Pliocene sediments. The limestone/sandstone aquifers of land system II provided the resources for much of the Cape Helles operation. However, as the Allies advanced up the slopes of land system I B groundwater became more difficult to obtain. Wells tapping this aquifer also existed around the southeastern margins of the Suvla Plain (Figure 4). Beach and alluvial sediments were found to be a good source of water by Beeby-Thomson (1924) and these provided some limited resources at Anzac Cove, and more abundant supplies at the eastern margin of the Suvla Plain. Salt water incursion was a problem within the central part of the plain itself occurring seasonally, and due to excessive water abstraction. Much of the Suvla Plain and Karakol water supplies were derived from Oligocene sandstone aquifers.

One of the biggest problems was a lack of reliable groundwater supplies at Anzac Cove (Figure 4). Here, the limestone aquifer of land system II was at depth, and seasonally supplies perched on impermeable strata, difficult to accurately locate, were the norm, together with some exploitation of seasonally dry rivers. These were ephemeral, however, and many of the wells dug in the surrounding gullies began to dry up with the approach of summer, and though others were found at depth. This meant that the daily ration from local sources in the 1st Australian Division was rarely more than one third of a gallon per man. Except on the extreme northern flank, wells were even scarcer in the area occupied by the New Zealanders, on the northern slope of land system IB. Most water supplies for these areas were imported.

Conclusions

The Gallipoli Campaign was doomed to failure primarily because of a lack of commitment to it from the Allied high commands in London and Paris. Too few men, too little planning, inadequate munitions, and indecisiveness together with woefully inadequate communications ultimately led to the stagnation and defeat of the allied troops. At the heart of the failure lies an inadequate understanding of the nature of the terrain. Although the strategic aims were well served by the initial plans, the terrible loss of life at Cape Helles was due to a commitment to landing on beaches which were clearly too narrow and commanded on both sides by easily fortified positions. The mistake at landing at Anzac Cove was costly, as after the initial failure to exploit early gains the positions became untenable. The vexed question of water supply was not clearly addressed and became a major issue, as was the provision of rest camps which were not overlooked by the Turkish artillery. Both of these the Turkish forces had in abundance.

The most important conclusion is a question as to why a landing at Suvla Bay had not been adequately explored, given the suitability of its landing beaches, the trafficability of the wide Suvla Plain, and the suitability of the plain in the provision of groundwater supplies.

References

Aspinall-Oglander, C.F. 1929. *History of the Great War. Military Operations Gallipoli. Volume 1. Inception of the Campaign to May 1915.* W. Heinemann Ltd, London

Aspinall-Oglander, C.F. 1932. *History of the Great War. Military Operations Gallipoli. Volume 2. May 1915 to the Evacuation.* W. Heinemann Ltd, London

Beeby-Thomson, A. 1924. *Emergency Water Supplies for Military, Agricultural and Colonial Purposes.* Crosby Lockwood and Son, London.

Crampin, S. & Evans, R. 1986. Neotectonics of the Marmara Sea region of Turkey. *Journal of the Geololgical Society of London* 143, 343-348.

Dowson, E.M. 1920. Further notes on aeroplane photography in the Near East. *Geographical Journal* 58, 359-370.

Doyle, P. & Bennett, M.R. 1999. Military geography: the influence of terrain on the outcome of the Gallipoli Campaign, 1915. *Geographical Journal* 165, 12-36.

Ekins, A. 2000. *Master of the battlefield; Terrain and the Gallipoli Peninsula.* Unpublished lecture presented at the University of Greenwich, January 2000.

Elmas, A. & Meriç, E. 1998. The seaway connection between the Sea of Marmara and the Mediterranean: tectonic development of the Dardanelles. *International Geology Review* 40, 144-162.

Erguvanli, K. 1957. Outline of geology of the Dardanelles. *Geological Magazine* 94, 47-53.

Flett, J.S. 1937. *The first hundred years of the Geological Survey of Great Britain.* HMSO, London.

Geographical Section, General Staff, 1908. *Gallipoli, Sheet 1, One inch to one mile.* War Office, London.

Hamilton, I. 1920. *Gallipoli Diary, Volume 1.* Edward Arnold, London.

Hogarth, D.G. 1915. Geography of the war theatre in the Near East. *Geographical Journal,* 65, 457-451.

Karatekin, N. 1953. Hydrological research in the Middle East. *Reviews of Research on Arid Zone Hydrology* UNESCO, 78-95.

Lee, J. 2000. *A Soldier's Life. General Sir Ian Hamilton 1853-1947.* Macmillan, London.

Masefield, J. 1916. *Gallipoli.* Heineman Ltd, London.

Moorehead, A. 1956. *Gallipoli.* Hamish Hamilton, London.

Mumford, I. 2001. *A Small War, Gallipoli, 1915. He Who is Above Sees All: an Early Use of Aerial Photography in Terrain Analysis.* Unpublished lecture presented at the University of Greenwich, January 2000.

Naval Intelligence Division, 1942 *Turkey, Volume 1.* BR507, Geographical Handbook Series, Naval Intelligence Division.

Pamir, H.M. 1953. Hydrogeological research in the basin of the Ergene. *Proceedings of the Ankara Symposium on Arid Zone Hydrology* UNESCO, 224-231.

Perincek, D. 1991. Possible strand of the North Anatolian Fault in the Thrace Basin, Turkey - an interpretation. *American Association of Petroleum Geologists Bulletin* 75, 241-257.

Prior, R. 1985. The Suvla Bay tea-party: a reassessment. *Journal of the Australian War Memorial* 7, 25-34.

Rhodes J.R. 1965. *Gallipoli.* Batsford, London.

Sengor, A.M.C & Yilmaz, Y. 1981. Tethyan evolution of Turkey: a plate tectonic approach. *Tectonophysics* 75, 181-241.

Steel, N & Hart, P. 1994. *Defeat at Gallipoli.* Macmillan, London.

Strahan, A. 1919. Introduction. Work in connection with the war. *Memoirs of the Geological Survey. Summary of Progress of the Geological Survey of*

Great Britain and the Museum of Practical Geology for 1918. HMSO, London, 1-4.

Ternek, Z., Erentöz, C., Pamir, H.N. & Akyürek, B. 1987. *1:500,000 Ölçekli Türkiye Jeoloji Haritasi. Explanatory Text of the Geological Map of Turkey. Istanbul.* Maden Tetkik ve Arama Genel Müdürlügü Vayinlarindan, Ankara.

United Nations, 1982. *Ground Water in the Eastern Mediterranean and Western Asia.* Natural Resources/Water Series No. 9. United Nations, New York.

Peter Doyle & Matthew R. Bennett
School of Earth & Environmental Sciences
University of Greenwich
Chatham Maritime
Kent, ME4 4TB

British, French and German Mapping and Survey on the Western Front in the First World War

Peter Chasseaud

> **ABSTRACT:** All belligerents entered the war supplied with small-scale, ungridded topographical maps of the expected area of operations. Soon, with trench warfare, it was realised that accurate large-scale gridded maps (1:20,000-1:25,000), with tactical intelligence plotted from air-photos, were essential for the planning and control of indirect artillery fire; such maps had previously only been prepared for the attack and defence of fortresses. Although their pre-war general staffs had sections responsible for maps and survey, all belligerents had to improvise field survey organisations to create and print large-scale maps, to conduct the necessary surveys, and to provide the essential firing data for the artillery. Meanwhile the existing national survey departments rapidly responded to the new requirements by producing enlargements of existing maps. All together, Britain printed 34 million war maps, France over 30 million, and Germany a staggering 775 million (including the Eastern Front). These unprecedented totals give some idea of the requirements of mass armies. This paper examines the approaches taken by the main protagonists to military mapping on the Western Front.

1. Introduction: a new battlefield geometry

For centuries the medium and small-scale topographical map has been invaluable to commanders as a model of the terrain, giving an image of the features vital for strategic and tactical operations. However, the lack of crucial details on these maps, such as swamps and other topographical details could lead to disaster. In 1914 all belligerent countries were equipped with topographical maps (1:60,000-1:126,000, and smaller-scale) for a war of movement and manoeuvre, showing the terrain in sufficient detail for formation commanders to plan, and unit commanders to execute, their operations. The purpose of this paper is to examine the approaches to mapping of the main protagonists engaged on the Western Front from 1914-1918[1].

The manoeuvre of mass armies with modern firepower which, together with earthworks, favoured the defensive, soon led to the stalemate called trench, siege or position warfare. The range and accuracy of modern weapons forced troops below ground and behind cover, and the artillery, required to destroy the enemy's barbed wire, entrenchments, machine-gun and battery positions, was forced to rely on 'map-shooting' and 'indirect fire,' often unobserved. In the course of the war this developed into the sophisticated method known as 'predicted fire.' Indirect fire had developed slowly in the late 19th century as technology enabled ranges to lengthen dramatically; guns were fitted with panorama or dial sights which enabled fine angular measurements to be made from aiming points such as church spires, and fire could be brought to bear on targets invisible from the battery position, either by observation from a forward or flank position or from the map.

A new type of map was required for artillery work, prefigured from 1903 in European siege plans and the Russo-Japanese war, and was developed in the field from 1914 onwards by all belligerents. This was an accurate, large-scale contoured and gridded topographical map, built on a dense network of fixed points, of angular accuracy (preferably on an orthomorphic projection), and on which detail of all offensive and defensive works — trenches, strongpoints, dugouts, machine-gun, trench-mortar and battery positions, signal stations, cable routes, command posts, etc. — could be plotted from air-photos. In addition, located enemy battery positions were plotted from data supplied by flash-spotting and sound-ranging sections — part of the newly emerging artillery intelligence and counter-battery organisations. Thus a new geographical information system had been created, providing a sophisticated three-dimensional fire-control database or matrix of the battlefield, and typically on the 1:20,000 or 1:25,000 scale for medium and heavy guns and howitzers, and 1:10,000 scale for field artillery. Even larger-scale (e.g. 1:2,000) plans were created for trench mortars, which used similar survey techniques for indirect fire, and for offensive and defensive mining. These indirect fire techniques were also applied to machine gun fire. In effect, the battlefield had been digitised; every point could be rendered in metric or yard rectangular co-ordinates.

The dense fixed point network was crucial as control for plotting detail from air-photos and for precise fixing of one's own battery positions, aiming points, observation posts (OP's) and sound-ranging microphones. When they, and the targets in enemy territory, were accurately fixed to the survey grid, this battlefield geometry enabled range and bearing to the target to be calculated and applied to the gun. Similarly, the angle of sight was obtained from relative heights of gun and target, given by the contours. The key elements of the artillery map were therefore accurate planimetry and contours; when these features were absent, fire missed the target and often fell on friendly troops. Such reliance on the map had to be reinforced by careful calculation of influences caused by atmospheric changes, gun-wear, and shell and propellant variances, through the development of meteorological and calibration services.

Air-photos were the main source of data for enemy tactical detail and assumed overwhelming importance. Where vital observation could not be obtained from high points on the terrain-hence mine warfare and costly battles for key points-the 'eye in the sky' was the only way of obtaining the

Figure 1. British Field Survey Company, RE at work; plane-table resection of a field battery position. (Reproduced by permission of the Royal Engineers Museum, Chatham)

trace and details of the enemy's trench system and, augmented by sound-ranging results, of plotting his battery positions (Chasseaud 1986, 1991, 1999).

Required accuracy of battlefield geometry
It was important to avoid survey discontinuities within a war-theatre, and this was partially achieved through the use of national high-precision primary triangulations, onto which were fitted the less accurate secondary and lower-order triangulations (Figures 1 & 2). However, serious discrepancies occurred along the junctions between national systems, and each system had its own origin to determine absolute position, its own base measurement affecting the scale, and its own azimuth or bearing along the basic side or sides of the triangulation. The 'common points' between two systems were often some metres apart, and triangulation values of neighbouring countries had to be brought into agreement before they could be used. Relative discrepancies of 5

Figure 2. British Field Survey Company, RE at work; triangulation using theodolite and heliograph. (Reproduced by permission of the Royal Engineers Museum, Chatham)

or 10 m caused large angular errors between points lying close together but belonging to different systems, which were needed by the artillery for fire-control and predicted shooting. The First World War proved the vital need for an integrated survey framework. An error of a few metres in the absolute position of a point was not disastrous if other points in the area had similar errors; it was relative positional accuracy between neighbouring points that mattered. Field survey organisations had to provide such points giving bearing accuracy within 2 minutes of arc. For points 1 km apart, this implied a relative positional accuracy of about 0.5 m. Modern national surveys easily achieved this standard, but in France and Belgium, as in other theatres, confusion between old and new systems, faulty methods of adjustment and trigonometric list errors produced relative displacement errors greater than this (Jack, 1920; Clough, 1952).

International developments in photogrammetry before 1914
Most photogrammetrical co-ordinate determination and plotting before 1914 had been done from terrestrial photos, the most recent developments having

been Fourcade's and Pulfrich's stereocomparators, Thompson's stereoplanigraph of 1907-8, and von Orel's stereoautograph (a development of the stereocomparator) of 1908, 1911 and 1914. The latter revolutionised plotting from terrestrial photographs, being the first stereo-plotter which could automatically plot planimetry and trace contours from stereograms, thus forming a key stage in the development of photogrammetry, despite being restricted to horizontal camera axes. The von Orel stereoautograph was later adapted for oblique air-photos (Hart, 1943), forming the initial basis of wartime plotting of points and heights from oblique pairs. Subsequent development in aerial photography introduced the problem of camera axes which were relatively fixed, leading to the invention of the spatial autograph, i.e. reconstruction of the intersection of rays in three dimensions. Some high-precision apparatus later proved capable of plotting terrestrial and (oblique) air-photos (Zeller, 1952, p. 62). Thompson modified his design to give automatic plotting, but it never received the mechanical development which it deserved (Thompson, 1974; Atkinson, 1980).

Despite Thompson's work, and a growing awareness of the German lead in technology and education, the British attitude to photogrammetry was remarkably sceptical in the period before, during and even after the war. It was summed up by a British reviewer of a German publication in 1913 who remarked that 'photo-surveying is more akin to an amusing game than to a useful art' (Anon., 1913). The British were in general only dimly aware of progress in Germany and Austria.

2. The British experience, 1914-1918

The growth of British survey and mapping work, and the development of the Field Survey Companies (FSC) of the Royal Engineers on the Western Front has been well-documented (Winterbotham, 1919a, b; Jack, 1920; O'Donoghue, 1980; Chasseaud 1986, 1991, 1999) and therefore need only briefly be dealt with here.

Maps in the 1914 campaign

Unlike the French and Germans, the British artillery had little experience of survey work, with the result that during the war nearly all artillery survey was done by the Royal Engineers (RE). As a result of a report on the French 1911 Chalons exercises, cooperation between artillery and aircraft was practised at British pre-war manoeuvres, and squared maps featured in these trials. Colonial wars had led to the creation of three survey sections, equipped for trigonometrical and plane-table survey in unmapped country. These were available in 1914, additionally equipped with photographic apparatus, but it was decided that as the theatre of operations was in an already well-mapped area, they would not be needed. A Printing Company RE was also available, with hand-litho, letterpress and photographic equipment.

Mobilisation supplies of topographical maps of France (1:80,000) and Belgium (1:100,000), and smaller scales, had been organised by the Geographical Section of the General Staff (GSGS) in London. The only survey officers initially with the BEF were Major E. M. Jack, from GSGS, who represented 'Maps GHQ,' an assistant who returned to London after setting up the map distribution and depot organisation, and Captain B. Wilbraham, commanding the Printing Company at GHQ. Jack's original function was to supply maps for a war of movement, and send print orders to England when necessary.

The new artillery and trench map
It was on the Aisne in September 1914 that the first of the new artillery and trench maps were tentatively produced, which developed into an accurate, gridded large-scale topographical base, with tactical overprints showing all details of enemy organisation as revealed by air-photos. This formed the cartographic database for the planning and execution of all operations in the next four years. This type of map was developed simultaneously by all the belligerents. The Aisne demonstrated the need for a reference squaring or grid system to indicate targets, and when the BEF moved to Flanders in October, the Ordnance Survey supplied squared maps at 1:40,000 and 1:20,000. It was soon realised that rectangular co-ordinates on a theatre grid were needed, so that range and bearing for artillery could be calculated trigonometrically rather than merely being measured with rule and protractor. This development markedly increased the accuracy of artillery fire. Once allied batteries had been fixed by the surveyors, and all German batteries and other targets had also been fixed to the theatre grid, located through the evolving methods of flash-spotting, sound-ranging and aerial photography, complex bombardment and counter-battery programmes became possible and surprise achieved.

An accurate large-scale topographical survey was already available for Belgium, but for those parts of France not covered by 1:20,000 fortress and frontier *plans directeurs* (artillery maps); there was only the old, inaccurate and uncontoured, 1:80,000 map, on which relief was shown only by hachures and inaccurate spot-heights. From the beginning of 1915, therefore, it was decided that mere enlargements of the 1:80,000 would not do. A new map had to be made — by trigonometrical and plane-table survey for the area in British hands, and by compilation of all available material (communal cadastral plans, railway and mine plans, etc.) revised from air-photos for the enemy area. New 1:20,000 survey was begun in January by the 1st Ranging Section RE, under Captain Winterbotham. German defences were plotted from air-photos and overprinted on special 1:5,000 and 1:10,000 sheets, and later regular series 1:10,000 and 1:20,000 sheets which often had additional overprinted information such as enemy battery positions or barrage lines. These were all essentially the same map, usually enlarged or reduced photographically. The larger scales were essential for trench warfare, while the field artillery used the 1:10,000 scale, and heavy artillery, with longer range, used the 1:20,000, and even the 1:40,000 scales for very long-range railway guns. For accuracy, the latter calculated range and bearing trigonometrically. To avoid map distortion and ensure accurate fire, gunners used rigid artillery boards, with the map pasted in small sections onto a scribed grid, fitted with a graduated arc and rule pivoting on the battery position.

Most artillery and trench maps in the early years of the war were shipped from the Ordnance Survey in the UK, from drawings sent back from France, only small editions or urgent maps being printed in France. At first the showing of British trenches was prohibited for security reasons, but small secret editions with British trenches in blue were allowed for staff in September 1915. In late 1916 British front and support trenches were permitted to be shown, followed in 1917 by the notation of British trenches to a depth of about 600 m, and finally in 1918 all British trenches were displayed (but not any features which could not be interpreted from an air-photo).

Like the French, the British relied on simple apparatus and graphic methods for vertical air-photo rectification for cartography. Wide-angle (6-inch) survey photos were only taken by the Royal Flying Corps (RFC), after much pressure in the spring of 1917, for mapping the new German Hindenburg Line positions in the absence of cadastrals, enabling mosaics of existing verticals to be accurately located; the resulting maps turned out to be more accurate than the enlarged 1:80,000 French war plans directeurs (MacLeod, 1919). Content with their cadastral compilations (apart from the poor relief), no British experiments were made with automatic stereo-plotting apparatus for air survey.

The 1915 campaign
The development of the set-piece battles that came to dominate the Western Front took place in 1915, and the British were engaged at Neuve Chapelle, Aubers Ridge, Festubert and notably Loos in September-October of that year.

At Loos, intense preparations on First Army front led to the development of a new edition of the regular 1:10,000 sheet, with German trenches corrected from air-photos to 25th August. Secret editions of all relevant 1:10,000 sheets were authorised, British trenches being overprinted on cropped regular sheets by hand-press. The First Army's Printing Section also produced insets for regular sheets, showing new German defence intelligence in green, and the rapidly developing German second line in red. Tracings were also prepared to show new British assault trenches, and drawn panoramas were printed. Three editions of some sheets appeared, as well as revision insets. From a wider survey viewpoint, the battle occurred at an unfortunate time, as this part of First Army area was being replotted more accurately on the Brussels meridian, there having been various positional errors in earlier sheets. Adjoining sheets were thus out of sympathy; tracings showing correct fixed-point positions were issued to batteries for evaluating misplacement of detail. By October 1915, new surveys had been completed, and a new edition on all three tactical scales, incorporating the new trigonometric work and detail and ground-form surveys, was issued in December.

The Battle of the Somme, 1916
In March 1916, Third Army relieved the French Tenth Army in the Arras-Vimy sector. In Arras, the 3rd Field Survey Company (FSC), RE 'looted' various hand-presses, litho stones and paper from derelict printing works. A new survey method was now used; checked and amplified French trigonometric points (many problems arose with the old French triangulation) were plotted, and cadastrals compiled, onto plane-table sheets which were sent for field-checking, revision and contouring. For the German area the method was as before. The result was an excellent new map of Third Army area, joining at Vimy Ridge with new First Army work, and south of Gommecourt with that of Fourth Army. Second Army, in Flanders, redrew and revised its sheets, and by early 1916 the British possessed accurate, linen-backed artillery and trench maps of their whole front, showing German defences in detail, and impressive in their drawing, printing and production quality. From now on, frequent new editions of the topographical base map, and of the trench plates, were produced.

The Battle of the Somme, beginning on 1st July, severely tested the cartographic organisation, the main lesson being the need for power-presses at every army HQ, preferably with photo-mechanical process apparatus. It was not until the spring of 1917 that the FSCs obtained all this equipment, and even then they could only print small sheets. New types of maps appeared for operations. Hostile battery maps for counter-battery work showed German gunpits by sets of green dots, daily situation maps were overprinted by 4th FSC, corps front trench maps were introduced, track maps were produced so that harassing fire could be directed onto enemy troops and transport, maps showing enemy counter-attacks enabled defensive barrages to be dropped at the right places, barrage maps showing the timed lifts of creeping barrages were duplicated by corps and divisions for batteries and attacking infantry, 1:5,000 special trench operation sheets were prepared, particularly by 5th FSC (Figure 3).

During the Battle formations often improvised with poor-quality duplicator maps. The lesson, was that not just the army Field Survey Companies, but also corps, should be able to supply rapidly quantities of all types of maps, and should therefore have sufficient reproduction resources. In the autumn of 1916, the 3rd FSC had formed corps topographical sections for the rapid production of tactical maps to cope with the ever-changing situation. Early in 1917, such corps sections were formed for each army, equipped with Ellam duplicators.

The battles of Arras, Messines, Third Ypres and Cambrai, 1917
The 1917 battles — Arras, Messines, Third Ypres and Cambrai — saw the further development of the 'hostile battery' or 'positions' map, which was essential for effective counter-battery fire, and of the barrage map for set-

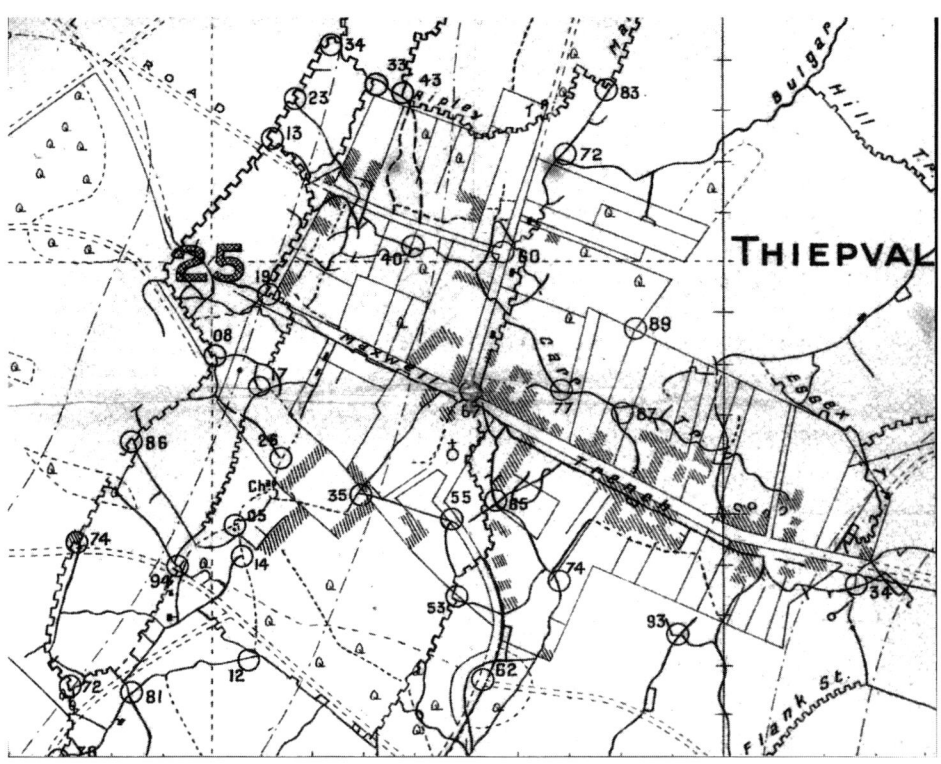

Figure 3. Section of a 1:5,000 scale British trench map of the village of Thiepval. A key sector of the Battle of the Somme in 1916, the buildings are all shown in ruins (Author's collection)

piece attacks. These were printed by the army Field Survey Company, or overprinted at corps on backgrounds supplied by the Field Survey Company or the Ordnance Survey itself. The First Army produced an excellent lithographed 1:10,000 barrage map for the capture of the Vimy Ridge on 9th April. Second Army's barrage map for the Messines battle in June was similarly produced (in several 1:10,000 sheets) on an army, rather than a corps, basis, thus ensuring proper co-ordination of the fireplan. The same pattern was followed by 2nd FSC in September and October at Ypres (Passchendaele). An ominous feature of this battle was the appearance of the 'going' map, prepared by the Tank Corps, indicating which areas in the Salient were liable to become flooded or boggy, and should be avoided by tanks. This did not prevent many tanks, as well as infantry, guns and mules, becoming 'bogged', some sinking without trace. Geological maps were also prepared to assist with tunnelling and dug-outs.

In the event the tanks found the right ground, and the right tactics, at Cambrai in November 1917 — a 'survey battle' *par excellence*. Surprise was achieved by a development of map-shooting; this was unregistered or predicted fire, in which range and bearing were carefully calculated and, after corrections for meteor and calibration, applied to each gun. No previous registration was permitted, and there was no preliminary bombardment to warn the enemy. Enemy batteries were suddenly neutralised at zero, while tanks crushed the wire and dealt with machine guns, and infantry advanced covered by a creeping barrage. An army barrage map (in two 1:20,000 sheet — one for each attacking corps) was prepared, by the 3rd FSC.

The German offensives and Allied advance of 1918
At the end of 1917 it was decided to improve map supply by establishing an Overseas Branch of the Ordnance Survey (OBOS) in France, and this became operational in March 1918, just before the German offensives. Map supply was therefore maintained while the Field Survey Companies were retiring (3rd FSC printed up to the last minute at Albert and lost a machine to the Germans) and were unable to provide continuous production. Back-area surveys were immediately begun in case of further German advances, and the resultant sheets were rapidly printed by OBOS, as were many other regular and special sheets. In the summer of 1918 the Allies decided to standardise their maps on the French Lambert projection and new Lambert sheetlines, and also to adopt the French colour scheme (red for own trenches and blue for enemy) and conventional signs. In fact, while new editions of existing sheets were redrawn along these lines, the new Lambert series did not see action.

The Allied attacks from the Battles of Hamel and Amiens (July and August) onwards benefited from the artillery survey experience of the Cambrai battle; predicted fire and surprise became the norm, except in the case of the deliberate bombardment of the Hindenburg line before the assault at the end of September. Equipped with one, and in some cases two, large printing machines, the FSCs (they became battalions in June) performed prodigies of map production in 1918, printing millions of maps in two, three or four colours. During the war the Ordnance Survey (including OBOS) and the War Office together printed something like 22 million maps for the armies in France and Belgium, while the FSCs printed another 10 million.

3. French maps and survey on the Western Front, 1914-1918

French maps and survey 1871-1914.
The *Service Géographique*, a General Staff branch, was created in 1887 from the *Dépôt de la Guerre*, to undertake all geodetic, survey and reproduction work and issue topographical maps. Its main geodetic work was the triangulation for the new 1:50,000 map, the replacement for the 1:80,000 (Frith, 1906). Although France had been the first country with a national survey, she was overtaken in large-scale mapping by others during the 19th century. The 1:80,000 was inadequate for many purposes, and in 1891 the Central Geographical Commission urged a 1:20,000 survey; this lagged because of insufficient funding, and covered only the fortress and frontier areas by revising existing *plans directeurs* (Arthur-Lévy, 1926). *Plans directeurs* were large-scale military maps, originally intended to support artillery.

A 1903 *Instruction*, 'on fortress and siege artillery fire organisation' stipulated fire-control using gridded large-scale plans, artillery boards, etc. Grid co-ordinates facilitated reporting, and the identification of points such as targets, battery positions, observation posts (OPs), aiming points, landmarks, and so on. This formed the basis for a 1914 Instruction extending the principles to medium and heavy artillery in the field (Bellot, 1936, p. 38). In 1908, new artillery siege survey methods were introduced for armies designated to attack the frontier fortresses lost in 1871; intelligence and survey officers were appointed for target location and fire-control, to fix their own batteries by resection, and plot all German defences, particularly batteries, ammunition stores and other targets, by cross-observation or aeroplane observation on battery commanders' large-scale plans. These *canevas de tir* and *plans directeurs de tir* for indirect fire showed all fixed positions and enabled accurate range and bearing to be given. The German 1:25,000 map of Alsace-Lorraine formed the basis of the *canevas de tir*, or artillery board, the topographical guide for indirect fire (Arthur-Lévy, 1926). In 1911 the Air Corps practised air observation, artillery fire control and infantry liaison, with pilots and battery commanders using identical gridded maps to identify positions by co-ordinates to within twenty m. Visual and wireless signalling systems were also tested. Both French and Germans were advanced in aerial photography experiments at this time (Raleigh, 1922; Albrecht 1969). It was soon found that the pre-war *plans directeurs*, on a polyhedric projection, were unsuitable for artillery work as bearings were distorted and it was difficult to plot new points or calculate range and bearing; similar problems arose with the 1:80,000 Bonne projection.

As in 1870, plans for map supply in 1914 were predicated upon an invasion of Germany. 1:80,000 sheets were printed at the *Service Géographique*, and during the war most *plans directeurs* were printed in Paris and the *Service's* outposts. Artillery survey was flourishing because of the need for large-scale plans of the frontier and other fortresses, with their associated surveys of gun positions, observation posts and fields of fire. *Brigades géodésiques* were to locate targets for the artillery in siege warfare (Winterbotham, 1918).

Artillery *groupes* soon each had a specialist artillery officer *orienteur*, trained by the *Service Géographique* and *groupes de canevas de tir* (GCT), to survey their batteries and supply firing data (Jack, 1920).

The pre-war view that the main wartime role of the Service would be to supply the existing topographical maps rapidly changed under the impact of trench warfare (Service Géographique, 1938). Throughout most of the war there were ten French armies on the Western Front, each with its *groupe de canevas de tir*. There was also, from August 1915 to February 1916 a GCT with the *Région Fortifiée de Verdun*, and the entrenched camp of Paris had its own GCT.

Trench warfare began on 16th August 1914, when extensive fieldworks were begun at Nancy. The stalemate following the 'race to the sea' led to siege procedures for field artillery, with a new survey organisation, by the end of October (Arthur-Lévy, 1926). What was initially required was a close network of trigonometrical points, whose precise co-ordinates could be given. The first army *groupes de canevas de tir* were formed on 30th October 1914 from the pre-war *brigades géodesiques*, led by officers of the *Service Géographique* and the Service hydrographique de la marine (Service Géographique, 1938).

Plans directeurs, close triangulation nets and artillery boards
In November 1914, the French Commander in Chief, Joffre stipulated that, for artillery use, all maps of the front including the 1:80,000 should carry the kilometre grid. From the beginning of the war, aerial observation was used to direct artillery fire, and grid co-ordinates were a crucial part of this process, being widely used in orders. Trench warfare codified this practice, and this *Instruction* extended fortress and siege artillery principles into the field. It envisaged:

1. Establishment of a *canevas d'ensemble*, that is a close triangulation amplifying the general triangulation.
2. Establishment of *canevas directeurs de tir* (artillery boards) for batteries, based on the *canevas d'ensemble*, consisting of 1:20,000 gridded plans marked with the exact position of the batteries' directing guns, landmarks or aiming points, and targets.
3. Use of 1:20,000 maps, based on pre-war 1:10,000 surveys, or on enlargements of the 1:80,000. All to carry the same grid as the artillery board *planchettes* and other topographical material used by the artillery.
4. Progressive improvement of the *canevas de tir* and its transformation into *plans directeurs*, giving a complete picture of the enemy trenches and other works.

The map distortion problem: Bonne Projection and grids
Before the war the *Service Géographique* had not used rectangular co-ordinates for its own survey work, and had abandoned the Bonne projection for its large-scale surveys and for the new 1:50,000 map, substituting the

polycentric system which could not be used for the general rectangular co-ordinates which facilitated the use of simple plane trigonometry for complementary triangulation calculations and the easy and precise designation of points for the artillery. The 1903 *Instruction* on fortress and siege artillery fire prescribed a simple conical projection system with easy conversion from geographical to rectangular co-ordinates. Unfortunately this system remained purely local, and could not extend beyond 30 km from the origin without generating appreciable distortions (SHAT, 3M 575).

In the first months of trench warfare it was realised that the projections used for the 1:80,000 map and the *plans directeurs* led to distortions sufficient to disturb artillery and survey work, and prevented the use of a theatre grid. One early solution was to divide the front into gores (segments based on central meridians). It was soon found that distortion around a point was noticeable at the edges of the gores. At each gore edge, despite solutions such as double-grids and plans overlapping onto the neighbouring gore, liaison between armies using co-ordinates and the *plans directeur* became very difficult, if not impossible, during operations (Bellot, 1936, p. 36). In early 1915 the situation was becoming very dangerous, particularly regarding the intended return to movement following a breakthrough of the German front. General Bourgeois, directing the *Service Géographique*, therefore planned the gridding of all 1:50,000 sheets beyond the front in the general Bonne system.

The Lambert orthomorphic projection was suggested, which preserved shape around points and consequently bearing, for the whole theatre. In June 1915 Lambert's conical orthomorphic projection was decided upon with two standard parallels, thereby reducing by half the maximum scale distortion. Armies received data in early September, but time was too short to use it on the new *plans directeurs* of the Champagne front, where the attack was to begin on 25th September. The battle was fought on the old projection and grid. In August, Joffre ordered that rectangular co-ordinates should be used for the identification of all targets. In November 1915 armies began to switch to the Lambert, and the Lambert grid was also used on 1:50,000 and 1:80,000 maps. In 1916, Verdun was fought on the Lambert, but the Somme was fought on the Bonne. The *plans directeurs* and *canevas d'ensemble* of the whole French front were established on the Lambert projection by the beginning of 1917 (Spilleux, 1917). By the Armistice a new Allied scheme of sheets, using Lambert projection, sheetlines and grid, was imminent.

Artillery survey
In the winter of 1914-1915, survey techniques were being extended to field batteries whose officers had little training in topographical methods, and there were great demands on the GCTs to survey batteries. There were often gross errors in line due to incorrect positioning of the target on the map and of the directing gun, and incorrect bearing from this gun to its aiming point; these errors led to the total disorientation of fire, which fell far either side of the target. This emphasised the need for the field artillery to apply strict survey principles.

Trigonometric work by the GCTs during the War
From November 1914 the GCTs established the fire-control triangulation. In the front areas the old triangulation was insufficiently dense and required amplification. Many of the old trigonometric points had disappeared, and it was necessary to fix new points, accurate to within 1 or 2 m; simple geodetic instruments and methods were used. The GCTs increased the density by theodolite resection of the new positions, combined with intersection from known points. Where it existed, the new (post-1870) triangulation was used with corrections to establish the concordance between the two networks. From the end of 1915, the GCTs gradually completed this *canevas d'ensemble*, using tacheometer traverses in covered or close terrain and in the battery and observation zones just behind the front (SHAT, 3M 582).

Air photogrammetry
Despite the pioneering French work in the 19th century, surprisingly little progress was made during the First World War compared with the Germans. In late 1914 and early 1915, Lieut. Sasportès of the Third Army GCT attempted to resolve the problems of plotting from air stereo-pairs, and his experiments on the Verdun front led to his 'Note on the graphic restitution of photo-pairs supposedly taken in a normal attitude' (February 1915). Rectification was at first done graphically using the four-point method — in 1915 he invented a graphic method called *'prolongement photographique.'* In March 1915 Captain Vavon, of Sixth Army's GCT, applied Laussedat's idea of a *'chambre claire'* through which a rectified virtual image of the air photo was seen and drawn on the plan. Some 330 were constructed during the war. In April and May 1915 Roussilhe, with the GCT of Paris, created an apparatus which projected an image of the rectified photo onto a plan; four fixed points marked on the plan were identified on the photo, enabling the photo image to be accurately placed. If the plan was replaced by photographic paper, the rectified image could be re-photographed for cartographic use. A little use was also made of a stereocomparator, but this appears to have been mostly restricted to the terrestrial photos taken for the Paris defences (SHAT, 3M 560; Bellot, 1936, p. 65).

The Sections undertook provisional rectification, various types of apparatus being designed by their officers — e.g. the *appareil de redressement Clerc*, and the *appareil de projection Demaria-Lapierre*. Ground forms were studied with the stereoscope. In July 1916 Capt. Schweisguth of the Seventh Army GCT invented a viewing apparatus which was widely used by GCTs and STCAs. Only verticals were used for plotting detail onto plans directeurs; obliques and high altitude panoramas were used to study details and give an idea of ground forms (SHAT, 3M 580).

The French relied on creating a close fixed-point net, onto which detail from verticals could be plotted using simple methods. Naturally these were not accurate where the control points were in doubt. In June 1917 the GCT of Fourth Army in the Champagne introduced a new method of planimetric restitution of their *plans directeurs* based on the identification of ruling points on the photos. Hitherto this had been done on a piecemeal basis as air photos came in from day-to-day. Now, however, the comparison of successive photos of the same terrain over six months enabled the *Groupe* to carry out a definitive restitution (Bellot, 1936, p. 121).

Development of plans directeurs
The *groupes de canevas de tir (GCT)* had the specific task of creating large-scale trench and artillery maps (*plans directeurs*) of their army areas, with German defences plotted from air photos, and dealing with related artillery survey matters. The first plan directeur specially printed during the war appears to have been on 31st October 1914 for a local operation. The November *Instruction* had charged each sector artillery commander with supervising the preparation of the *plan directeur*, the work to be done by his officers helped by those of the *GCT*, but it was quickly recognised that corps artillery officers had insufficient topographical training to create a *plan directeur* and therefore in a 20th November circular on GCTs, it was stated that among their work was the establishment of a *plan directeur de tir* for the artillery command. On 17th November, Joffre issued a *Note* prescribing the closest collaboration between the Flying Service, which undertook photographic missions, and the *GCTs* which used the photos to create *plans directeurs* (Bellot, 1936, p. 72).

Those armies which were in areas for which large-scale material was available were fortunate. Others had to enlarge from the 1:80,000 and amplify and revise from air photos. In all cases, German defences had to be plotted from air photos. The planimetry and relief of the enlargements was inaccurate, but were gradually improved as air-photo interpretation became more sophisticated. These provisional enlargements were soon replaced by more accurate sheets compiled from cadastrals made between 1817 and 1850 which had been stored at the *Service Géographique*, coal mining, railway and canal plans, etc. (Arthur-Lévy, 1926). GCTs at first produced monochrome plans, but in early 1915 began to add German defences in blue and contours in bistre. Larger scale (1:10,000 and 1:5,000) plans were soon produced, better depicting the increasingly formidable German defences. At first the GCT of each army evolved its own sheetlines, numbering or naming, and grid, and the solution to this last problem has already been noted.

The Second Army created the first complete 1:20,000 *plan directeur*, the 3rd edition (12th February 1915) of the Albert sheet, with very full planimetry, setting the pattern for the future. Signs were large and legible; trigonometric points were carefully plotted, with special signs, thus eliminating the need for a separate *canevas d'ensemble*. Bistre 10 metre form-lines were taken from the 1:40,000 '*minutes,*' and were gradually improved. German defences were shown in blue, with signs for each feature. Key points and targets were specially numbered to avoid confusion, with a special series for those in the German zone. This model was soon adopted for all armies, and underwent little modification throughout the war. From April 1915 onwards, German batteries were designated by their four-figure co-ordinates. On 19th April Second Army published such a sheet of the Somme area which was immediately standardised for all armies. From 1915, for areas just behind their front where large-scale surveys did not exist but were of great importance, some GCTs undertook simple tacheometer traverses, plotted on the 1:20,000 scale, to rectify the planimetry (Bellot, 1936, p. 75).

Plans directeurs for the Third Battle of Champagne (September 1915)
For the Third Battle of Champagne, the GCTs of Second, Third, Fourth and Fifth Armies were placed on 16th September under commandant Bellot of Second Army's GCT, to achieve co-ordination of survey and maps. For the attack of 25th September, three main types of plan directeur were produced:

1. 1:20,000 (artillery) to designate targets and measure firing data, forming the large-scale map of the whole front; a common framework for all other plans and sketch maps.
2. 1:10,000, for general staffs to study offensive projects and direct operations, giving a complete picture of all German defensive organisation; special secret edition also showing French organisation (in red).
3. 1:5,000 infantry sketch maps for attacking the first German position, giving a comprehensive picture of all details and information needed to guide and help advancing units.

These series had the same elements of planimetry, enemy organisation and designations, and an identical grid and numbering system to avoid misunderstandings. Full instructions ensured that maps and plans conformed to a rigorous specification (Bellot, 1936, p. 77).

Maps and survey at Verdun, 1916
The Battle of Verdun lasted from February 1916 to the end of the year, with further important operations in 1917. The area was covered by the good 1:20,000 pre-war fortress *plan directeur*, surveyed at 1:10,000 on a local projection. In 1914 the Verdun defence force was served by a *bureau topographique* attached to the fortress artillery, which supplied topographical data to the fixed artillery of the forts and flanking batteries, and to mobile units.

In August 1915 the former fortress of Verdun became part of the new *Région Fortifiée de Verdun* (RFV) (FOA, 1916). Lallemand was appointed to command the *GCT* of the newly created RFV and began to revise and update the existing 1:20,000 *plans directeurs* (Bellot, 1936). New defences on the left bank had to be surveyed and mapped. In December 1915 the GCT abandoned the polyhedric projection and local grid of the old fortress plan directeur, for the Lambert projection and grid. The old base map (*plan directeur* with detail and contours in black) was retained.

Early in 1916 there were indications of attack, and German artillery ranging began on 12th January. Bad weather prevented much French aerial photography, essential for mapping and intelligence purposes, until 17th January. This was particularly important for the plotting of new German batteries, from which the enemy's intentions could be divined. It was said that there was no officer on the RFV staff who could interpret air-photos, though certain officers of the GCT had become very experienced at precisely this. An unnamed 'expert', sent on 17th February, identified the sector of attack.

On 24th January, de Castelnau ordered the strengthening of the right bank defences and the construction of a new intermediate line (Horne, 1962). This was begun so late that it was not shown on the German assault maps. The offensive began on 21st February. The German artillery concentration was unprecedented; they surveyed-in 1,200 guns and howitzers (Conrad & Laspeyres, 1989), a big proportion of them heavies. The attack frontage was only 6 km on the right bank, on which over 850 guns gave a great density to the brief, devastating bombardment (Horne, 1962). It was impossible to locate individual guns or batteries; gun flashes were so dense that French airmen gave up trying to pinpoint them all.

Pétain's Second Army was hastily sent to Verdun, finding on the 25th that the Germans had captured Fort Douaumont, the keystone to the defences. Joffre suppressed the RFV on the 26th. Pétain established his HQ at Souilly, where his GCT's Advanced HQ (the rest was at Bar-le-Duc) under Courtier immediately took over officers, survey and mapping from Lallemand's RFV Groupe. The old 1:20,000 *plans directeurs* were immediately recast into two new standard sheets (*Verdun B* and *Étain A*, the first editions of which were both dated 11th March). On 6th March the Germans extended the attack to the left bank, capturing Côte 304 and the Mort Homme in May. On 15th April, therefore, the GCT of Third Army was placed under the command of the OC GCT Second Army to achieve a single technical control over maps and survey on both banks of the Meuse; Third Army's *Groupe* recompiled and revised the left bank 1:20,000 *plans directeurs*, ensuring that these were accurate and up-to-date (SHAT, 3M 561).

The *GCTs* provided crucial survey and mapping support. Map drawing was carried out under unrelenting pressure, closely directed by Courtier. The continual changes of the front and new trenches, battery positions, tracks, etc., meant very frequent editions in 1916: e.g. 1:5,000 *Fort du Vaux*, 12 editions; 1:10,000 *Douaumont*, 20 editions; 1:20,000 *Étain A*, 22 editions. Key editions coincided on all three scales. The Second Army distributed 365,000 *plans directeurs*, comprising 161 editions, between 15th June and 12th September 1916, about 100,000 of these in the first half of August (SHAT, 3M 561).

An enormous effort was undertaken to develop the army triangulation on which depended the accurate fire of the artillery. Capt. Pellet of Second Army's GCT organised *brigades topographiques* to push the triangulation into the battery zone. By the end of 1916 they had executed 1 500 km of tacheometer traverses to fix battery positions, aiming points and lines of fire.

In the shell-swept forward and battery zones, all trig points had been destroyed and resection of battery positions was impossible until new points had been established by traverses pushed up from the rear. On the battlefield, the region of Froidterre, Fort Souville and the approaches to Fort Vaux were resurveyed on the ground as the German were pushed back, and then the Hauts-de-Meuse towards St. Mihiel, and the second position (SHAT, 3M 561).

On 24th October, after much thorough survey and mapping, Douaumont was recaptured. On 2nd November, the Germans evacuated Fort Vaux. The new creeping barrage technique, based on careful survey preparations and map-shooting, had been developed since 1915 by the British and French, and during 1916 it became firmly established. Pétain's successor, Nivelle, was convinced that this was a war-winning formula, and based his set-piece operation to recapture Douaumont on it, over 650 guns being surveyed-in. Two new 400 mm railway guns came into position and were given line very accurately to bombard the fort, each shot being plotted on a large-scale plan (Conrad & Laspeyres, 1989). Mangin's Corps studied 1:5,000 relief models (the French long utilised *plans-reliefs*), constructed by Corporal Arnold who built 1:5,000 and 1:20,000 relief models, as accurate as *plans directeurs*. Arnold delivered 30 models, for briefing, per month (SHAT, 3M 573). A full-sized replica of the battlefield, including Fort Douaumont, on which the assaulting units practised, was also laid out (Horne, 1962).

The Nivelle Offensive, April-May 1917
General Nivelle believed that his creeping barrage with the surprise use of tanks would create a breakthrough on the Aisne. The Germans, however, adjusted their tactics accordingly, to defence in depth with a lightly held forefield and increasingly heavy counter-attacks. The offensive failed due to over-confidence and poor security; the survey preparations however were most thorough. Fifth Army's GCT on the Chemin-des-Dames front was significantly reinforced. Lieutenant Feuardent studied the battlefield closely from a special OP, using an astronomical telescope with x75 magnification, and provided precious information. Other sources such as air reconnaissance confirmed this observation and led at the end of 1917 to the GCTs collaborating with Intelligence in operating many more such posts, 12 officers being specially trained and equipped with telescopes of x100 magnification. On the right, Fourth Army's CGT supplied survey support for the attack on the Monts de Champagne. On the left, Third Army's GCT, based at Noyon, surveyed the ground abandoned by the Germans in their March retirement to the Hindenburg Line, and prepared new maps of the forward zone and the new German defences for its attack on 16th April. The *plans directeurs* were of their usual high standard but could not indicate the recent changes in German defensive tactics. Tenth Army was on the Chemin-des-Dames from 21st April, attacking the Craonne Plateau, and from 8th May organised the captured position. Its GCT resurveyed the Aisne valley and used the caves of the Chemin-des-Dames ridge as shelters during a very large-scale survey of 'an

absolutely special nature', presumably in connection with defence works and the establishing of OPs to observe towards Laon (SHAT, 3M 583; Bellot, 1936).

4. German maps and survey on the Western Front, 1914-1918

Peacetime organisation and preparations for war
The Prussian State Survey, directly under the General Staff, co-ordinated state civil and military mapping. Before 1914 surveys were mainly concerned with completing the 1:100,000 Staff map of Germany, based on 1:25,000 or 1:50,000 surveys, but contoured 1:25,000 plans (*Planmaterial*) were produced for the artillery from 1904, artillery boards, fire-control plans and plotting boards at this scale being created for fortresses and coast defences (Anon, 1915; Schimrigk, 1920, p. 82). Field Survey Sections were first used about 1904 in South West Africa. A Photogrammetric Section was formed in 1908, enlarged in July 1914 to a Company. At the outbreak of war the Survey comprised Trigonometry, Photogrammetry and Cartography Departments, Artillery *Planmaterial* Section, Colonial Section and Map Room, 911 personnel in total. A further 350 more joined annually as field reinforcements (Albrecht, 1969).

On mobilisation, most Survey personnel went to fighting units or staff, though the Cartographic Department remained in Berlin for drawing, reproduction and administration. In mid-1915 all war survey was placed under a central directorate at GHQ. For war, sets of the 1:100,000 map, the obsolescent hachured 1:80,000 French map, and the Belgian contoured 1:40,000 map reduced to 1:60,000, were issued. Issues for the Schlieffen Plan only covered Belgium and north-east France, the break-of-scale at their frontier causing problems. The advance rapidly swept through this area, and maps soon ran out (Boelcke, 1921).

German artillery survey and photogrammetry in 1914
In 1901 the Survey adopted terrestrial photogrammetry for fortress and siege war, especially for artillery map-shooting, a stereoscopic method using Pulfrich's photo-theodolite provided the first practical results. In 1904 it established sections for fortress artillery surveys and to produce artillery *Planmaterial* — a dense, gridded 1:25,000 trigonometrical and topographical framework including fixed points of the fortress area, battery emplacements and observation posts-for German fortresses and attacks on enemy fortresses (Albrecht, 1969). After the start of army aerial photography in 1911, enemy battery positions, etc., were plotted from air-photos. The Photogrammetric Section arranged for survey practice in fortresses. Two Survey Detachments (*Vermessungsabteilungen*), including Photogrammetry Sections, were set up in 1912 and practised terrestrial and air photogrammetric cooperation (from airships, balloons and aeroplanes) at Wahn and Thorn. In December 1912, Moltke issued an instruction 'Reconnaissance and Survey in Siege Warfare,' Fortress Survey Detachments (*Festungsvermessungs-abteilungen* or FestVAs) being formed in 1912-13 for frontier fortress artillery photogrammetry using

balloon and aeroplane photos. In March 1914 three, created from the Photogrammetry Department, existed at Cologne, Metz and Strassburg (BA-MA PH9/XX, PH9V/98 & 99). The Survey's photogrammeters were transferred to the FestVAs on mobilisation (Albrecht 1969). Much of the German lead in photogrammetry can be traced to this development, for the FestVAs became the nuclei of the field survey units (VAs). Unlike British and French practice, air photogrammetry featured in German survey units from the start.

The Germans were ahead in air photogrammetry, notably with the Scheimpflug multiple-lens camera used in Zeppelins. In 1909 Zeiss developed a survey camera with which gun-flashes, shell bursts, enemy positions, OPs etc. were photographed from high ground or balloons. Plotting was initially done by hand, and later with the Stereoautograph. Soon after flying sections were created in 1911, the first vertical and oblique air surveys were made (Boelcke, 1921; Albrecht 1969).

Survey units move into the field
On mobilisation, eight further FestVAs were established (Albrecht, 1969), and two trigonometric and two topographic surveyors were provided per heavy artillery brigade (PRO WO 106 1529). The three original FestVAs followed the advance, FestVA1 from Cologne surveying the 42 cm mortars to bombard Maubeuge (August) and Antwerp (October), and intersecting targets and hits. Aerial photogrammetry was used at Antwerp to determine by graphic resection the air station of the camera and to plot positions (Fleck & Jacob 1916, p. 23). The FestVAs from Strassburg and Metz pushed forward their triangulations to provide artillery control for attacking the French fortresses between Nancy and Verdun. The eleven FestVAs proved inadequate for the position warfare artillery surveys which soon developed. They gradually became mobile and worked with the field artillery, the units which advanced into the field being called *Feldvermessungs-abteilungen* (FeldVAs, or simply VAs). Three more FestVAs were formed during the winter (Albrecht, 1969).

Trench warfare forced field artillery to adopt the heavy gunners' map-shooting techniques, necessitating accurate surveys. From December 1914, emergency Field Survey Sections *(Vermessungstrupps)* were formed to provide survey support for the artillery (Landmann, 1996, p11), undertaking extensive surveys in the battery zone for large attacks. Further VAs were created in the spring of 1915, including Wurtemburg and Bavarian units, the former having 71 personnel, with cameras, stereo-comparator, Finsterwalder photocartograph, photo-reduction apparatus, power-press, etc., far stronger than British and French units (Jack, 1920, pp. 7, 190).

Maps situation in the first few months
The plotting of trenches, gun-pits, etc., from air-photos, needed detailed, accurate large-scale plans as a base, which were only available for fortress areas (Boelcke, 1921). Trench warfare forced most armies to enlarge the

French 1:80,000 and revise from air-photos. Belgium was covered by a good 1:20,000 survey, and French frontier fortress areas by pre-war *plans directeurs*; those in the area of operations, obtained by the Germans, formed the base of 1:25,000 *Planmaterial* (for Belgium at 1:20,000). A provisional 1:25,000 artillery and trench map of the Western Front was created in early 1915 by 50 Air-Photo-Plotting Sections formed by flying squadrons. These produced large-scale sketch-maps of Allied defences with backgrounds and trenches drawn from unrectified mosaics; lacking a trigonometric framework they were seriously distorted and useless for artillery work (Boelcke, 1921).

War survey organisation (Kriegsvermessungswesen), *and field survey detachments (VAs)*

By mid-1915, mapping of varying accuracy was being done by the VAs, the *Vermessungstrupps*, Artillery Survey Sections and Flying Units. The first three worked from fixed points, but although the flying unit maps were unreliable, staff and troops believed they were accurate; a conflict developed between the flying and survey units which the latter lost, unable to deliver such rapid and tempting results. The experience of the first months indicated the need for a central survey directorate. Realising that the existing units could not cope with the growing demands for survey and mapping, armies demanded better support (Boelcke, 1921).

In July 1915, therefore, a War Survey Directorate was created at GHQ under Major Siegfried Boelcke *(Kriegsvermessungschef* or *KVC),* for ground and air survey (and geology from 1916), cartography, map-printing, raising and equipping new units, selecting personnel and issuing technical instructions (KVC, 1918a). He reorganised existing units into a flexible, mobile organisation of VAs, each army having one (some two), with attached map-printing section. Prussia, Bavaria, Saxony, and Wurtemburg created their own VAs (Boelcke, 1921).

Air photography and tactical photo-interpretation was designated an Air Force task; air survey and plotting tactical material for mapping was a War Survey Organisation task. Photogrammeters and draughtsmen were transferred from the Air Force sections to the VAs, which were the only units permitted to produce maps. In practice, competition between Air Force and Survey Organisation continued until the end of the war, notably in precision air photogrammetry for which both created research establishments, because of the impossibility of making a clear distinction between intelligence and topographical information on photos. This led to duplication and delays, particularly after November 1916 when the Air Force became independent (General Staff, 1918, p. 107) and set up a Photographic Inspectorate and an Air Survey Unit.

In 1916 frequent transfers of officers were made between France and Russia to disseminate best practice (SHAT, 3M 576). By 1918 there were 29 VAs, averaging 300 personnel (Albrecht, 1969).

VA Artillery Work in 1916
Their main artillery tasks were battery surveys and the production of *Planmaterial* for battery boards, plotting boards and fire control plans. Battery boards had the gridded plan pasted down, marked with the directing gun, zero line and aiming points, and fitted with a pivoted rule and graduated arc. For long-range guns it was advisable not to take the shooting data (range and bearing) from their huge boards, but to determine them trigonometrically. Plotting boards for flash-spotting and sound-ranging had the four survey posts carefully marked on the gridded plan, and a graduated scale. Intersections onto an enemy gun-flash pin-pointed the target. Similar boards, usually prepared at 1:2,000 from rectified air-photos, were used by trench mortar survey sections *(Minenwerfer Messtrupps)* (KVC 1918b).

The geodetic situation, trigonometric work and grids
Each army created its own triangulation, map-series and sheetlines and report and trig grids, thus preventing the adoption of uniform artillery fire-control; artillery cooperation on army boundaries was difficult when the gun was on one grid, target on another, and aiming point on a third. Difficulties in reconnaissance and confusion in reports and referencing also occurred. There were nine army grid systems in the West, based on French fortress grids; report grid numbering was completely different. These problems led to the creation of the War Survey Directorate in July 1915, but Boelcke was not permitted to rationalise the geodesy, standardise sheetlines and grids, and introduce an orthomorphic projection, and the problems persisted. A partial rationalisation occurred towards the end of the war (Boelcke, 1921).

To provide a rigid framework for artillery surveys, mapping and air-photo control, a geodetic base was created, initially provisional frontier fortresses local systems. In early 1915 ex-cadastral surveyors hastily created army Soldner (Cassini) co-ordinate systems; the projection led to angular distortion, giving bearing and range accuracy up to 60 km either side of the astronomically-fixed central meridian (Boelcke, 1921). It did not provide integrated geodetic data. In the autumn of 1915 Boelcke ordered the creation of an integrated co-ordinate system, based on the Gauss-Kruger Conformal (Transverse Mercator) projection and grid, for the whole Western theatre, but armies refused to abandon existing systems, believing they would soon advance (Albrecht, 1969).

In 1917 three test bases were measured to determine the scale of individual networks (Werkmeister, 1923). The original provisional work led to wasteful patching and amplifying; an accurate, integrated theatre co-ordinate and levelling system, initiated at the outset, would have providing basic data for all subsequent orders of survey and supported the topographical surveys for the 1:25,000 map. This accuracy was vital, as the Paris shooting at ultra-long ranges showed (Albrecht, 1969).

German photogrammetry during the War
'Germany, by its profound connection of science and technology, led the world optical industry' (Eckert, 1921, p. 241). The first air-photos taken at the start of trench warfare were immediately used for air photogrammetry, crucial for creating the planimetry of the 1:25,000 map. Systematic serial-photo missions were flown from 1915 and control points, including points in the enemy area, fixed through ground and air survey. In active sectors, enemy trenches and batteries were photographed daily; the photos, tactically annotated, were passed immediately by the Corps Photo Section to the corps and divisional topographical sections where they were plotted and the results overprinted on base maps. At army HQ, Air Force Staff Photo Units worked closely with the VAs (KVC, 1918a; Albrecht, 1969). Balloons were used as 'high OPs,' becoming an established part of German flash-spotting, and also as camera-platforms for oblique photogrammetry, a significant element of German survey (GHQ, 1918). Over one million air-photos were plotted photogrammetrically during the war (Boelcke, 1921).

The accurate, heavy Stereoautograph apparatus was kept at Zeiss, Jena, photo-survey material being sent there for plotting. Designed for terrestrial photogrammetry, it was later adapted for high obliques, but was unsuitable for low obliques and verticals. The Stereocomparator was used in broken terrain for determining co-ordinates of points for planimetry and heights from terrestrial photos. The Survey and Air Force air-photo research establishments at Berlin and Stuttgart, and the Bavarian Survey at Munich, undertook similar experimental work. In 1918, after two years of research, armies were being equipped with Prof. Dr. Carl Hugershoff's *Bildmesstheodolit* (photogoniometer) proto-autocartograph for accurate point-fixing from obliques (KVC, 1918a). Stereoscopic air survey for plotting contours made progress but was not fully developed during the war. The Gasser double projector of 1915 was neglected (Gasser, 1925, 1953). Relief was usually taken from the poor French 1:80,000, occasionally amplified through local tachymetric and air surveys (Albrecht 1969).

Many points in and behind the enemy line were rapidly fixed by terrestrial stereophotogrammetry, periscopic cameras developed from pre-war photo-theodolites being used to obtain intersecting views to fix points up to 16 km away, providing reference points for artillery survey sections and control points for air-photos. In the Champagne, 6,000 points were fixed stereophotogrammetrically in the enemy area as air-photo controls. Enemy trench systems were surveyed in this way, and contours of enemy territory were also accurately drawn with the Stereoautograph from terrestrial photos. Terrestrial stereophotogrammetry was most productive in hilly country, as in the 1:25,000 contoured plan of the Meuse Valley north of Verdun produced in 1915, and in mountainous areas where new topographical maps were made, as in Macedonia where planimetry and relief were drawn in 1915 with the Stereoautograph (Boelcke, 1916; Albrecht, 1969).

In 1914 Oskar Messter developed a successful semi-automatic strip camera taking 250 exposures, and by May 1915 tested a fully-automatic one. At 2,500 m altitude, a strip of ground 2.4 km by 60 km was photographed in one flight without changing the film. A strip overlap of 25% was aimed for. Success depended on factors outside his control — aircraft which couldn't fly on a straight and level course, lens quality and film problems. Oskar and E.O. Messter developed and built 241 serial-photo air-cameras for reconnaissance mapping, which photographed over 7 million km^2 (Karlson, 1941; Baring, 1963, p. 125).

Air photogrammetry
Plotting was first done by graphic and analytical methods, amplified by stereoscopic examination (Eckert, 1921, 1925). If the focal length was known, a photo could be accurately plotted from only 3 points by the Hugershoff pyramid method. The pyramid method was used widely for accurate air photogrammetry before and during the war. Several other methods developed from the pyramid and four-point methods were used. In 1917 von Rudel developed a graphic method of radial triangulation (Schwidefsky, 1954, p. 132).

Photogrammetrical experimental work was done in the Survey Organisation's Reserve and Experiment Section in Stuttgart by Hermann Cranz in 1917, and later by Traugott Fischer, including 'the drawing of contour maps from air-photos on the basis of a fixed-point network, with investigations into accuracy' (Landmann, 1996). Similar work was also carried out in the Air Force's Air Survey Department in Berlin under Hugershoff, who had worked on photogrammetry before the war and, as a *Feldphotogrammeter* at Posen from 1914 to 1916, invented and developed revolutionary photogrammetric instruments and methods. In 1916 he went to the Air Force Photographic Inspectorate, and in 1917 commanded its Air Survey Department. In 1918 Cranz joined Hugershoff, reporting in March an attainable accuracy for point-fixing of ± 0.8 to 1.2 m by *(Bildmesstheodolit)* photogrammetry, ± 2 to 3 m by forward intersection of new points, and ± 3 m for height-fixing (Albrecht, 1969).

In 1915 E.O. Messter joined the flying troops as a cinematographer, and soon moved to their testing ground. For plotting accurately from air-photos, Pulfrich's stereoplotting methods were applied from 1915, and later led to contact with Hugershoff. At the end of 1916, Messter went to the Air Force Photographic Inspectorate, developing his ideas and contacting experts, especially Hugershoff, with whom he discussed the automation of air-photo mapping. The strip-photo technique was extended to stereo-pairs; Boelcke noted: 'With the Messter *Reihenbildner*, sectors of ground are surveyed to scale. The *Doppelreihenbildner* (stereo-pair strip camera) provides the stereo survey from which the general ground-forms can be recognised. To create an accurate true-to-scale map, the Hugershoff [*Bildmesstheodolitû*] method is used and can survey at 1:10,000, surpassing the accuracy of the Prussian 1:25,000 Survey.' The apparatus' quality was confirmed by operational use. The first experimental model of the Hugershoff Autocartograph appeared towards the end of 1918 (Karlson, 1941; Baring, 1963, p. 125).

Dr. Max Gasser's revolutionary double-projector, patented in 1915, provided an optical method of plotting planimetry and contours for reconnaissance mapping. It was the forerunner of the Multiplex (Gruber, 1932; Hart, 1943). By 1915 Gasser had produced a prototype and solved the orientation problem, but a military committee (including his rival Hugershoff), was appointed to assess it. Its 'incorrect judgement . . . prevented his method producing accurate mapping for precise gun-fire etc., so his prototype remained unused' (Gasser, 1923, 1953; Meier, pers. comm., 2000).

The Hugershoff proto-autocartograph
During the war, Hugershoff and Pulfrich revolutionised photogrammetry by making direct angular measurements on oblique air-photos; their mechanical plotting first used the modified Stereoautograph and then Hugershoff's new photogoniometer *(Bildmesstheodilit)* proto-autocartograph. Hugershoff and Cranz provided the theoretical basis and achieved operating efficiency for optical-mechanical automatic plotting apparatus (Hugershoff & Cranz, 1919; Krebs 1922; Eckert 1925). In the absence of firm co-ordinates of three ruling-points per photo, the trigonometers selected prominent points on the air-photos whose position and height could be surveyed from the nearest fixed points. Oblique survey photos provided many control points, and were combined with vertical serial photos to obtain detail and contours. The method was mainly used for monocular point-determination, and constructing contours from individual points on the map was laborious (Eckert 1921, p. 255).

The Battle of Verdun, 1916
Verdun was the only large German offensive in the West between 1914 and 1918. Fifth Army maps and survey were undertaken by VA3 and VA15, and corps and divisional *Kartenstellen*. VA3 comprised about 100 personnel. The pre-war 1:25,000 Verdun-Belfort series, from *plans directeurs*, was revised from air-photos with French defences overprinted. This omitted the new French intermediate line on the right bank begun in January 1916 (Horne, 1962). Terrestrial stereophotogrammetry had created an accurate, contoured 1:25,000 plan of the Meuse Valley north of Verdun (Boelcke, 1916), ensuring that heavy guns were accurately plotted relative to targets. Battery commanders were increasingly converted to survey methods, and by the end of 1916 they all demanded the fixing of their batteries. One trigonometrical officer was attached to corps artillery for local triangulation, and two topographical officers for fixing foot artillery batteries and observation posts. During 15 days of preparations, and during the attack, they fixed advanced positions for each battery. VA3 prepared 150 battery boards. Corps sent daily to the VA a list of French batteries located by *Artillerie-Messtrupps*, balloons and aircraft (SHAT, 3M 576).

Before the attack, large numbers of 1:5,000 (each infantryman had a sketch of the French defences — in insufficient depth) and 1:25,000 maps, and fewer 1:10,000 maps, and 1:2,000 plans for *Minenwerfer*, were supplied. Relief plans were supplied on demand. 20,000 copies of new 1:25,000 sheets were printed in a train at Metz. VA3's own printing train produced 1:5,000 sheets. The 1:25,000 was the most carefully prepared map, from which all others were enlarged, though photogrammeter officers rectified air-photos at 1:5,000 using precision methods and instruments, and took panorama photos for the artillery. They used a method of reconstitution similar to the French *'chambre claire'* to provide control; other ranks fitted detail onto these points. 1:25,000 revision slips were printed overnight during the attack. Troops often complained of lack of maps, and many revision slips were not stuck on by the users. New editions appeared monthly. Regimental sketch maps, showing their new front line, were sent with an overview sketch by corps to the VA, the front being marked by ground panels visible on air-photos. Photogrammetry officers rapidly rectified them, consulting the corps sketches (SHAT, 3M 576).

The Battle of the Somme, 1916
This was the largest defensive battle fought by the Germans in the West until 1918. On the first day, 1st July, Second Army, with VA12 as its survey and mapping unit, had three corps. On 3rd July it was reorganised into three 'Groups' *(Gruppe),* and on the 19th its troops north of the Somme became a new First Army, served by VA23. The Battle highlighted certain inadequacies in German mapping. General von Arnim noted poor quality, map shortages, and the paucity of points for fixing German batteries on the maps. Detail was inadequate for map-shooting. Later, the VAs fixed and plotted many more points. Subsequent map supply was also deficient despite the provision of printing trains. British air supremacy, fine photos, accurate cadastral compilations, impressive gunnery and precise map-shooting data disturbed the Germans; without air observation, and under very different conditions than previous shoots, British batteries accurately hit small targets.

German offensives and Allied counteroffensives in 1918
By March 1918, survey personnel totalled 9,200 (Boelcke 1921). 'Artillery Trigonometers' fixed battery and gun-positions, zero lines, observation posts and sound-ranging and flash-spotting survey posts. Integration was crucial: in June, Boelcke stated: 'It is essential that they connect the Flash, Sound and Trench Mortar Survey Sections' local surveys with the lower-order trigonometric and topographic fixings, and those with the main trig grid' *(KVC* 1918b). Accurate map-shooting depended on them. They surveyed and gave line to the long-range railway guns. In the summer of 1918 a glowing report appeared on their work in the Marne attack (Albrecht, 1969, p. 32).

For the March offensive, Bruchmüller insisted on no registration, only permitting a few ranging shots. Surprise was achieved by a short, devastating preliminary bombardment. Captain Eric Pulkowsky, the Meteorological Chief, trained the artillery in the new predicted methods, which were successfully adopted despite opposition from many senior gunners; meteor and calibration corrections were applied, a methodical system being developed for transmission and application of data; ranges were accurately measured, and reliable maps provided for this. Eighteenth Army's VAs surveyed and providing battery boards for hundreds of batteries. Targets located by sound-ranging, flash-spotting and aerial photography were plotted with great accuracy (Ludendorff, 1919). The Germans deployed 10,000 guns and mortars, a

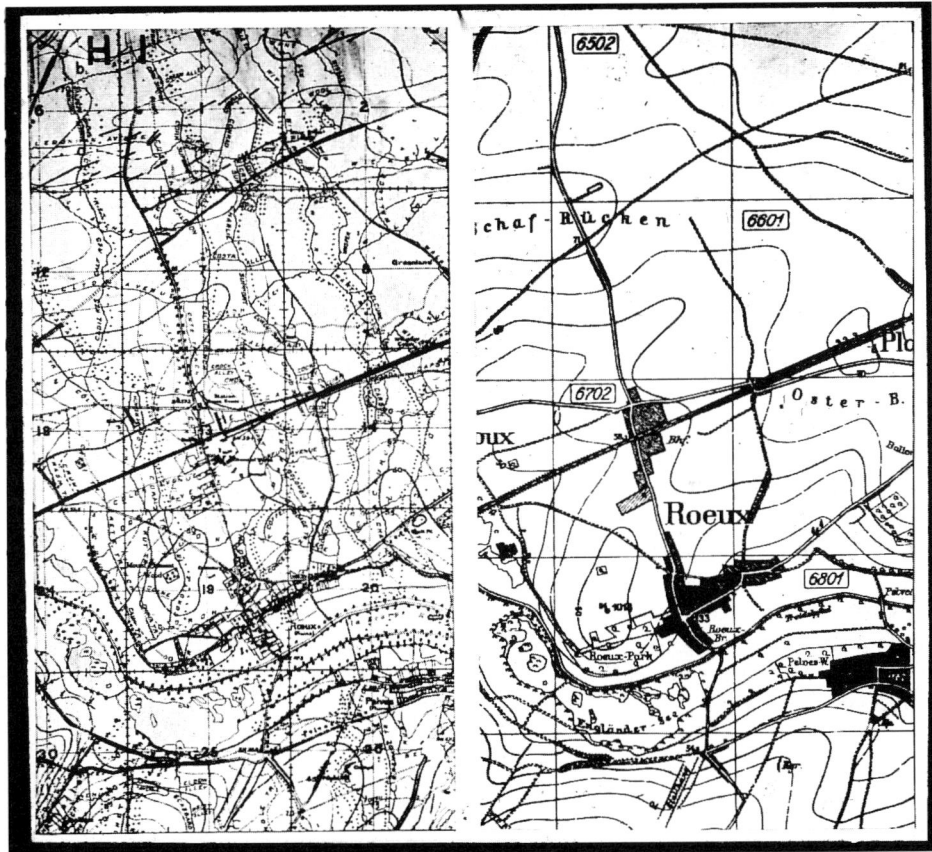

Figure 4. Comparison of British (left) and German (right) mapping of the same area, Roeux, east of Arras in northern France. Reproduced from a contemporary GHQ lantern slide

prodigious survey effort. The bombardment stunned the British, who praised the accurate German map-shooting; battery positions, defended areas, redoubts, posts, HQs, telephone exchanges, buried cables and cable-trench junctions, wireless and buzzer installations, railway junctions and other centres being pinpointed from air-photos and accurately shelled.

German infantry and accompanying artillery, using compasses and 1:10,000 'strip maps' with panorama drawings in the margin for orientation, sped across no man's land and through the British defences in the fog. All arms had contoured (for the artillery) 1:50,000 sheets, enlarged, and generally redrawn, from the 1:80,000, extending to the coast. Valleys, identified for infiltration, were in green. 1:25,000 sheets of the front area were also used, particularly as artillery maps. Some sheets carried three overlapping grids! (PRO WO 153 981). Similar tactics and maps were used for the later offensives. On 12th April, of the Western armies, six had one VA each, five had two each, and one (Fourth Army, currently attacking on the Lys) had four (five if the Marine VA is included). On the Vosges front, two army-detachments had one VA each and one had two *VAs*. There were seven VAs on other fronts (BA-MA, PH3/462).

Some advancing German forces only had enlargements from old 1:80,000 material, creating a serious gunnery problem; detail was often 300 m out, and much new detail had to be added (BA-MA, PH3/52). Trigonometric data became unobtainable in the advance to the Marne; triangulation from the old battle-zones was overstretched, temporary surveys being created from known trig values in the new artillery zone. Lacking fixed points, a short base was measured between points located on the map. Onto this, a reliable connection was created between the map and trigonometrical survey; acoustic and light-flare surveys by the artillery survey sections were combined with triangulation and traverses (Boelcke, 1920b, p. 14). In August a new pamphlet, *Der Punktplan* (Point-Plan), dealt with the lack of an accurate map in mobile operations; it emphasised the need for all survey units-VAs, Divisional *Kartenstelle*, flash-spotting and sound ranging sections, Air Force (oblique and vertical photos), trench mortar Survey Sections and Artillery Trigonometers-to work together to fix points in the operational area, and stressed that a fixed point network, carefully plotted on a 1:25,000 gridded board was superior to the map in accuracy and formed a reliable basis for indirect fire (KVC 1918b).

The Paris guns (Wilhelmgeschütze)
The accurate fire of the Paris guns in 1918 was a survey triumph. Such ultra-long-ranges (c. 120 km) had never before been achieved, and new problems had to be solved. Accurate preparations were crucial; a bearing error of 1° would create a 2,000 m lateral error at the target. Krupp provided most firing data, including meteorological corrections and bearing corrections for the earth's curvature and rotation during the time of flight (Bull & Murphy, 1988). The zero line was adjusted for the influence of lateral winds on the shell. Geographical co-ordinates of the target were obtained from the map and the gun-position grid co-ordinates converted into geographical co-ordinates, referred to a refined geographical zero-meridian; range and bearing were then trigonometrically determined (Albrecht, 1969). The surveyors marked the gun

bedplates with the finest divisions, the zero-line being astronomically orientated to run through central Paris. The first shots indicated perfect line and range (Bull & Murphy, 1988). Seismic location methods, already used in 1915-16 (Kaiser, 1916), were used to pinpoint the fall of shells in Paris; British and French also investigated this method, but the Germans took it

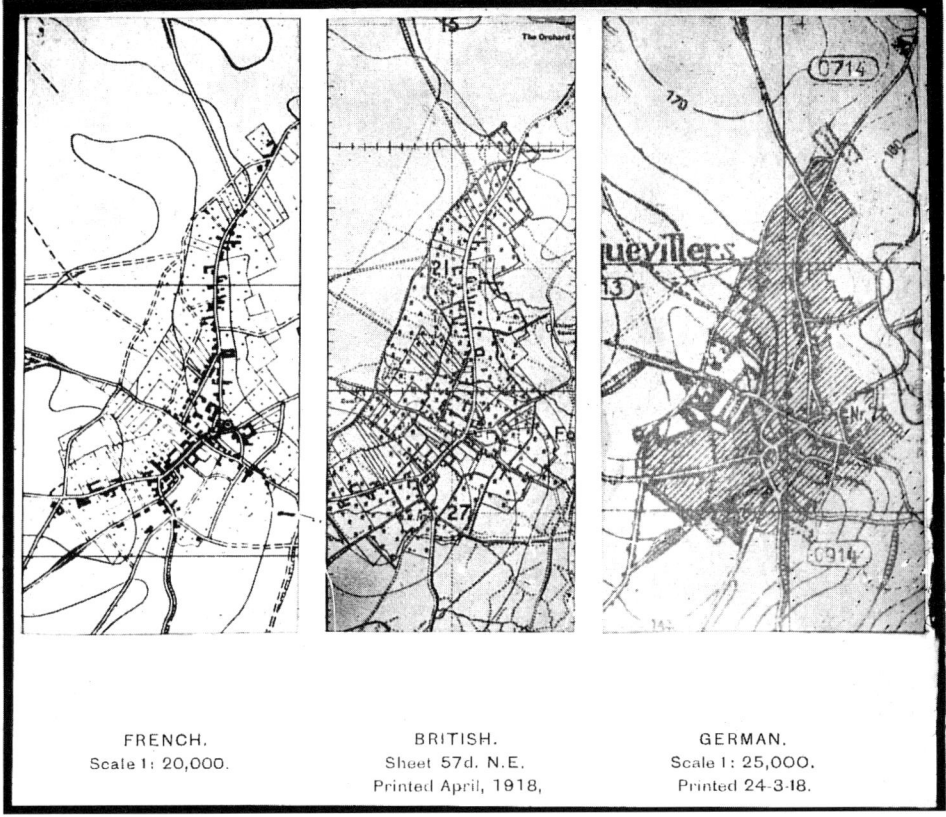

Figure 5. Comparison of French, British and German mapping of the same area. Fonquevilliers, Somme, northern France. Reproduced from a contemporary GHQ lantern slide

furthest. In April 1918 there were five officers and scientists working on 'seismic reconnaissance during the Wilhelm shooting' (BA-MA, PH3/507).

The Allied counter-attack and the German retreat to the Armistice Line
From July 1918, the Germans were continually withdrawing. Whenever new defence lines were taken up, hundreds of battery positions had to be changed; there were insufficient personnel in the West to undertake the heavy survey,

cartographic and printing tasks for these rear positions. In the crisis, survey work in Russia was terminated, and personnel were sent West as reinforcements (BA-MA, PH3/ 465). It was too late. The last feasible defence lines were breached by the British in early November.

5. Conclusions

It is important not only to compare the maps themselves, but also the survey methods used by the protagonists in order to consider how this has contributed to the war effort (Figures 4 & 5). French mapping and survey undoubtedly suffered from insufficient resources for pre-war domestic mapping, particularly the lack of a reliable large-scale topographical map such as existed in Britain, and to a lesser extent in Germany and Belgium. From this unpromising beginning, the French rapidly established a strong geodetic and cartographic base, adopting standard sheetlines and the Lambert projection and grid by 1916-17. They had the great advantage of having immediately available a mass of trigonometric and cartographic data-in particular the results of their old but invaluable 1:2,500 cadastral survey. The British produced the clearest and most accurate maps (given the projection caveat).

The Germans were the most technically advanced in optics, were ahead in air survey and photogrammetry, and directing by refined survey methods the most accurate artillery fire on Paris at ultra-long range. They were also ahead in using mobile and fixed power-presses, with process equipment, in the field, particularly at corps level; nearly all German large-scale maps were printed in the field, whereas most British and French equivalents were printed far from the front, in Southampton or Paris. German trigonometric beacons, often specially erected for the artillery, were much admired by the British and used whenever rapid advances captured them intact. The Germans created for themselves the problem of separate army trigonometric and report grids; each army had a different origin, and some sheets included three grid systems.

The British may have been ahead of the French in compilation from cadastrals, and both were ahead of the Germans (who, however, did not have access to most of the major archives). The German survey extended, both in breadth and in depth, much further than the British, and in some ways was more integrated. Greater breadth resulted from survey work being conducted by VAs, corps and divisional topographical sections, flash-spotting and sound-ranging sections, the Air Force (obliques and verticals, stereo-pairs and serial photos), trench mortar survey sections and artillery trigonometers. Depth was provided by much of this work being done at divisional level whereas in the British service it was not done below corps. The British did little mortar survey.

Both French and Germans were ahead of the British in the use of artillery boards and large-scale gridded maps, and also in flash-spotting for artillery location. The French were first in the field with automatic sound-ranging apparatus, but the British developed sound-ranging to its most efficient level;

by 1918 British organisation and methods of flash-spotting and sound-ranging were ahead of French and Germans, and obtained a greater number of accurate locations. French and German heavy gunners were ahead of the British at the beginning of the war in the application of survey methods, but this gap had almost closed by 1918. The French system of training in 'topographic organisation of fire,' and of training artillery *officiers orienteurs*, was in advance of the Germans and British. The British 'bearing picket' system of giving line to the guns was considered more accurate and efficient than the French or German systems. All developed astronomical methods of artillery survey, the French having an excellent prismatic astrolabe. All were moving in the same direction with predicted fire, but the British were first to open a battle with a fully predicted barrage (Cambrai). All developed the application of meteor and other corrections to determine the error of the moment, the Germans very systematically with the 'Pulkowsky method' in 1918.

The most important failures were those of Germany in suppressing the Gasser Double-Projector for air survey, in the retarded development of automatic sound-ranging apparatus, and in refusing to adopt an orthomorphic projection and theatre grid. They also suffered from the conflict between Air Force and Survey Organisation over photogrammetry. The British also failed as far as projection and grid were concerned, only agreeing to standardise on the French system in June 1918, but the situation was more serious for the Germans with their proliferation of trigonometrical and reference grids. Though German cartography was uneven, it displayed some excellent features, notably the clear plotting of various orders of fixed points which was of great value to the artillery.

In almost every aspect of war survey and mapping, the three national armies developed remarkably similar organisations and methods, suggesting that problems were clear and solutions obvious. As Winterbotham said on the compilation of large-scale maps, 'the procedure stood out as large as Gladstone's nose'. There were, however, distinct differences in the ways that these solutions were adopted, reflecting national cultures; years after the war, the British looked askance at new developments in scientific photogrammetry, saying that it was not in keeping with the national inclination to adopt anything so theoretical or mathematical (Air Survey Committee, 1923, p. 17). No nation was so far ahead in any respect that it made a strategic difference. The Allied advances of the final battles in 1918 were undoubtedly greater blows for the quality of the survey support that enabled devastating surprise fire to be laid down on successive German positions. Bruchmüller had taken advantage of similar good survey earlier in the year to hammer the Allies back with his stupendous barrages, but not to the point of collapse. The war was won by the Allies for quite other reasons.

Acknowledgement

Research for this paper was completed during the tenure of a University of Greenwich student bursary, held at the School of Earth & Environmental Sciences.

References

Air Survey Committee 1923. *Report of the Air Survey Committee, No.1.* War Office, London.
Albrecht, O. 1969. *Das Kriegsvermessungswesen wahrend des Weltkrieges 1914-18.* Deutsche Geodatische Kommission, Munich.
Anon. [EHH] 1913 [Review of] Die Geschichtliche Entwicklung der Photogrammetrie und die Begründung ihrer Verwendbarkeit für Mess- und Konstruktionswecke, by Max Weiss, Stuttgart, Strecker & Schröder, 1913. *Geographical Journal* 42, 189.
Anon. 1915. *Zu D.V.E. Nr. 197.K., Teil I, Anfertigung von Batterieplänen, Vom 11. März 1905.* Reichsdruckerei, Berlin.
Arthur-Lévy 1926. *Les Coulisses de la Guere-Le Service Géographique de l'Armée 1914-1918.* Berger-Levrault, Paris.
Atkinson, K.B. 1980. Vivian Thompson (1888-1917): not only an officer of the Royal Engineers. *Photogrammetric Record* 10, 5-38.
Baring, R. 1963. E. O. Messter siebzig Jahre. *Bulletin der Sonderheft* 1963, 124-138.
Bellot, A. (Ed.) 1936. Rapport sur les Travaux Exécutés du 1er Aout 1914 au 31 Décembre 1919. *Historique du Service Géographique de l'Armée Pendant la Guerre (Raport de Guerre).* Service Géographique de l'Armée, Paris.
Boelcke, S. 1916. *Die Tätigkeit der Vermessungs-Abteilungen - Zusammenfassung der Arbeiten: Verfügung vom 24-3-1916 (Druck),* GrHQ 1916.
Boelecke, S. 1920a. *Die Engländer und das deutsche Kriegsvermessungs-Wesen.* Petermanns, Berlin.
Boelcke, S. 1920b. *Kriegsvermessungen und ihre Lehren.* E. S. Mittler & Sohn, Berlin.
Boelcke, S. 1921. Das Kartenwesen. *In:* Schwarte (Ed.), *Der Grosse Krieg 1914-1918, Bd. 8: Die Organisation der Kriegsführung.* Leipzig.
Bourgeois, G. 1919. Le Service Géographique de l'Armée et la cartographie de guerre. *La Géographie* 32, 463-484.
Bruchmüller, G. 1922. *Die deutsche Artillerie in den Durchbruchschlachten des Weltkrieges.* Mittler, Berlin.
Bull, G.V. & Murphy, C.H. 1988. *Paris Kanonen-The Paris Guns (Wilhelmgeschütze) and Project Harp.* Mittler, Herford & Bonn.
Chasseud, P. 1986. *Trench Maps-A Collectors Guide.* Mapbooks, Lewes.
Chasseaud, P. 1991. *Topography of Armageddon-A British Trench Map Atlas of the Western Front 1914-18.* Mapbooks, Lewes.

Chasseaud, P. 1999. *Artillery's Astrologers-A History of British Survey and Mapping on the Western Front 1914-1918.* Mapbooks, Lewes.

Clough, A.B. 1952. *Maps and Survey, The Second World War, 1939-1945, Army.* War Office, London.

Conrad, P. & Lespeyres, A. 1989. *La Grande Guerre 1914-1918.* EPA Editions, Paris.

Eckert, M. 1921. Luftbildaufnahme und Kartenherstellung. *Geographische Zeitschrift* 27, 241-260.

Eckert, M.. 1925. *Kartenwissenschaft.* Berlin.

Fleck, J. & Jacob, O. 1916. *Kriegsvermessungswesen. Das Lichtbildverfahren und Seine Verwertung.* Hermann Brücker, Berlin-Friedenau.

Frith, G.R. 1906. *The 'Service Geographique de l'Armee' in The Topographical Section of the General Staff.* School of Military Engineering, Chatham.

Gasser, M. 1923. *Über die Unparteilichkeit von Gutachten und Kritiken.* Kalkberg.

Gasser, M 1953. *Die Eroberung des Luftraumes für die Kartographie durch die Photogrammetrie.* Verlag Mayer, Günzburg/Donau.

GHQ 1918. *German System of Squaring Maps, Ia/48518a 1-6-18* [British GHQ, Operations], SS 715, PRO WO 297 61.

Gruber, O. von 1932. *Photogrammetry-Collected Lectures and Essays.* London

General Staff 1918. *Handbook of the German Army In War. April, 1918. S.S. 356.* Issued by the General Staff, London

Hart, C.A. 1943. *Air Photography Applied to Surveying,* 2nd edition. Longmans, Green & Co., London.

Horne, A. 1962. *The Price of Glory, Verdun 1916.* Macmillan, London.

Hugershoff, R. & Cranz, H. 1919. *Grundlagen der Photogrammetrie aus Luftfahrzeugen.* Stuttgart.

Jack, E.M. (Ed.) 1920. *Report on Survey on the Western Front.* War Office, London.

Krebs, H. 1922. Die Hugershoff-Heydesche Auto-Cartograph. *Zeitschrift für Feinmechanik* 4, 9.

KVC 1918a. *Sonderbestimmungen fürs Kr. Verm. W., 15. Juli 1918.* Druckerei des General-Gouvernements im Brüssel, Brussel.

KVC 1918b. *Der Punktplan. Vom 1. August 1918. Chef des Generalstabes des Feldheeres.* Druckerei des General-Gouvernements im Brüssel, Brussel.

Landmann, J. 1996. *Das Militärische Karten- und Vermessungswesen in Südwestdeutschland-Ein Beitrag zur Geschichte im 20. Jahrhundert.* Militärgeographisher dienst Der Bundeswehr, Stuttgart.

Ludendorff, E. GEN. 1919 *My War Memories 1914-1918,* Volume II. Hutchinson & Co., London

MacLeod, M.N. 1919. *History of the 4th Field Survey Battalion.* Unpublished typescript copy in Defence Geographic Centre, Tolworth.

O'Donoghue, Y. 1980. The Ordnance Survey 1914-1918. *In:* Seymour, W.A. (Ed.), *A History of the Ordnance Survey.* Dawson, Folkestone.

Raleigh, W. 1922. *The War in the Air,* Volume.1. Oxford University Press, Oxford.
Schimrigk, W. 1920. *Grundlagen der Geländekunde. Handbuch für Offiziere und Offizieranwärter sowie zum Selbstunterricht.* Ernst Siegfried Mittler und Sohn, Berlin.
Schwidefsky, K. 1954. *Grundriss der Photogrammetrie,.* Stuttgart. English edition, Pittman, London, 1959.
Service Géographique 1938. *La Service Géographique de l'Armée, Son Histoire, Son Organisation, Ses Travaux.* Ministière de la Guerre, Paris.
Spilleux, LT.-COL. 1917. *Conférences sur le Tir de l'Artillerie Lourde.* 3rd Edition. Lafolye Freres, Vannes.
Thompson, E.H. 1974. The Vivian Thompson Stereo-planigraph. *Photogrammetric Record* 8, 81-6.
Winterbotham, H. 1918. *Survey on the Western Front.* Maps GHQ, London.
Winterbotham, H. 1919 a British Survey on the Western Front. *Geographical Journal* 53, 253-276.
Winterbotham, H. 1919b Geographical Work with the Army in France. *Geographical Journal* 54, 12-28.
Zabecki, D. T. 1994. *Steel Wind; Colonel Georg Bruchmüller and the Birth of Modern Artillery.* Praeger, London.
Zeller, M. 1952. *Text Book of Photogrammetry.* H. K. Lewis & Co., London,

Notes
1. This work has made extensive use of maps and archived literature in the following repositories: Bundesarchiv-Militärarchiv, Freiburg-im-Breisgau, Germany, (BA-MA); Public Record Office, Kew, London (PRO); Service Historique de l'Armée de Terre, Vincennes, Paris. (SHAT). These abbreviations are used throughout the text to denote original documentary sources.

Peter Chasseaud
School of Earth & Environmental Sciences
University of Greenwich
Chatham Maritime
Kent, ME4 4TB

Terrain and the Messines Ridge, Belgium, 1914-1918

Peter Doyle, Matthew R. Bennett, Roy Macleod & Louise Mackay

> **ABSTRACT:** An analysis of consecutive trench maps for the Messines Ridge in Belgium, from 1915-1918, was carried out as a pilot study to examine the role of GIS as a historical tool for considering military operations. The project involved digitising original trench maps held at the Public Record Office, London, and comprised three components: a time-series analysis of consecutive trench positions; an analysis of geological suitability of trench position; and the development of a three-dimensional digital elevation model (DEM) of the battlefield. Analysis of position supported the concept of three intervals of trench development: a static period up to June 1917; the battle of Messines of 7th June 1917; and the war of movement in 1918. Geologically, the German line was disadvantaged by wet ground, and this led to the development of alternative strategies for protection from heavy artillery fire, with extensive use of surface concrete shelters (MEBU or 'pillboxes'), and ultimately a defence-in-depth strategy. The DEM represents a significant tool in visualising the battlefront.

1. Introduction

The Great War of 1914-1918 was a largely static war. Although initially a war of movement — with German advances developed on a timetable of progressive movement to surround Paris and cut off troops from the coast — the war in the west soon became restricted to a narrow strip from the North Sea to the Swiss Frontier (Figure 1). From late 1914 onwards the armies in the west occupied this narrow strip, and although minor adjustments followed battles, the dominance of this strip remained until early 1918 (Figure 1). This framework of a static war provides a unique opportunity to study the relationship between the battle lines drawn by both sides and the intimate nature of the terrain, including details of topography and underlying geology.

The historiography surrounding the development of the Western Front is vast, and encompasses all aspects from the personal life of the soldier through to detail of the major battles. In this literature, however, there has been a distinct absence of an objective assessment of the role that terrain has played in battle outcome. In most cases terrain is described with reference to simple topography, a concept that is particularly difficult to access in simple terms on the British sector of the Western Front, where the topography is that of a simple mature fluvial landscape. Geology in particular has a distinct role to play in the development of battles, and up until recently the relationship of battle to geology had not been examined in any detail. The positioning of trenches relative to geology and topographical features has been recently discussed in overview (e.g. Doyle & Bennett, 1997; Doyle, 1998; Doyle et al., 2000, 2001), but this work has been hampered by the lack of empirical data. The present project was proposed in order to determine the nature of the relationship between battle and terrain (topography and geology) using the vast database of maps and other data in the Public Record Office and other archives.

The project was conceived therefore to provide an objective assessment of the role of terrain in battle outcomes of the Great War, using the vast storehouse of military maps and employing the enabling technology of a commercially available Geographic Information System (GIS). One relatively constrained area of the Western Front, that of the Messines (Mesen) Ridge near Ypres (Ieper) in Belgium was selected for detailed study. This paper sets out the methods, approaches and results of the research.

2. The Messines Ridge

The Messines-Wytschaete Ridge is a topographical component of the Passchendaele Ridge, a low (<50 m high) ridge that bounds the eastern margin of the Belgian town of Ypres (Figure 2). In this report the convention of maintaining the French spelling of Flemish towns and map features is retained, as this is consistent with wartime practice, and with features as mapped on the contemporary trench maps. This ridge formed for the major part of the Ypres Salient, a shallow arc that surrounded the eastern margins of the town for the majority of the war. The Salient allowed the encumbents of the ridge top to produce enfilading fire — literally firing upon the troops in the Salient on three sides — which meant that the British troops took many routine casualties on a day-to-day basis.

From the early stages of the war the tactical advantage of inhabiting the high ground was not lost on the opposing sides. From late 1914 the German armies occupied positions that provided them with the opportunity to observe the British armies and their activities which were broadly confined to the clay plain below. In late 1914 and mid 1915 the Germans launched offensives in the Salient in order to break through the line, and this was reciprocated in 1917 by the British in their offensives at Messines in June, and along the Passchendaele Ridge in late Summer and Autumn. The appropriate volumes of the British official history, *Military Operations France and Flanders*, published between 1922-48, provide further details of the conduct of these operations.

The Messines-Wytschaete Ridge extends from Messines to the village of Wytschaete on the ridge crest, running broadly N-S in orientation. It is characterised by mature vegetated and cultivated slopes, and is drained by

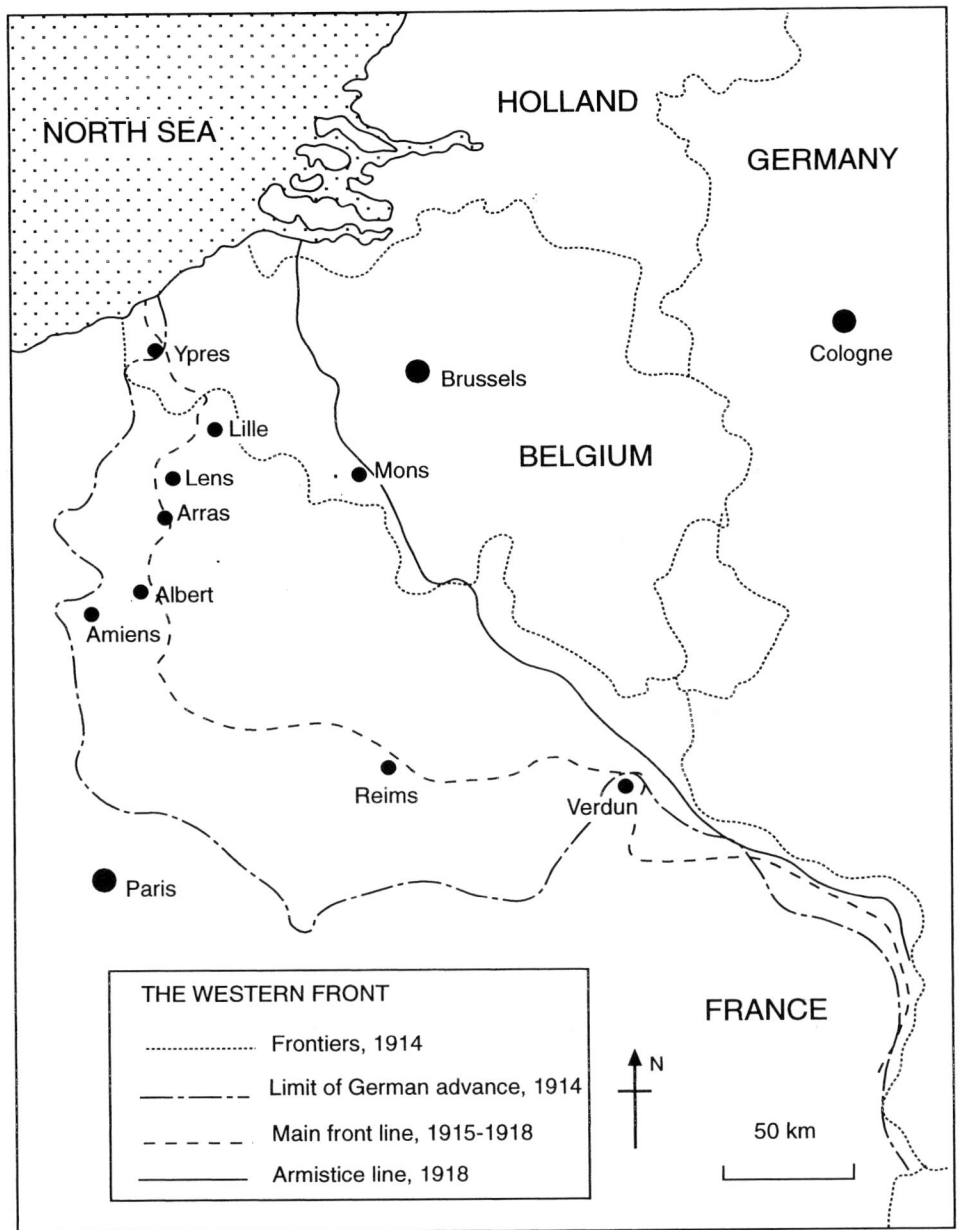

Figure 1. The Western Front, 1914-1918. Messines lies just to the south of Ypres (now Ieper)

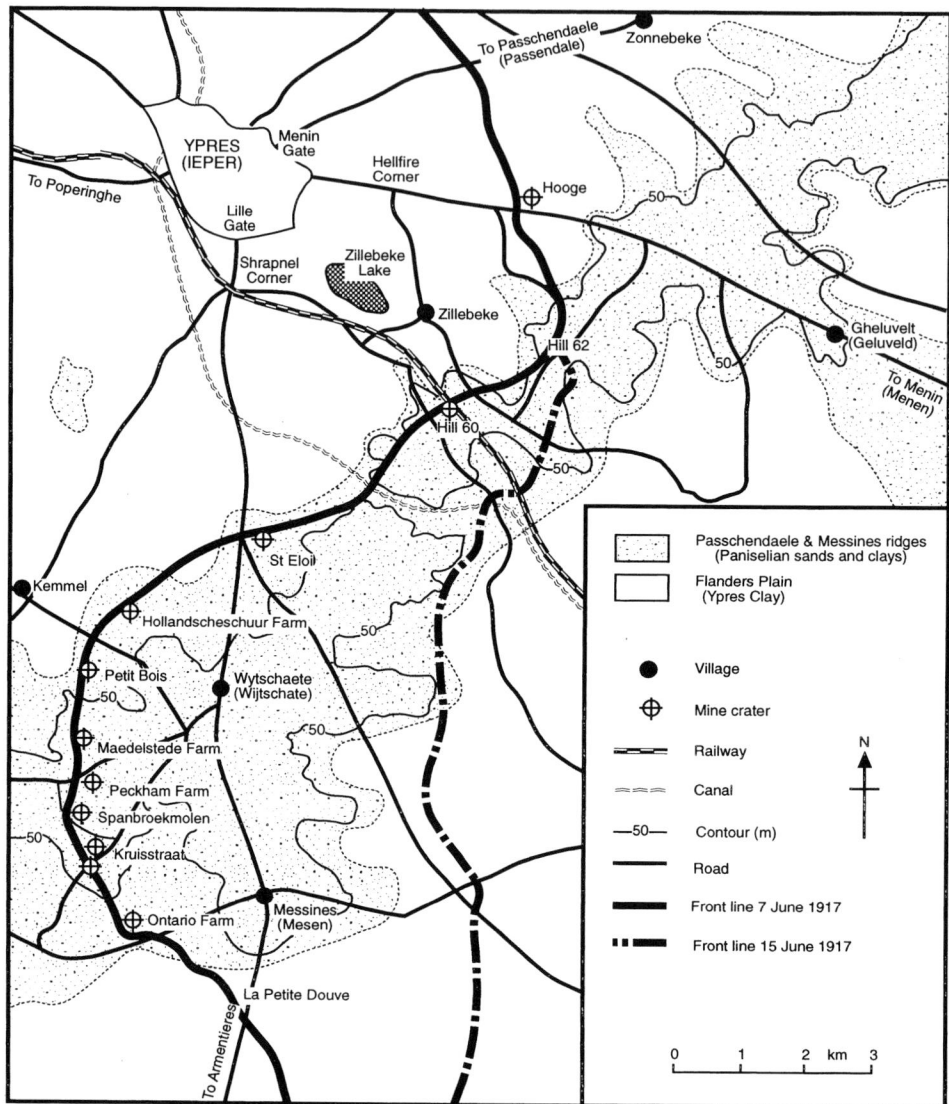

Figure 2. The Passchendaele Ridge in relation to Ypres. The Messines Ridge trends approximately N-S from Wytschaete to Messines itself

several minor streams (beeks). Geologically, in common with the rest of the Passchendaele Ridge system, it comprises interbedded Palaeogene fine sands and silty clays which have been previously grouped into the Paniselian Formation, overlying the Ypres (Ieper) Clay Formation (Figure 2). The details — both stratigraphical and geotechnical — of these units have been discussed elsewhere (Doyle, 1998; Doyle *et al.*, 2000, 2001).

The Messines Ridge was famously the scene of one of the most important British successes on the whole of their sector, when the ridge was carried by an assault opened with the blowing of nineteen mines beneath the ridge itself. The conduct of the battle has most recently been described by Passingham (1998). Prior to this battle, however, and in common with other parts of the Salient, the British had been disadvantaged by the German occupancy of the high ground. After the battle, the British were to be swept off the ridge again in the German offensives of 1918.

The Messines Ridge was selected for this study for the following reasons:

1. The ridge forms a compact and tangible feature on the Flanders landscape that can be constrained by maps, and that has sufficient topographical relief to provide a useful base with which to interpret geology, topography and trench movements;
2. Operations on and around the Ridge were known to have involved both static periods of trench warfare and more dynamic periods of offensive action; and
3. The trench maps existing for this sector of the Salient were sufficient to provide a comprehensive review of the role of terrain throughout the war.

3. Trench maps and archive resources

The Great War saw an unprecedented number of maps produced at all scales, and it was a war that saw the development of the first aerial photography in any useful way (Chasseaud 1999). The cartographic basis for this mapping exercise is currently under study, and detailed catalogues already exist (Chasseaud 1986). Large archives of trench maps exist in both public and private hands, and these provide a unique opportunity to examine the terrain as surveyed by the mapping establishment, based on the principles of the Ordnance Survey. Details of that establishment exist in both contemporary reports (e.g. Winterbotham 1919), and in more recent analytical studies (Chasseaud 1999), and is the basis of ongoing research (Chasseaud, pers comm.).

The British mapped their sector of the Western Front using a combination of existing map surveys and their own specially commissioned surveys. This included both traditional survey methods and also increasingly the use of aerial photographs. A full history of this approach is given by Chasseaud (1986, 1997, 1999). Trench maps were intended both for artillery use and infantry use, and were produced at a variety of scales. Arguably the most useful were the 1:10,000 scale trench maps, which plotted the positions, in successive editions, of the German trenches (printed in red initially, and by the end of the war, reversed to blue). British trench positions were at first not plotted, then plotted in outline (the front line only), and finally plotted in full, as arguments over the security of these maps receded as the war progressed. Secret maps were, however, produced which showed the British positions in detail. For continuity and ease of examination the study concentrated predominantly upon the German lines, which are plotted in some detail on all maps. It should be noted that the nature and extent of the trench systems changed with time. In 1915-16 for example, both sides were employing trench systems composed of mutually supporting parallel trench lines, while by 1917-18 this had been modified to include a lightly held front position only with many concrete fortifications (Doyle & Bennett, 1997).

The Ypres Salient was encompassed in various 1:10,000 scale sheets of the 1:40,000 scale topographical sheet 28. The complexities of the grid referencing system is fully discussed in Chasseaud (1986). The main 1:10,000 sheets covering the Messines Ridge were Ploegsteert (28SW4) and Wytschaete (28SW2), and after the battle of Messines in June 1917, four extra sheets: Bailieul (28SW3), Bas Warneton (28SW4 & SE3), Kemmel (28SW1) and Kortewilde (28SW2 & SE1), were required to encompass the positions of the trench lines. The Public Record Office was taken as the largest archive in selecting the maps, although other archives, such as the Imperial War Museum were also consulted. The final number of sheets selected was a function of the need to have temporal continuity between the trench lines on adjacent sheets, and as such only those sheets that were closely related in time were used. A full listing of the maps used in compiling the database is given in Tables 1 and 2. These maps are contained within WO153 files at the Public Record Office, London.

Table 1: Trench maps for Wytschaete 28SE2 1:10,000

Edition	Lines	Number
27 July 1915	German	GSGS 3062 762
26 October 1915	German	GSGS 3062 764
16 October 1916	German/British front	GSGS 3062 768
3 July 1917	German/British front	GSGS 3062 771
1 November 1917	German/British front	GSGS 3062 777
19 September 1918	German/British	GSGS 3062 784

Table 2: Trench maps for Ploegsteert 28SW4 1:10,000

Edition	Lines	Number
26 July 1915	German	GSGS 3062 797
19 October 1915	German	GSGS 3062 798
7 December 1916	German/British front	GSGS 3062 800
18 July 1917	German/British front	GSGS 3062 803
6 March 1918	German/British front	GSGS 3062 804
19 September 1918	German/British	GSGS 3062 809

4. Methodology and approach

Methodologically, two stages were identified in the production of the analysis, these were:

1. Selection of the maps and digitising them for manipulation in Arc/info; and
2. Analysis and manipulation of the data layers in order to quantify relative trench movement

These aspects are discussed in turn below.

Selection and digitising
Selection of the maps from the Public Record Office collection (listed in Tables 1 & 2) was subject to two constraints. In essence maps were selected that would:

1. Encompass the geographic range of the Messines-Wytschaete Ridge and provide for the construction of a digital elevation model; and
2. Provide a sufficient temporal interval to demonstrate potential range of variation of trench position relative to topography and geology.

The 1:10,000 scale trench maps of the Ploegsteert and Wytschaete areas listed in Tables 1 and 2 were digitised into the ArcInfo GIS for dates from July 1915 to September 1918. This was to allow a safe margin around the major battles of the summer of 1917, the German advances of Spring 1918 and the subsequent advance to victory of the Allied troops by the end of the year. The majority of the trench maps used were standard (i.e. not secret) editions, and as such, provided only consistent information about the German lines, with the British lines mostly being represented by simple dashed lines or similar skeletal information. German front lines were identified through recognition of primary and secondary fire trench lines, generally the first two parallel trench lines, although often it was denoted by the presence of saps or small trench systems running from out into No Man's Land.

The digitising produced a series of data layers, specifically topography, communications, soil and dugout suitability; and trench position These aspects are discussed in turn below.

Topography. The Messines study area is represented topographically here with the initial front of October 1915 (Figure 3). This provides the simplest understanding of the relationship between trenches and terrain, as the morphology of the front line so closely follows the contours. The use of 80-year-old maps with a unique projection system, questioned the quality of the data that could be used. A 1991 Belgian map, with a Lambert projection, was used to geo-correct the trench maps with visible ground features, and final transformation produced a low RMS error. Tiling of final map sheets was an easy process waylaying fears of map shrinkage

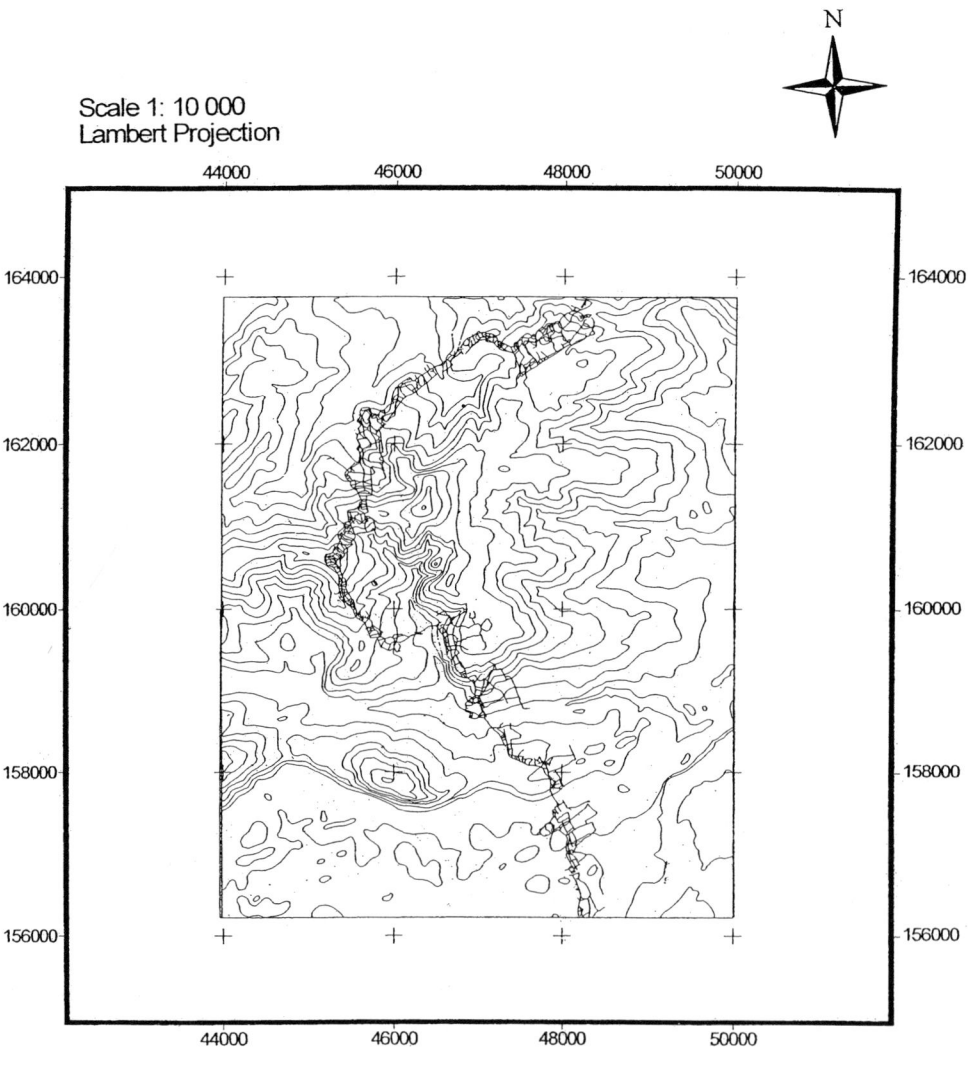

Figure 3. Terrain of the Messines Ridge with German front line of October 1915

Communications. A descriptive map of the study area, concentrating on the main towns, A-roads and waterways. A buffer around these features of 50m, 100m and 200m produced no significance with respect to trench position. Completion of the database was the addition of the viable communications to the front, with road systems and major towns and villages providing a backdrop to the tactical and strategic objectives of the opposing armies.

Soil and dugout suitability maps. A soil map of the locality and its inferred suitability for dugouts, provides an opportunity to model trench position with, the most important parameter gleamed from the pilot study as discussed below.

Trench position. Relative trench position was noted from the trench maps series held in the Public Record Office. Tied to the topography this allows for the direct comparison of trench systems relative to topography and soil maps.

Analysis
The data layers produced during the process of digitising were then overlaid in order to demonstrate percentage movement of the lines and their relationship with the underlying topography. This analysis provided the means for direct comparison of trench position with topography and geology. Three analytical layers were developed from the database: a trench position time series; the relationship of soil suitability to trench position; and a Digital Elevation Model (DEM) of the battle front. These are discussed below in turn.

Trench position time series. Representing net gain associated with trench movements for the period. This allowed for a direct comparison with topography and ground suitability. The static period (1915-17) was taken as the comparator, to which the movements associated with the final stages of the battle (July 1917) and the return to the war of movement (Spring 1918) could be compared.

Relationship of soil suitability to trench position. Representing the direct relationship of the front line position with the soil type and suitability index developed by the British Expeditionary Force (BEF) in 1917. The dugout suitability maps used by the BEF — essentially geological maps produced in 1917 in order to exploit the newly captured terrain for permanent dugout excavations — provided an extra data set. These maps were the result of detailed borehole investigations (see Doyle *et al.*, 2001), and provided an accurate assessment of the nature of ground conditions, colour coded red for dry and successive shades to blue for wet.

Three-dimensional visualisation (Digital Elevation Model). Representing the three-dimensional representation of the battlefield in order to visualise the development of strongpoints and the development of the battle. The Digital Elevation Model (DEM) was constructed from the 5 m contour lines and this provided a suitable backdrop of Messines Ridge for the analysis of the data.

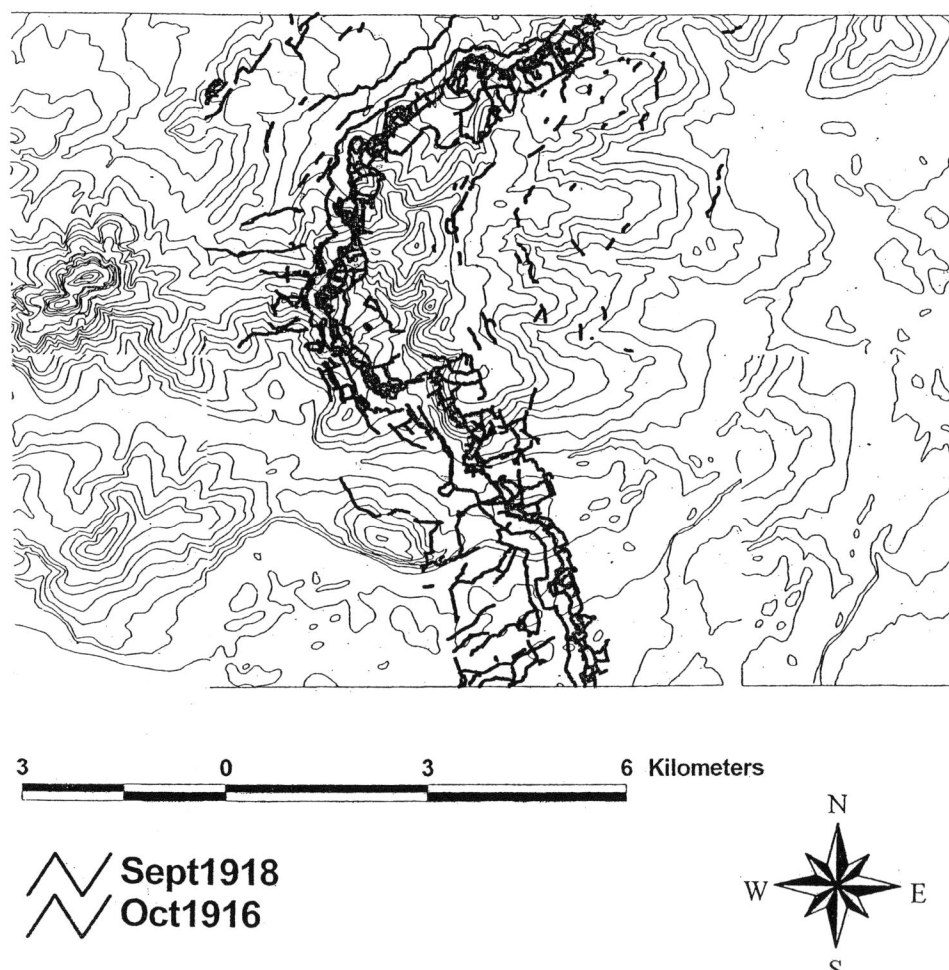

Figure 4. October 1916 and September 1918 trench lines

5. Results

Trench position time series

A time series of trench movement, in this example only two dates were used to ease map clarity, in this case the fragmented nature of the later date shows that the terrain element was no longer of importance, or not capable of modelling at this extreme stage (Figure 4). The trench movement series were plotted using a comparison of several maps. The longest occupancy, from 1915 up to the British offensive of 1917 was taken as the base, and this was compared with the maps before and after the battle, and during the German advances of 1918.

A comparison of all the trench lines is given in Figure 5. This demonstrates the central position of the static lines occupying the high ground of the Messines Ridge, and encompassing both spurs of the Steenbeek Valley. This central zone demonstrates the static nature of the trenches between 1915 and early 1917 (shown by positions for October 1915, June 1916 and October 1916). As expected, significant movement was experienced only during the June 1917 offensive (trench lines for July 1917 on Figure 5) and the war of movement in mid to late 1918 (trench lines for July 1918 and September 1918 on Figure 5). These aspects are further discussed below.

Static period, 1915-May 1917. The Messines Ridge is a subset of the main Passchendaele Ridge system, composed of sands, silts and clays overlying the clay plain of Ypres. The Passchendaele ridge comprises an arcuate low ridge, and acts as a watershed, with rivers draining from its apex onto the clay plain below. The Steenbeek is one of the most significant of these minor rivers and it flows to the south southeast from the ridge crest below the village of Wytschaete, while the Wytschaetebeek flows in an opposite direction SW-NE off the Passchendaele Ridge. To the south, the Steenbeek valley is relatively steeply incised, and its position marks the western slope of the Messines Ridge itself, with the main transport and communication routes following its line. West of the Steenebeek is another spur, bounded to the west by the route of the Stuiverbeek, which also flows to the south from the main ridge crest. South of these spurs, and paralleling the main Passchendaele Ridge, is the spur referred to by the British as Hill 63, which rises to 63m. This spur and those forming the Messines Ridge are either side of the broadly W-E flowing La Douve River, of which the Steenebeek and Stuiverbeek are tributaries.

The main German frontline positions, established early in the war, parallel the crests of these two NNW-SSE trending spurs of the Passchendaele Ridge (Figure 5). Central to the defensive system is the strongpoint of Spanbroekmoelen, situated on the W-N trending ridge top at just over 70 m, and forming the axis of the salient arc of the German defensive position. From this point the trenches curve in a broad arc, following the line of the Steenebeek to the SSE, and to the north, following the line of the Wytschaetebeek that flows to the northeast (Figure 5). To the north of

Spanbroekmoelen, the line follows the range of spurs associated with the right bank of the Wyschaetebeek valley. Strongpoints, created as minor salients, follow wooded areas and fortified farms, consistent with the pattern to the south, discussed below. Petit Bois, Ontario Farm and Bois Quarante are representative of these, with the line following the contour of these spurs. To the south of Spanbroekmoelen, two other strong points are at Kruisstraat, following the line of a small spur, and Ontario Farm, significantly representing the southern-most extent of the German line on the spur forward of the main Messines Ridge, before the German system switches back to the main Messines Ridge in front of the village of Messines itself. Finally, on the opposite side of the valley from Hill 63, the German line again hinges on the La Petit Douve Farm, adjacent to the old Messines to Ploegsteert road.

During this static period, the German line occupied ground that was in essence around 10 m higher, significant in a mature fluvial landscape with greatest relief of around 70-75 m. The lines in Figure 5 show no significant change during this period, and this is therefore taken as the optimum position of strength for the German line, with progressive strengthening of fortifications and with a series of strong points approximately 100 m apart in order to provide arcs of covering fire. However, as discussed by Passingham (1998), the defensive strategy established by the Germans from 1917 onwards was to create defence-in-depth, with a flexible or plastic approach that meant that the attacker had to take each line in turn, with the troops being flexible enough to shift position in order to avoid the inevitable artillery bombardment prior to offensive action. Clearly the front line, with its strong redoubts would break up any attack and allow the main defensive troops to counterattack the divided attackers. Importantly, fundamental to this strategy was the establishment of a strong defensive line, on a reverse slope, with observers and machine-gunners on the forward slope. The rationale was to enable the German gunners to engage the enemy as they advanced over the open ground (Passingham, 1998, p. 74). This defensive line was established as the Oostaverne line, which linked the villages of Messines and Oostaverne with the Menin Road. It was placed on the reverse slope of the Messines ridge, effectively the dip slope of an asymmetrical plateau and as such the defensive strongpoints described were in the forward position, providing the machine gun and artillery observation cover already described. This system developed throughout the static interval was clearly appropriately sited with respect to the topography of the battlefield, apparent from the trench maps.

The Battle of Messines, June 1917. The Battle of Messines opened on 7th June 1917 with the explosion of 19 mines excavated under the German front, and specifically under the strong-point positions that formed the basis of the front line of defence (e.g. Anon., 1917; Ball, 1919; Institution of Royal Engineers, 1922b; Edmonds, 1932; Mullins, 1965; Pennycuick, 1965; Passingham, 1998; Figure 6). As is well known, preparation for this offensive had been intensive, and as such it is considered to be one of the best prepared offensives of the war.

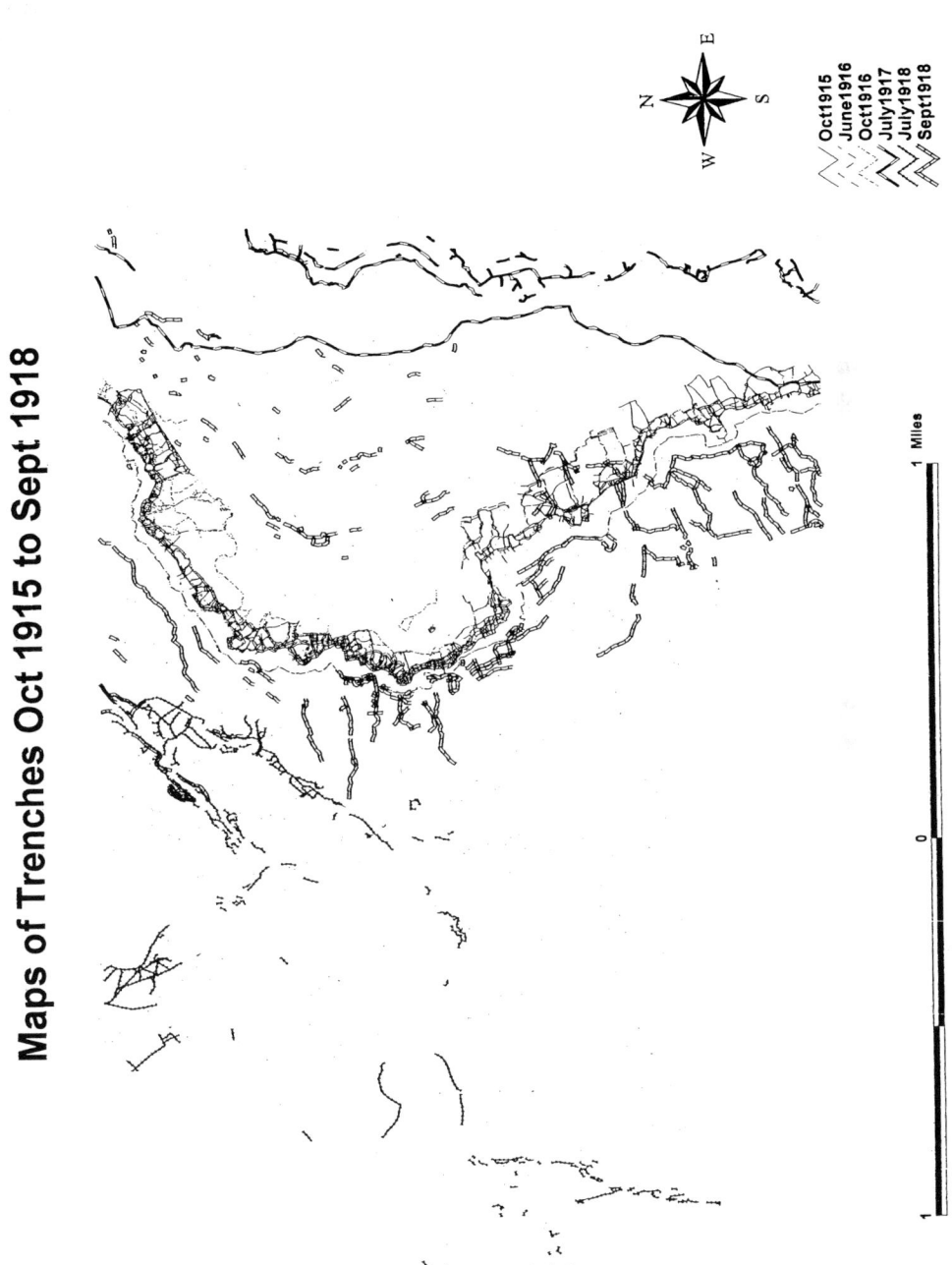

Figure 5. All trench positions for the Messines Ridge, 1915-1918

The combined fire power of the mines and a well-conceived artillery barrage launched on a narrow front carried the line. Examination of the trench lines relative to those of other intervals demonstrates that the static line could be carried with a concerted effort, and the British troops were able to overrun the German positions and carry the ridge top. It is clear from this that the topographical elements comprising the front were overcome, and the trenches were carried to a maximum depth of 100 yards (just under 100 m). This positioned the British troops on the original reverse slope, with the British front established at the German final line of defence, the Oostaverne line.

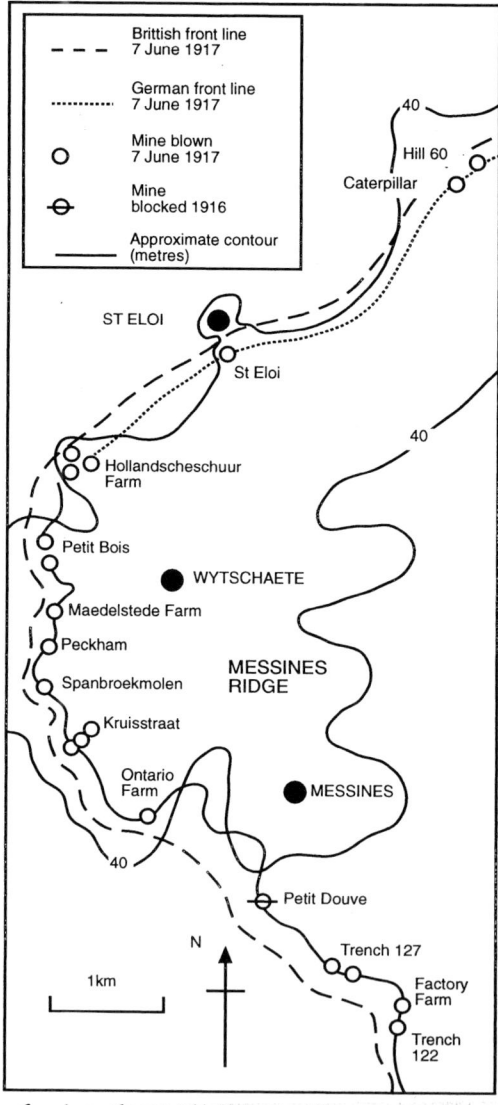

Figure 6. *Spacing of mines beneath the significant strongpoints in the German frontline. These mines were blown on 7th June 1917*

Mobile period, Spring-Autumn 1918. Following the British success at Messines in June 1917, the German offensives of Spring 1918 were able to regain the ground lost and take the ridge in its entirety. The trench lines for July 1918 show that the Germans were able once again to sweep up the dip slope of the Messines Ridge, and carry the British lines as far west as Mont Kemmel, the most significant high ground in the region. Significant in this was the position of the La Douve river, and the new German line was established at the head of the river, maintaining the high ground forming the plateau from which the hill of Mont Kemmel rises. This line was to be untenable, however, as the Germans were dislodged and swept back in the Summer and Autumn of 1918. The time series map for this interval demonstrates the establishment of fragmentary trench lines for September 1918, with a return to the ridge top as the most significant feature, with a almost identical comparison between the trench lines of 1916 and those of late 1918 (Figure 4).

Relationship of suitability to trench position

The percentage of trench occurring within as opposed to without a suitable area provided a means to ascertain if there was a relationship between trench position and soil type, and these results are presented in Table 3. These results represent the relative frontage of the German front line, the standard as represented on British trench maps for the interval studied, in relation to the soil type and basic geology

Table 3 *Statistical summary of trench dugout suitability cross-tabulation for the three acceptable trench dates*

Stratum	Suitability	October 1915 Length (m)	%	June 1916 Length (m)	%	July 1917 Length (m)	%
1. Alluvials	Bad	916.6	0.9	155.5	1.7	542.6	3.4
2. 'Wytschaete Sands'	Good	7424.4	7.6	7493.1	8.2	-	-
3. Sands/clays	Fair	23375	23.8	21113.3	23.2	-	-
4. 'Kemmel Sands'	Bad	29386.8	30	26560.1	29.1	9955.0	62.4
5. Green sands	Fair	200.7	0.2	14981.3	0.2	328.4	2.1
6. sandy clay	Fair	18288.2	18.7	4048.8	16.4	-	-
7. 'Ypres Sands'	Bad	2225.3	2.3	4048.8	4.4	2119.4	-
Total (m)		98006.6		91151.5		15965.1	

The results in Table 3 show the percentage of trench line positioned within each zone, although only the first two dates were feasible and this affected statistical quality. The cross tabulation analysis between suitability zone and trench occurrence was only applicable to earlier dates where the front was compactly structured. At later dates the lines were highly fragmented and positioned in older trench lines, historically the terrain at this stage had been greatly affected by the ravages of the preceding war years.

This analysis reinforces the conception that the German front line trench system was actually disadvantaged with respect to geological conditions. It is well documented that the 'Kemmel Sands' represent the poorest conditions due to high porewater pressures (e.g. Doyle & Bennett, 1997; Doyle et al., 2001), making this unsuitable for deep dug-out construction. As seen from the analysis, the greatest single percentage of front relative to a particular geological conditions (30 %) is found within the worst ground conditions, and that this percentage did not materially change through time. As could be expected from the relatively small surface area of exposed and well-drained Wytschaete Sands capping the ridge top, the percentage of frontline trench in such advantaged conditions is low, around 8 %. This is not the case, however, with respect to the main defensive line (the Oostaverne line) that was dug into rather more suitable and more readily drained sands lying stratigraphically beneath the Kemmel Sands.

These conclusions are, in themselves unremarkable, but reinforce the direct influence of terrain, and more specifically geological conditions in determining the defensive policy adopted by the Germans. Ultimately, the capture of high ground was of the greatest significance to the Germans in a war that depended so much on artillery and therefore on artillery observation (see Prior & Wilson, 1996 for a discussion of artillery in the Third Battle of Ypres). As was illustrated by the extensive programme of borehole investigations in developing suitability maps for dug outs (Institution of Royal Engineers, 1922a; Doyle et al., 2001), relatively few if any of the geological units of the ridge were suitable for really deep dugouts, and therefore, as the need for greater head cover to combat heavy howitzer and mortar shells increased as the war progressed, so the German strategy had to turn to the development of concrete shelters (MEBU or 'pillboxes'; Oldham, 1995) at the surface. Ultimately this led to the development of a defensive strategy based around the spacing of such pillboxes for co-ordinated firepower, and in turn, contributed to the development of the defence-in-depth system. Simply put, the large percentage of front line associated with geological conditions unsuitable for deep underground construction meant that a new defensive system of concrete shelters had to be developed; a strong indication of the importance of geological conditions in determining both tactical and strategic policy.

Once the static warfare had been broken in 1918, then the long-term influence of the geological and ground conditions was similarly reduced, although it is clear that dugout construction at mid altitudes on the Ridge would always be compromised by the presence of the wet Kemmel Sand unit.

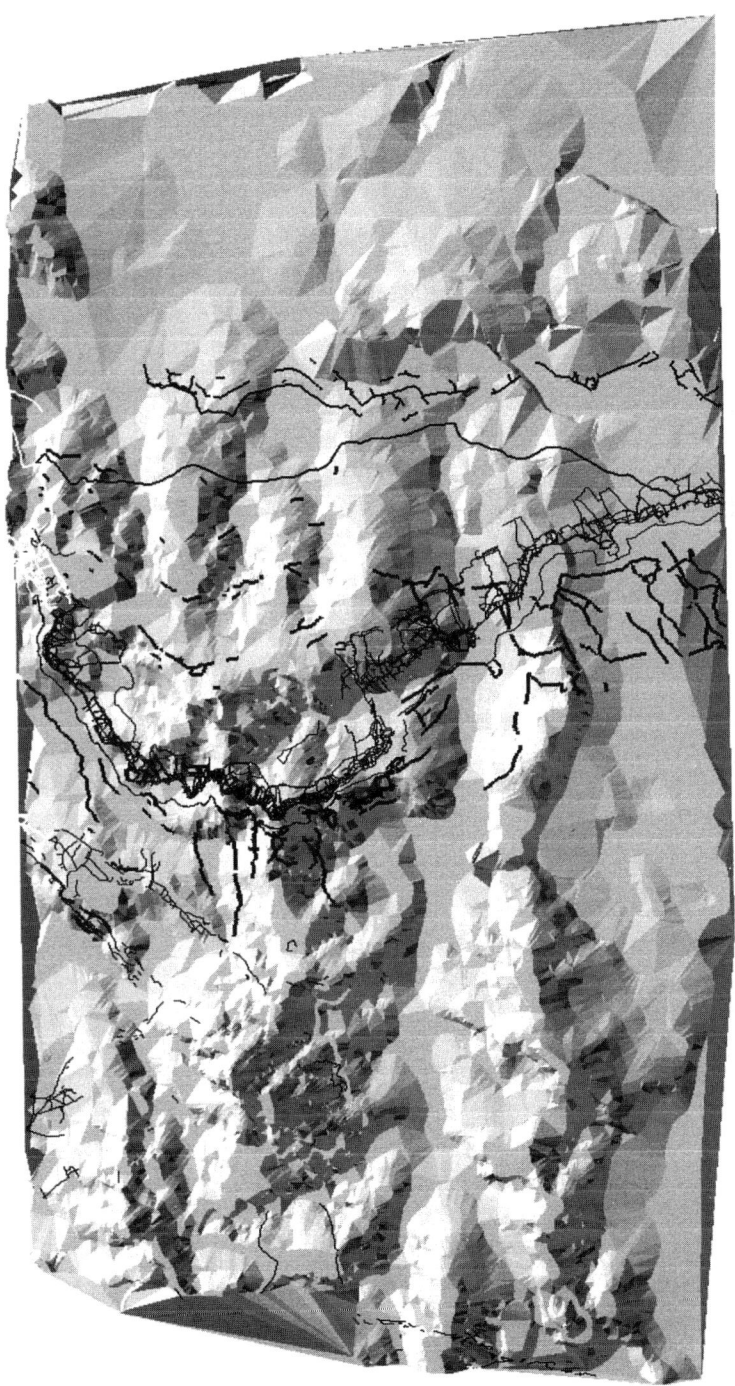

Figure 7. Digital Elevation Model of the Messines Ridge, with overlaid trench positions

The greater mobility and volatility of the front was also a factor that undermined the importance of the system of pillboxes, reducing overall the validity of the geological determinant of strategy. Increasingly, however, the deployment of tanks by the British towards the end of the war required a full understanding of the 'going' characteristics of the surface.

Three-dimensional visualisation (Digital elevation model)
A 5m resolution DEM representation of the Messines study area was constructed, with trench lines overlaid (Figure 7). The purpose of the DEM was to provide a visual representation of the trenches through a three-dimensional model of the topography.

The Messines Ridge (Figure 7) shows in more detail than can be expressed in a trench map the association of the German front line with the ridge crest, and the employment of parallel spurs of the Steenbeek valley. The strongpoints later undermined by the British are also clearly demonstrated by the DEM, with the strong E-W orientation of ridges and valleys created by the fluvial landscape being used to great effect, obviously intended to break up the attacking forces into separate packets, that would them be swept by artillery fire as they crested the ridge. The DEM also demonstrates the importance of Kemmel Hill — the most prominent feature in the region (Figure 7) — in providing an observation position over the German lines, as well as the obvious attack routes for the Germans in retaking the lines in Spring 1918.

6. Conclusions

This project was developed as an exercise in analysing the Western Front through a detailed examination of one small but identifiable area, that of the Messines Ridge in Belgium. The database of British trench maps — plotting understandably only the German defences in any kind of detail — is a valuable one, and one that can be manipulated using GIS in order to accurately plot trenches and defensive positions in relation to topography and geology. In fact, the accurate production of the database lends credence to the quality of map production and survey at the time, and the comparison with current Belgian survey maps is favourable. The use of a GIS in this instance has provided an opportunity to visualise the Western Front in a unique manner. Most appreciations of trench maps having been used in a passive role as objects, rather that in an active role in ascertaining the course of a battle or the development of the defensive positions in relation to topography or geology. It is in this aspect that the use of a DEM has proven a great success to aid visualisation of terrain effects for ground -based warfare.

This pilot project has identified that the methodology is sound, and that it has provided a valuable tool in the examination and analysis of military maps. For example, it has provided a mechanism whereby the development of trench systems through time can be accurately and quantifiably compared.

At Messines, what is clear is that the static trench systems of the early part of the war were effective, but that the advantages of altitude were countered by the disadvantages of ground, with 33.2% of the front in 1915 being in bad (i.e. wet) ground, and only 7.6% being in good (i.e. dry) ground, the remainder being only fair (Table 3). It is primarily for this reason that the German strategy of using surface concrete shelters with overlapping arcs of fire, and ultimately the development of the defence-in-depth concept; a key example of the role of terrain and particularly geology in influencing policy in warfare. This analysis has reinforced the already accepted view that the British operations in June 1917, using mine warfare and co-ordinated artillery, were both cleverly conceived and necessary in tackling the appropriately sited strongpoints constructed from concrete fortifications.

Finally, the development of the Digital Elevation Model of the battlefield has created an opportunity to examine the line of sight of both attacker and defender, and this has thrown up a clearer understanding of the deployment of troops in both the static interval leading up to June 1917, and the more mobile phase that followed. As a visualisation tool the DEM developed is an incredibly valuable resource, again underlining the strong relationship that the defenders of the ridge top had developed during their tenure.

The pilot project has demonstrated the potential as well as the pitfalls of the use of historical maps, albeit accurately surveyed ones. It represents the first step in the characterisation of the Western Front as a whole, and ultimately a comparison with other fronts engaged in trench warfare during the 1914-1918 world conflict. The demonstration of the role of geology and terrain in the establishment of German concrete positions, and ultimately the defence-in-depth system is illustrative of the power of this analytical tool.

Acknowledgement

This research was funded by Leverhulme Trust grant F/345/E to the authors at the University of Greenwich. Ben Holmes provided GIS support at the University of Greenwich, for which we are grateful.

References

Anon. 1917. *Artillery operations of the Ninth British Corps at Messines, June 1917.* USA War Office. War Department Document 647, Government Printing Office, Washington.

Ball, H.S. 1919. The work of the miner on the Western Front, 1915-1918. *Transactions of the Institute of Mining and Metallurgy* 28, 189-248.

Chasseaud, P. 1986. *Trench Maps. A Collector's Guide.* Mapbooks, Lewes.

Chasseaud, P. 1997. Field survey in the salient: cartography and artillery survey in the Flanders operations in 1917. *In:* Liddle, P. H. (Ed.)

Passchendaele in perspective: the Third Battle of Ypres. Leo Cooper, London, 117-139.

Chasseaud, P. 1999. *Artillery's Astrologers. A History of British Survey and Mapping on the Western Front 1914-18.* Mapbooks, Lewes.

Doyle, P. 1998. *Geology of the Western Front, 1914-18.* Geologists' Association Field Guide 61, Geologists' Association, London

Doyle, P. & Bennett, M.R. 1997. Military geography: terrain evaluation and the British western front, 1914-1918. *Geographical Journal* 163, 1-24.

Doyle, P., Bennett, M.R. & Cocks, F.M. 2000. The influence of geology on the British Sector of the Western Front, 1914-1918. *In:* Rose, E.P.F. & Nathanail, P. (Eds), *Military Geology in War & Peace,* Geological Society, London, 179-235.

Doyle, P., Bostyn, F., Barton, P. & Vandewalle, J. 2001. The underground war 1914-18: the geology of the Beecham dugout, Passchendaele, Belgium. *Proceedings of the Geologists' Association* 112, in press.

Edmonds, J.E. 1932. *Military operations in France and Flanders, 1917.* Volume 1. HMSO, London.

Institution of Royal Engineers 1922a. *The work of the Royal Engineers in the European War 1914-1919. Military Mining.* W.J. Mackay, Chatham.

Institution of Royal Engineers 1922b. *The work of the Royal Engineers in the European War 1914-1919. Geological Work.* W.J. Mackay, Chatham.

Mullins, L.E. 1965. The mines at Messines. *Royal Engineers Journal* 79, 286-292.

Oldham, P. 1995. *Pillboxes on the Western Front.* Leo Cooper, London.

Passingham, I. 1998. *Pillars of Fire. The Battle of the Messines Ridge, June 1917.* Alan Sutton Publishing, Stroud.

Pennycuick, 1965. Hill 60 and the mines at Messines. *Royal Engineers Journal* 79, 388-397.

Prior, R. & Wilson, T. 1996. *Passchendaele, the Untold Story.* Yale University Press, New Haven

Winterbotham, H.S.L. 1919. British survey on the Western Front. *Geographical Journal* 54, 253-271.

Peter Doyle, Matthew R. Bennett & Louise Mackay
School of Earth & Environmental Sciences
University of Greenwich
Chatham Maritime
Kent, ME4 4TB

Roy Macleod
Department of History
University of Sydney
Sydney
New South Wales, 2006

Zero Hour: Historical Note on the British Underground War in Flanders, 1915-1917

Franky Bostyn

> **ABSTRACT:** This note reviews the work of the Tunnelling Companies of the British Royal Engineers in developing mine warfare in Flanders from 1915-1917, culminating in the Battle of Messines in June 1917. Mine warfare had been carried out by both sides for some time, but the detonation of 19 mines in opening the Messines offensive is arguably the pre-eminent example of the use of mine warfare. The note examines in turn the developments leading to Messines, including: the explosives used — gun powder, gun cotton and ammonal; the construction and design of mine chambers; the importance of the closure (tamping) of these chambers; the methods of preparing detonators; the effect of the explosions themselves; and finally the form of the craters.

1. Introduction

The Battle of Messines in June 1917 is generally considered to be one of the most successful and best prepared offensives of the First World War. The opening of the battle by the detonation of 19 deep mines caused an artificial earthquake and total confusion of the German defenders, as a result of which the ridge, previously considered to be impregnable, fell quite easily into the hands of the British (Passingham, 1998). The digging of the deep galleries allowing these explosions to take place was the work of specialist Tunnelling Companies of the Royal Engineers, whose work has been described by Grieves & Newman (1936), Barrie (1962), Bostyn (1998), and Barton (1998). This paper reviews the work of the tunnellers in Flanders, and more specifically investigates mine warfare itself, examining the explosives, mine chambers, tamping, detonators, explosions, and finally the craters formed. The article is based mainly on published first-hand accounts of ex-tunnellers and contemporary technical analysis and instructions.

2. Ammonal

For their first mines in the Ypres Salient, the British used black gunpowder and gun-cotton (Institution of Royal Engineers, 1922). The latter was by far the best but the demand, especially from the Artillery and the Navy, was so immense that the tunnellers did not have enough at their disposal.

Therefore, for the undermining of Hooge on the Menin Road in 1915, the 175th Tunnelling Company used a new explosive charge, ammonal, which had been known about for 15 years but had never been used for military purposes. After a series of misunderstandings and difficulties, including confusion with ammonol, a drug used in suppressing sexual excitement (Grieves & Newman, 1936; Barrie, 1962), the ammonal finally reached its destination and on 19th July, 1915 the biggest charge so far was detonated (Barrie, 1962). Ammonal was found to be 3.6 times more powerful than black powder and combined both the lifting power of black powder and the shattering effect of gun-cotton. Its composition is ammonium nitrate 65%, coarse aluminium 16%, T.N.T. 15%, charcoal 3% and fine aluminium 1% (Davis, 1919).

Initially the explosives arrived in 100 lb. (45kg) bags, but, since these were difficult to manipulate, it was packed in 50 lb. (23kg) tins (Institution of Royal Engineers, 1922). Because the explosives sometimes had to be left for several days underground — and in the case of Messines in 1917, several months — before they were exploded, the question arose about how to protect them against the increasing water seepage. This was especially important since ammonal had a high absorbency level and exploded irregularly when it contained more than four percent water (Barrie, 1962). Initially, this problem was solved by packing the explosive in bags made out of gun-cotton, which also improved the ignition quality. However, due to a constant shortage of these bags, four gallon petrol tins (resistant oil tins with a volume of 18 litres) were also used. These tins could be completely waterproofed by using pitch. For 16 out of the 19 mines exploded at Messines in 1917, the explosives had been delivered in special explosive tins; in the case of the other three mines, petrol tins had been used[1].

Next to the placement and dry storage of the explosives, came the problem of transportation. The same road was used to transport both soldiers and materials: first by train or truck to control depots, then by horse or small Decauville-rails to the advanced material dumps and finally by the infantry, attached infantry or the tunnellers themselves, to the shafts (Pennycuick, 1965). For the mine war of 1917 this was certainly a tremendous amount of work. 'Getting this into position was a task of some magnitude as, for the whole show, nearly 20,000 boxes had to be carried up to the shafts, all by night. We had some exciting times doing it, but luckily ammonal requires high detonation and we frequently had tins burst by shell fragments or pierced by bullets. We all had our little troubles, but I do not think any Company had any serious mishap while loading and wiring' (Anon., 1930). In respect of this Lieutenant Murray (1st Canadian Tunnelling Company) noted the following anecdote: 'We went into the line in the famous Plugstreet sector for instruction with the 171st Tunnelling Company. While we were there, we received very excellent instructions on how to carry ammonal up to the frontline at night. Our boys went at it with a will, but before long one of the English miners told them that if one of the cans was hit with a bullet, it would explode, so I had a full fledged Cape Breton strike on my hands at a very short notice. Visions of court martial, firing squads, etc. were very vivid until I had one of the cans placed on a mound of sand bags a short distance from the trench and fired at it with my revolver. It did not explode, so the strike was off. As the men were moving off, a very angry officer hailed me, wanting to know the reason for the firing behind the line. I told him and if the can did not explode, he did! The mound of sandbags was his dugout and he was in at the time. He examined the can, said it was good shooting, asked me to have a drink and that finished it. We got the explosives up the line in good time' (Murray, 1929, p. 116).

3. Mine chambers

It is generally considered that almost 500,000 kg of explosives were laid under the Messines Ridge in preparation for the battle in 1917. Individual charges were large, such as at Sint-Elooi where 43,400 kg of ammonal was laid[2]. Such large charges were not placed at the end of tunnel galleries, but were placed in large specially constructed chambers, sometimes referred to as 'large living rooms' (Anon., 1930, p. 19). These 'rooms' were constructed for any charge over 1,000 lb. (Woodward, 1920). A second, but no less important, reason to build these mine chambers was to be found in the necessity that in order to obtain the best explosion, the explosives should be as concentrated as possible. Contemporary mining notes thus prescribed an extremely high cubic construction within a slight incline in order to keep the explosives as dry as possible[3]. In 1915 the two largest mines near Hill 60 were situated in cubic chambers of 1 m^2, while the third (500 kg) was placed at the end of the gallery (Institution of Royal Engineers, 1922). In 1917 the typical explosion chamber of the Kruisstraat No 3 was a mine and had a height of 1.9 m, 4.2 m in length and 1.7 m in width, intended for 30,000 lb. of ammonal (13,620 kg)[4]. With other deep mines, the different chambers came from a central tunnel, e.g. below Hill 60 and Caterpillar with 3 and 6 charging rooms over a total length of 9 and 14 m. At Sint-Elooi and Ontario, these mine chambers had remarkable shapes as well. The latter had to be deviated under a double square angle due to problems with sandy soil and measured a final length of 9 m. Difficult and unequal explosions were the consequence of this[5].

4. Detonation

Technically, the detonation of a mine is dependent upon the actual detonators attached to or in the explosives, the exploders to make the detonators explode and the wires connecting both to each another. As all charges were fired electrically, only one type of detonator was used; No. 13 Mark III, containing 43 grains of fulminate of mercury. The detonator was placed in a one ounce dry gun-cotton primer. A wooden cover was placed beneath and on top of the primer. These were recessed to take eight commercial detonators which then surrounded the gun-cotton primer. The whole was then bound together with fine wire. The detonator, thus reinforced with the gun-cotton primer and eight detonators, was placed in a 50 lb. box of gun-cotton, forming the main primer (Institution of Royal Engineers, 1922). The reason is that gun-cotton ignites twice as quickly as ammonal and is less sensitive to water (Frayling, 1988). In extremely wet conditions — which was often the case in the Salient around Ypres and Wijtschaete — the explosives were frequently put into waterproof rubber bags or, as with Peckham, in bottles hermetically sealed with tar and pitch (Grieves & Newman, 1936). 'A general rule may be laid down for large mines that when the charge is over 40,000 lb., it should have not less than one priming charge per 1,000 lb. of explosives per circuit. Smaller mines should have four priming charges per circuit. Priming charges must be evenly disturbed throughout the charge, both as regards each individual circuit and as a whole. This ensures good priming even if only one circuit can eventually be fired'[6]. Since the beginning of the mine war, each charge was equipped with two detonators containing mercury, each connected by two wires to two exploders, which consists of a small series wound dynamo, operated by a handle[7] (Figure 1). The first wire of each detonator was connected to the first exploder and the second wire to the second exploder. If these worked properly, each charge thus had four chances to explode. In case both detonators should fail to explode, the explosive itself also had two ordinary fuses (Institution of Royal Engineers, 1922).

'I also placed two shorter instantaneous fuses to be ignited in case of a double electrical failure and reflected ruefully at the time that if I was unlucky enough to have to light them, I would almost certainly be buried before getting clear', according to Lieutenant Cassels (175th Tunnelling Company) in placing the mine at Hooge (reported by Barrie, 1962, p. 75). In 1916 more attention was paid to the impact of simultaneous explosions and the trouble was actually taken to link each set of explosives. The detonators still had two wires, but whereas the first was connected to an exploder, the second was connected to a generator delivered by the Australian Electrical Mechanical Mining and Boring Company. Regardless of the number of charges, there were three exploders and one generator[8]. However, only half of the Messines mines in 1917 were exploded by a generator, whereas, for the other half, ordinary exploders were used according to the first principle[9]. The exploder of one of the Ploegsteert mines is to be found at the Royal Engineers Museum in Chatham, whereas the Imperial War Museum displays one of the three mentioned switches of Hill 60.

The wires were frequently marked with coloured wool, which were attached to the wooden galleries with hooks and were even pulled into a tube to protect them from extreme moisture (Coulthard, 1919; Institution of Royal Engineers, 1922). In 1916, when the mine battle of Messines was postponed for one year, the tunnellers were faced with two problems. Not only did they have to keep the explosives dry, but they also had to maintain the wires to ensure that they remained in good working order (Grieve & Newman, 1936; Barrie, 1962). Because they often suffered from the water which seeped after bombardments and counter-mines, small amounts of electricity were sent through several times each day. The amount of electricity was too small to cause an explosion, but was strong enough to check if all the measuring instruments still worked. The checks, though, were not limited to the wiring alone. The electrical testing was divided into: (1) testing the firing apparatus; (2) testing the detonators; (3) testing the conducting wires; and (4) testing the complete circuit after everything is ready for firing (Woodward, 1920). It is therefore no surprise that accidents happened. As reported by Oliver Woodward (1st Australian Tunnelling Company) on 25th April 1917: 'On arrival at the Hill [Hill 60], I found that the disaster arose from a premature explosion when a charge was being prepared for a mine in 'D' left gallery. The only one of our officers who escaped, had returned to our Headquarters dugout to get a measuring tape'. He noticed that a 50 lb. box of gun-cotton was being prepared as a primer for the charge and overheard the remark, 'We did not test the detonators for continuity'. Evidently it was decided to test the detonators while they were in the primer and by a thousand to one chance there must have been a supersensitive detonator which had exploded when the testing current was put through the circuit. The whole of the dugout was wrecked and it was only by the purest good fortune that the four officers survived[10].

Figure 1. British Exploder Mark III, 1914

5. Tamping

Once the explosives had been placed, the load was 'tamped' or blocked off. Explosions always follow the direction of least resistance which means that the explosives would usually explode sideways, causing little disturbance on the surface and much damage to their own galleries, much more damage than to the enemy. Therefore, the immediate surroundings had been made stronger than the soil on top. Sandbags were most commonly used and these would be filled with loose clay as well, as in some cases, wood (War Office, 1915). To make absolutely sure, it was the rule to make it two and in some cases three times the distance (Coulthard, 1919). The tunnellers also established that different sandbags placed at various intervals resisted the shock better than one great mass (Grieves & Newman, 1936). At Hill 60, during the spring of 1915, the 171st Tunnelling Company alternated 3 m of sandbags with 3 m of open space until they reached 15 m in length (Institution of Royal Engineers, 1922). In 1917, the deep mine at Sint-Elooi was blocked off with 1800 sandbags over a distance of 180 m alternating between 9 m of reinforcements and 3 m of open space (Murray, 1929). After the explosions of 7th June 1917, Captain Woodward (1st Australian Tunnelling Company) noted that the mining system below Hill 60 was left practically intact from about 100 ft (approximately 30 m) from the face of the tamping. There were alternate layers of 27 m of filling, 30 m open soil, 55 m sandbags, 60 m open and again 18 m reinforcement with sand bags (Woodward, 1920).

Figure 2. German photograph of a British mine explosion in Flanders, 1916

6. Zero hour

Once the chamber had been constructed, the mines placed and tamped and the exploders and detonators connected, everything was ready for 'zero hour'. The best description of an explosion was found with the Inspector of Mines, General Harvey: 'First, there is a powerful earthshake and then appears over the site of the mine as it were an enormous mushroom which slowly grows till it is about 50 feet in height. Then it breaks outwards with a dull roaring 'cruummp'. Immediately after, all the gases are liberated and rush as a pillar of flame and smoke heavenwards, reaching a height of about 200 feet. Forty-five seconds later all fragments of debris have fallen and all that remains in view is a cloud of smoke in the sky and a gaping crater over the surface of which flicker like 'will-of-the-Wisps' the pale blue flames of remaining combustible mine gases'[11] (Figure 2).

A remarkable note here is that, according to Captain Harry Trounce, a clear pattern arose at the time of most underground explosions. Quite a large number of mines were fired at 'Stand-to' at dawn or dusk, average time 4 AM or 7 PM, depending on the time of the year. The 'Stand-to' times were common because the infantry of both sides occupied the front line trenches at these times as a protection against surprise attack (Tounce, 1918). It is interesting to examine the pattern of timing during an interval of seven months, when 1,025 camouflets (small counter mines) and mines (499 British and 526 German) were fired (Institution of Royal Engineers, 1922). During this interval, the German

Figure 3. One of the big 1917 craters, captured and photographed by the Germans in April, 1918

line suffered most detonations between 5 p.m. and 9 p.m. and the large number also recorded around about dawn. Detonations beneath the British line showed a wider range of operations in the evening and early night, with the frequency of these explosions around or after sunset was very pronounced. In both instances, the interval from 7 a.m. to 3 p.m. was shown to be undisturbed (Institution of Royal Engineers, 1922).

Thirty months of mine warfare in Flanders was characterised by a growing number of explosions and a gradual increase in the quantity — and thus the force — of the explosives. During the first half of 1917, in the Ypres area, 59 German and 57 British mines and countermines were fired[12]. After Messines and a series of German explosions near Railway Wood and Bellewaerde, the underground war in Flanders came to an end[13]. Finally a few figures for comparison. In 1915, three Hill 60-mines contained 4,300 kg of gun powder and 450 kg gun-cotton (Institution of Royal Engineers, 1922). A year later, the six loads near Sint-Elooi contained nearly 33 tons of ammonal which represented an explosive value more than 20 times that of Hill 60 (Barrie, 1962). In 1917, the Messines Ridge was provided with nearly one million pounds of explosives, but the earthquake caused by this amount was still 40 times smaller in force than each of the atom bombs dropped on Hiroshima and Nagasaki in 1945 (Clifford, 1977).

7. Craters

The effect of mines depended on the quantity of explosives used, the depth of the charge below the surface and the nature of the soil, and pre-war formulae were used to calculate the surface effect of such underground exposions (War Office, 1915). Unfortunately, these formulae were based on sandy soil and the explosive value of gunpowder. Moreover, in the initial phase of the underground war, the tunnellers did not know the exact depth of their excavations due to a lack of appropriate measuring instruments.

In describing the craters produced, attention was given to its diameter and depth. For example, where the diameter equals depth, the crater was referred to as a 'one-line crater'; when the diameter was double the depth, it was a 'two-lined crater', and so on. Mines charged to produce two-lined craters were called 'common mines'; those more heavily charged 'overcharged mines'; those with smaller charges 'undercharged mines' and those with charges so small compared to their depth as to produce no craters were called 'camouflets'[14]. In developing this practice turned out to be the best teacher and, by means of 'Mining Notes', the tunnellers were kept informed of the latest findings. From 1916, the Mine Schools behind the front line and at military bases in Britain also carried out experiments with explosives so that, from time to time, new tables were published with more accurate formulae. For example, at the School of Military Engineering Fortification School at Darland, a replica of the trench system around Hill 60 was built, below which, on 15th February 1917, a deep test mine of 5,000 lb. ammonal was detonated. Such activities improved accuracy and prediction: after 7th June 1917, Captain Woodward noted the close agreement of the calculated and actual dimensions of the craters found: Hill 60, 70 yards (actually 68 yards); and Caterpillar 90 yards (actually 91 yards) (Pennycuick, 1965).

Of other explosions, the resulting craters may be described as remarkable at the very least, and some basic data for some of these is given in shown in Table 1, with illustrations in Figures 3 and 4.

Table 1. Basic geometry of some of the Messines mines

	Caterpillar	Ontario	Petit Bois no.2
Charge	31780 kg	27240 kg	13620 kg
Depth to charge	30 m	31 m	17 m
Depth of crater	16 m	nil	14 m
Diameter of crater	79 m	61 m	66 m
Total diameter	116 m	67 m	127 m

The most eye-catching crater is certainly the one near Ontario Farm. In the official comments on the explosions of 7th June, it was noted that the detonators, which were well-placed, were too few a number at this location. The heavy overburden of saturated sand also affected the result (Woodward, 1920). The British ammunition expert D.R, Whitaker associated the result of the explosion first and foremost with the shape of the mine chamber, which he envisaged as a tunnel running quite some distance and with a small cross section. Pennycuick (1965) has suggested that the peculiar shape of the Ontario mine was due to two completely different factors. Firstly, the Ontario mine was relatively isolated, whereas the other loads were always detonated in twos, and secondly, the charges were placed only two days before zero hour, whereas others closeby were waiting for eight days (Caterpillar) and ten days (Petit Bois No.2) to explode. Comparison may be made with the Sint-Elooi mine which caused a crater of only 5 m deep, despite the fact that the gallery was 38 m deep and 43,400 kg of ammonal was used. It was also an isolated, deep mine, which was only charged a number of days before the explosion. Officially, it was considered that the fact that the detonators were placed along the gallery at Sint Elooi and were thus not central, and that the mine gallery was in Ypresian clay, which was not compensated for in the formulae, was the reason for the shallow crater formed (Pennycuick, 1965, p. 397). Probably the most straightforward explanation for negligible craters was given by Mullins (1965), who stated that for mines exploded in soft, water-saturated areas, little or no crater was left as the sides flowed into the holes in the manner of quicksand.

In conclusion, eight factors may be summarised which were important for the shaping of craters in Flanders following underground explosions: the quantity and the nature of the explosives; the condition of the soil, (sand, clay, water-saturated); gallery depth; the way in which the load was blocked off (tamped); the construction and the position of the mine chambers; the location of the detonators in the charge; the presence of other mines nearby in case of simultaneous explosions; and the time interval between the placing and the detonation of the explosives.

Figure 4. Spanbroekmolen, one of the Messines craters with an original width of 75 m and depth of 12 m, resulting from a regular explosion

References

Anon. ['Tunneller'] 1930. Messines. *Tunnellers Old Comrades Association, Bulletin* 5, 16-25.
Barrie, A. 1962. *War Underground: the Tunnellers of the Great War*. Frederick Muller, London.
Barton, P. 1998. *The Underground War*. Parapet Productions, screened 11th November 1998, Channel 4.
Bostyn, F. 1998. *De vergeten oorlog onder de Salient: bijdrage tot de geschiedenis van de Tunnelling Companies in Vlaanderen, 1915-1918*. Unpublished Licentate thesis, University of Louvain.
Clifford, N.D. 1977. Early history of Sapper tunnelling. *Royal Engineers Journal* 91, 263-276.
Coulthard, R.W. 1919. Tunnelling at the Front. *Transactions of the Canadian Mining Institute* 22, 444-461.
Davis, A.W. 1919. Tunnelling reminiscences. *Transactions of the Canadian Mining Institute* 22, 475-481.

Frayling, B. 1988. Work in the 171 Tunnelling Company. *Royal Engineers Journal* 102, 170-176.

Grieve, G.W. & Newman, B. 1936. *Tunnellers*. Herbert Jenkins, London.

Institution of Royal Engineers 1922. *The work of the Royal Engineers in the European War 1914-1919: Military Mining*. Institution Of Royal Engineers, Chatham.

Mullins, E. 1965. The mines at Messines. *Royal Engineers Journal* 179, 286-292.

Murray, R.R. 1929. Tunnelling in the Ypres Salient. *Journal of the United Services Institute of Novia Scotia*.

Pennycuick, J.A.C. 1965. Hill 60 and the mines at Messines. *Royal Engineers Journal* 79, 388-397.

Trouce, H.D. 1918. *Fighting the Bosche Underground*. New York.

War Office, 1915. *Mining and Demolitions*. War Office, London.

Woodward, O.H. 1920. Notes on the work of an Australian Tunnelling Company in France. *Proceedings of the Australasian Institute of Mining and Metallurgy* 37, 1-54.

Notes

1 Public Record Office (PRO) WO153/909 Second Army Offensive Mines 7/6/17. Four-gallon petrol tins were used near Hill 60, Caterpillar and Kruisstraat No.3.

2 PRO WO153/909 Second Army Offensive Mines 7/6/17. Actual 95,300 lb. ammonal + 300 lb. gelignite.

3 19RECL Mining Note 128 19/1/1918. Mining Note 128 is entitled 'Firing Arrangements for Large Mines', contains four pages and is a kind of recap on 7th June, 1917.

4 19RECL Mining Note 128 19/1/1918.

5 PRO WO153/909 Second Army Offensive Mines 7/6/17.

6 19RECL Mining Note 128 19/1/1918.

7 Australian War Memorial (AWM) (940/48194) The War Story of Oliver Holmes Woodward, 1933, p. 95. The extremely interesting memoirs of Captain Oliver Woodward, 1st Australian Tunnelling Company, were typed in 1933 in Adelaide for private use, 171pp.

8 19RECL Mining Note 128 19/1/1918. Captain Woodward wrote about the switch scheme of the mines under Hill 60 and Caterpillar that the three double pole throw-over switches were closed to the dynamo circuit and the single pole firing switch was left open. Thus, to fire the mines, all that was necessary provided the dynamo was in operation, was to close the single pole firing switch. Had the dynamo failed, the double pole throw-over switches would have been thrown over the exploder circuit and the mines fired with the exploders. (AWM 940/48194 The War Story of Oliver Holmes Woodward, 1933, p. 95).

9 PRO WO153/909 Second Army Offensive Mines 7/6/17. Hill 60 & Caterpillar, Petit Bois (2), Maedelstede, Spanbroekmolen and

Kruisstraat (3) were exploded by use of a generator, the others by exploders on one or more circuits. In the report, one can read about the explosion of the left deep mine under Trench 127 with good result, good priming, but firing with one exploder on one circuit is risky.

10   AWM 940/48194 The War Story of Oliver Holmes Woodward, 1933, p. 88.
11   Harvey, R.N. British Military Mining 1915-1917 In: *The Military Engineer*, 1931, 513. The article of the Inspector of Mines is more of a superficial summary of the activities of the Tunnelling Companies and is based mainly on the official Military Mining, published in 1922.
12   These figures are calculated from the Weekly Summaries of Mining Situation (PRO WO 158/133): Railway Wood-Bellewaerde (31D 16E); Hooge (0D 0E); Sanctuary Wood-Tor Top (2D 0E); Mount Sorrel (1D 2E); Hill 60 (10D 9E); The Bluff (2D 8E); Sint-Elooi (5D 6E); Hollandse Schuur (2D 3E); Petit Bois (0D 2E); Maedelstede (1D 1E); Peckham (3D 1E); Spanbroekmolen (0D 1E); Kruisstraat (2D 3E); Ontario (0D 1E); Petite Douve (0D 0E); Trench 127 (0D 2E). It can be noted that in the 57 British charges, the 19 Messines-mines are included and that the Germans from February until the first week of March 1917 had exploded no less than 15 charges. It can be clearly seen how bad things were around Bellewaerde.
13   On 13-14th June, 1917 the Germans detonated two charges near Tor Top and six near Railway Wood. These were proclaimed as a big victory over the British. However, all of them were indicated in the Summaries of Mining Situation of 16th June as camouflets. Evidently the moral effect of the Messines mines was felt in Berlin and this last effort was probably intended to convince the home front that their own men could blow mines also. With the Third Battle of Ypres, the frontline began to move again, by which time the static mine systems were losing their function.
14   Mining and Demolitions 1915, 26-27. Here reference should be made to Section 5: Charges and Effects of Mines. The craters vary from the small ones, about 70 or 80 ft in diameter and 12 - 20 ft deep, to larger ones to such dimensions as 300 ft in diameter and up to 120 or 130 ft deep. The size, of course, depends on the charge of high explosives used, the depth of the mine galleries and the soil one springs the mine in (Trounce, 1918, p. 45).
15   Institution of Royal Engineers (1922, p. 22). An example is Mining Note 59 of September 25, 1916 (RECL), from which can be taken some figures from the 'Radius of Rupture' (R.R.) as the distance within any normal timbered gallery would be destroyed. Outside this distance, there may be all kinds of damage which are beyond determination by formulae.
16   Davis (1919, p. 477) also indicates the long time scale between the placement and the explosion of the charges. In a few cases, the craters were shallow, due either to slight undercharging or a possible deterioration of the charges, some of which had lain in place for a year.

Franky Bostyn
Astridlaan 6, Zonnebeke B-8980
Vlaanderen

Mud, Blood, and Wood: BEF Operational and Combat Logistico-Engineering during the Battle of Third Ypres, 1917

Rob Thompson

ABSTRACT: The First World War was the first mass industrialised war, creating the basic 'blueprint' for large-scale warfare this century, yet its historiography has been dominated by the 1960's perception of it as a 'futile' war of suffering from which nothing can be learned. The apogee of suffering was the Battle of Third Ypres, a 'Dantean inferno' where men drowned in liquid mud. Revisionist military historians have challenged this perception lately producing work that demonstrates the huge strides made by the British Expeditionary Force (BEF) in terms of combat effectiveness. Key to understanding the dynamics of the First World War is the significance of terrain, which in the static environment of the Western Front dictated the type and positioning of trenches and communications, whilst possession of high ground defined observation. Military historians have long studied the operational effect of terrain, but have done so from a narrow perspective, defining 'effect' in terms of battle. This is a mistake as modern mass warfare is defined not by notions of martial prowess, but by its mechanistic nature that utilises the systems based ethos of late 19th and early 20th century industrialism for the most efficient application of modern firepower. Viewed from this perspective therefore the key elements that defined operational success are engineering, logistics and administration. This paper seeks to redress that balance and explores the relationship dynamic between terrain and logistics, engineering and administration, and its impact on operations by examining the provision of road communications during the Battle of Third Ypres 1917.

1. Introduction

'Probably few commanders or staff officers realised...to how large an extent the success of every operation on the Western Front depended on engineering, how much the campaign was, in the popular phrase, "an engineers war".
Cyril Falls

On October 9th 1917 the British 66th and 49th Divisions, supported by Australian troops, assaulted the village of Passchendaele as part of General Haig's plan to force a strategic breakthrough of the German lines, ending three years of stalemate and possibly winning the war for the allies. This attack, known as the Battle of Poelcappelle, followed on the heels of three previously successful operations launched by General Plumer, the Second Army Commander. It was a complete and costly failure, which together with an equally disastrous assault launched three days later by the New Zealand and 3rd Australian Divisions, equated to the end of Haig's strategic designs and the failure of the Flanders campaign upon which so many hopes had been pinned.

The accepted reason for the failure of the campaign is quite simple — the battle bogged down in the mud occasioned by unseasonably bad weather and heavy British shelling. The British Official History cites the 'unanimous' opinion of corps and divisional engineers that '... up to 4th October there had been no serious difficulty in maintaining communications to the front, weather and ground conditions being tolerable and damage done by the enemy being readily repairable. Some even say this was the case until the 12th.' (Edmonds, 1948). After this date light railway locomotives were up to their boilers in mud, guns embedded themselves in the mud after firing a few rounds, lorries simply skidded off roads, and any movement of artillery and men became virtually impossible.

Consequently the fierce debate surrounding this controversial battle has been defined according to ground and weather conditions with the axiomatic assumption that 'mud stopped play'. The debate therefore has almost exclusively focused on the choice of Flanders as the most suitable area for operations; the paucity and shortcomings of strategic, operational, and tactical method and command in the face of such conditions; and whether the weather conditions obtaining in the autumn of 1917 were foreseeable.

It is the purpose of this paper to challenge the central assumption that mud and poor weather conditions were the primary reasons for failure in Flanders and propose an alternative thesis that identifies logistic and engineering practice and development in the BEF as the main reason for failure, and that the battles of the 9th and 12th October were in fact a logistic and engineering 'step too far'. This paper also suggests that this was indicative of a wider pre-existing logistic and engineering problem within the BEF savagely exposed by the Battles of Poelcappelle and Passchendaele.

2. Third Ypres

The ancient Flemish town of Ypres, situated on the Flanders Plain, lies approximately 35 km inland east of Calais and is surrounded by an arc of low ridges to the east (20-60 m high) that form a natural 'amphitheatre', bounded by the Messines Ridge to the south and Passchendaele Ridge to the north (Figure 1). Consisting of heavy Ypresian clay, subject to intense cultivation and lying at sea level, the entire area was dependent on a complex system of irrigation and drainage that was soon destroyed by shellfire, leaving the area subject to flooding where the main irrigation streams had broken down.

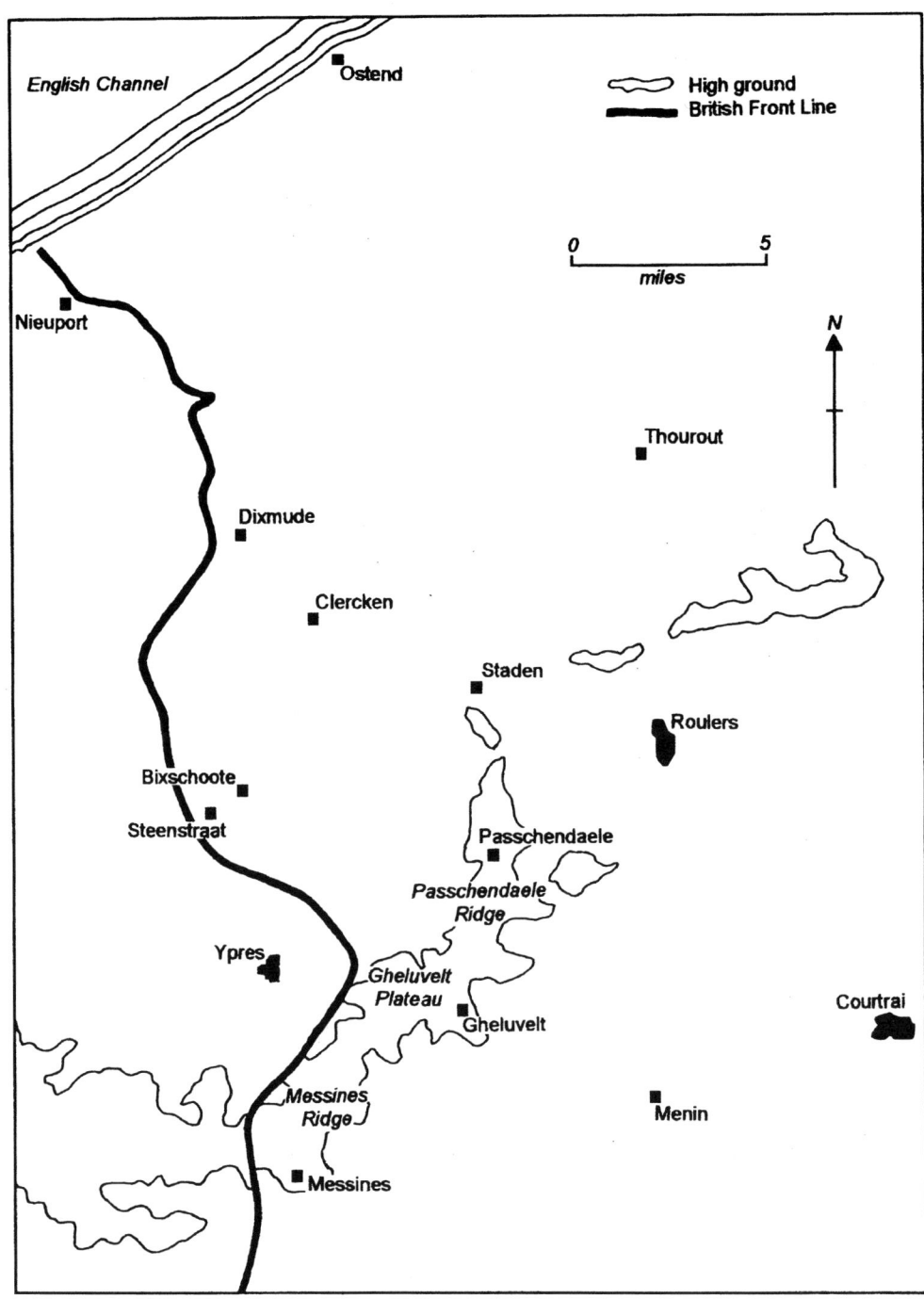

Figure 1. The Flanders Front 1917

Although the BEF initially occupied the ridges, by 1915 they had been forced back onto the plain around Ypres forming a salient 15 km wide at the base and up to 3 km in depth, giving the Germans unparalleled observation of Ypres from three sides. The combination of ridges, mud, the inability to dig deep defences (caused by the high water table) and the fact that the enemy could shoot from either the front, side, or rear meant that 'Wipers' as the area was known to the British 'Tommies' was universally loathed.

3. Why Ypres?

It is outside the scope of this paper to discuss the wisdom of aiming at such grand strategic designs in the late summer of 1917, despite all previous experience. However, there were a number of very sound logistic reasons to launch an attack from this sector, and a brief outline of the plan and the rationale underpinning the campaign is appropriate.

The first stage of the plan was to capture the tactical prize of the ridges, followed by a drive east and north-east to capture the strategic rail junctions of Roulers and Thourout. Simultaneously an amphibious operation to capture Ostend and drive eastwards was planned in conjunction with an assault from the coast at Nieuport. This would capture the German-occupied Channel Ports creating a strategic pincer movement to form a flank around which the allies could eventually attack the Germans in the rear.

Whilst half the British army's men and military supplies moved into the French ports of La Rochelle and Le Havre, these supplies had to be transported north along underdeveloped lateral road and rail communications. The increase in matériel demanded by large-scale industrial warfare in conjunction with the proximity of the Channel Ports to Great Britain meant that these ports were becoming of paramount importance.

In order to advance anywhere along the line in 1917 vast quantities of men and material were required and Flanders, with its highly developed road and rail system, was the only realistic place from which to supply an offensive. The Flanders sector was the only part of the British line that contained any worthwhile strategic targets and where a major assault had not yet been attempted: the Somme battlefields were devastated; Cambrai had no strategic targets of any significant value, and the La Bassee-Givenchy sector was described as 'unattackable and indefensible' (Guggisberg, 1917). Ypres had long been the 'pivot' of Flanders and represented the last realistic line of defence against German attack. Any retirement west of Ypres would immediately expose Calais and Boulogne to German artillery fire, threatening the British First and Belgian Armies with encirclement, as well as the nightmare scenario of creating a flank around which the Germans could manoeuvre and attack the allies in the rear[1]. Furthermore any German advance that imperilled the continental Channel Ports also threatened the British Channel Ports. As Sir John Cowans, Quartermaster General at the War Office put is, somewhat melodramatically: 'If we lose the Channel Ports

I shall make for the Welsh hills. I shan't wait to do the goose step down Whitehall behind Kaiser Bill' (Fay, 1937).

The loss of the Continental Channel Ports also equates to the loss of some of the main transit ports for supplies to Mesopotamia, Salonika, and Palestine as well as supplies of coal to Italy and France, thus imperilling subsidiary theatres as well as French and Italian industry.

4. The first two months

The Battle of Third Ypres began on 31st July 1917 under the control of Fifth Army Commander General Sir Hubert Gough, and was preceded by the Battle of Messines, conducted by Plumer's Second Army to the south. This ingenious and successful set-piece assault was designed to secure the southern end of the ridge and guard the right flank of Gough's planned assault.

However, a gap of nearly two months between operations allowed German engineering expert Col. Von Lossberg to create a new triple line of defence systems six to seven miles deep (Prior & Wilson, 1996). Consisting of successive lines of mutually supportive concrete pillboxes, trenches, and garrison posts, utilising every terrain advantage the ridges had to offer, these represented a formidable obstacle to the BEF.

Gough's initial attacks, which were strategic in concept and envisaged the capture of Passchendaele as a first day objective, were unsuccessful. A combination of rushed preparation, lack of aerial reconnaissance, unseasonable rain, and a failure to secure the heavily defended Gheluvelt area (which lay directly to the east or right of the main axis of attack) prior to shifting the assault north-east towards Passchendaele, caused horrendous losses for little apparent gain. Despite repeated assaults during August Gough made little headway and operational responsibility was transferred to Plumer's Second Army.

Plumer realised the importance of the right flank and the folly of aiming for distant strategic objectives, and proposed a series of 'step-by-step' limited assaults utilising overwhelming amounts of artillery. Under Gough assaulting troops aiming for distant targets tended to overextend their advance, thereby leaving them too weak and under-supplied to withstand the inevitable German counter-attacks. With Plumer the depth of the assault was defined by the range of the artillery, thus providing maximum protection for the assaulting troops and allowing them to consolidate ground taken. Once consolidated the artillery could be moved forward and the whole process begun again, creating a series of 'hammer' blows that would 'inexorably' push the Germans off the ridges.

Beginning on 20th September with the Battle of Menin Road and followed by Polygon Wood (26th September) and the spectacularly successful Battle of Broodseinde (4th October) it appeared that Plumer had found a 'system' to defeat the Germans. However, this entire system relied upon pushing the artillery forward as quickly as possible to minimise German recovery time between assaults. Thus the provision of roads for artillery and ammunition and tracks for mules and troops became of paramount importance.

Whilst this system appeared inexorably successful it held within it a number of inherent and fundamental problems. The shallow nature of the attack and the overwhelming use of artillery meant that troops always attacked over shattered ground. Consequently, pre-existing roads and communications were always destroyed and had to be rebuilt, cumulatively slowing the advance over time. The geological and terrain circumstances obtaining in the Ypres Salient combined with a shattered drainage system and unseasonable rain further compounded the problem.

The preparation time required between each operation was defined by the rapidity with which the guns could be moved forward, which was deemed to be between three and six days depending upon the calibre of the gun (Maxse, 1917). Indeed guns could be moved that quickly but only on good quality pre-existing roads. Whilst roads did still exist in the Ypres Salient they were for the most part beyond repair or reclamation and subjected to constant accurate shellfire afforded by the superior observation enjoyed by the Germans.

As the depth of the devastated zone increased so the difficulties in building and maintaining communications magnified. Advances rapidly outpaced the preferred method of tactical transportation, light railways, which required an immense amount of material in the form of grading, ballasting, and rail, and which was unsuitable for such boggy conditions. This left road-dependent motor transport, pack animals, and troops as the main means of transportation, supplemented by light 'push cart' tramways, which required no ballasting and were quick to lay and repair.

Consequently artillery movement did not define the pace of the advance and road construction was hostage to the movement and supply of material, and the amount of material required to build even a wooden plank road was prodigious. As the length of the road increased so more transport and men were required to deliver materials to the road head, increasing the logistic, engineering, and manpower load, and decreasing the pace of construction. Under reasonably good circumstances double-track road capable of handling Field Artillery could be pushed forward at a rate of 150-200 m a day. Given the conditions in the Salient: repairs necessitated by shell fire, maintenance, and the 'swamps' caused by destroyed drainage in the areas crossed by the irrigation streams, there was no chance that anything like this rate of construction could be maintained without massive amounts of additional motor transport, matériel, and manpower. In turn these required extra parks, petrol facilities, workshops, spares, and skilled personnel, all at a time when breakdown rates were increasing from 10% to, in some cases, 30% (Shelton, 1917).

Of course there is only a finite amount of transport and manpower available, and since the Royal Engineers lacked their own organic labour and transport, or a centralised organisation similar to the artillery to deal with the problem on an operation scale, Divisional CREs (Commander, Royal Engineers) were at the mercy of the preoccupations of Divisional and Corps commanders who had little grasp of the problem. Consequently infantry working parties provided the bulk of the labour. Since these were men who had often come out of (or were about to go into) a hard fight, they took little interest in the work. Dedicated Labour Companies consisting of medical category 'B' men and Chinese contract labourers were of poor quality and unable to cope with the rigours of the forward zone. In addition the Royal Engineers had to compete for labour, materials, and transport, as well as having to deal with the myriad other tasks allotted to them at a time when their needs were paramount.

The last successful operation, Broodseinde, undertaken mainly by Australian and New Zealand troops was a huge success. Ironically this success was due principally to a change in German defensive tactics that went disastrously wrong, and the fact that the Germans were preparing a major assault themselves. When the Australians and New Zealanders went forward they found trenches packed with Germans who were effectively unable to defend themselves. The success of the attack also caused further problems as the sheer volume of returning prisoners and wounded meant that further road and track construction for the next phase of the operation was seriously hampered due to congestion. With such an extensive victory the Plumer 'system' was seen as a success and the next phase of the attack, two rapid pushes to take place on 9th and 12th October respectively, was implemented. The Broodseinde success masked the reality of a forward communications system that was wholly inadequate to the task and dangerously close to collapse. In addition the relatively dry weather that had accompanied Plumer's assaults broke, torrential rain set in, and by 9th October key areas of the battlefield had become a quagmire requiring an even greater focus on communications. Whilst the decision to go ahead was reviewed in light of the weather, both Plumer and Haig were confident of success, as was II ANZAC Corps Commander A. J. Godley and his Chief Engineer, E. A. Panet, both of whom were responsible for planning and executing the assault. Consequently the attack was entrusted to 49th (West Riding) and 66th (2nd East Lancashire) Divisions, both poor quality 'junior' Divisions, with the 66th Division in particular lacking experience.

5. The battles of Poelcappelle and Passchendaele (9th and 12th October 1917): preparation prior to the assault

The north-eastern sector of the Ypres Salient contained the heavily fortified village of Passchendaele and was serviced by two roads (Figure 2). Panet (s'Gravenstafel) Road ran north-east along the s'Gravenstafel Ridge, bypassing Passchendaele to the north, whilst Zonnebeke Road ran from Ypres eastwards via Potijze and the Frezenberg Ridge to Zonnebeke and thence northeast into Passchendaele. Three other roads (Oxford and Cambridge Roads, Godley Road and Kansas Cross-Zonnebeke Road) laterally connected the two main roads, allowing materiel to be transported from the Ypres-

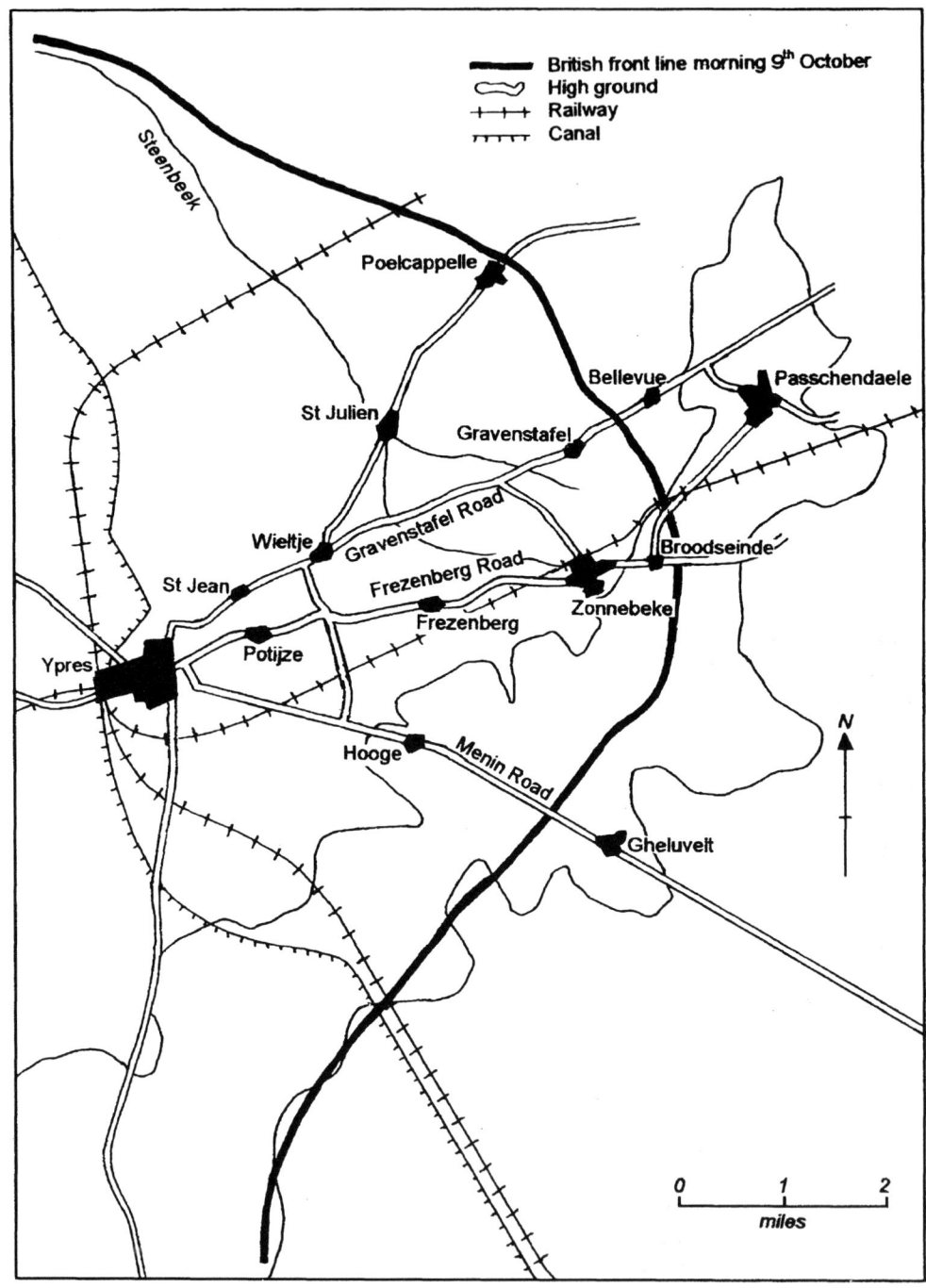

Figure 2. Ypres Salient on the morning of 9th October 1917 showing the general area of assault and main road and rail communications

Roulers railway (which ran southeastwards or away from the main axis of attack) to the north-eastern sector. Panet and Zonnebeke Roads represented the main battlefield arteries down which all material flowed and had been effectively obliterated during previous fighting. They were replaced with wooden plank roads consisting of 3.5 m beechwood slabs on an elevated hardcore base. These two roads were in full view of the Germans who, despite elaborate camouflage attempts by the BEF, had every single yard registered by their guns. The enemy knew the importance of these battlefields arteries and subsequently they were shelled around the clock necessitating the creation of 'Rapid Reaction' repair squads stationed every few hundred metres along the length of the road. The construction of forward roads was now so important that on 7th October, a mere two days before the assault, all traffic except that relating to road construction was halted for half a day in order to allow the Royal Engineers to push the roads forward (Williams, 1917). In addition the whole area was criss-crossed by duckboard (wooden planking) tracks that allowed some degree of troop movement across the lip-to-lip sodden craters. Both the roads and tracks were the responsibility of II ANZAC Corps who made two crucial mistakes when preparing for battle. These are outlined below:

1. Lacking manpower and unaware of the magnitude of the problem that faced them Panet concentrated on roads at the expense of tracks, his overriding concern being to push the artillery forward as quickly as possible. On the day of assault the infantry struggled along inadequately maintained duckboard tracks, which eventually petered out into waist-deep mud some distance from the jumping off point. Consequently the troops were totally disorganised and exhausted even before they entered battle (Adlercron, 1917). 'The average distance from the reserve positions to the assembly position was three and a half miles of which about two and a quarter miles was along duckboard track. The forward end of each track was unboarded and with heavy traffic the mud became very deep.' (Perceval, 1917). Duckboard tracks for men were not the only problem. Mules, the primary means of transportation in the leading edge of the forward area would not use the standard 'mule mat' trails, which lacked a firm footing. They also refused to move down any single-track road that contained motor traffic, thus requiring their own dedicated miniature plank roads to be built or the construction of twin-track roads for use of both classes of traffic (Irving, 1917). One consequence of failing to attend this was a 50% shortfall in ammunition for the 49th Divisional machine gun barrage and hasty re-arrangements had to be made to cover the assault front (Lewis, 1917).
2. The overwhelming need to push the artillery forward meant that single-track rather than double-track roads were constructed in order to facilitate speedy construction. However, the provision of single-track roads with passing places ignored the fact that such roads were easily and fatally prone to blockages caused by shelling and breakdowns, and

could only be used by either mules or lorries, but not by both. 'In the event of heavy shelling transport could not be got ahead out of the danger zone and the road blocked for 24 hours or more' (Lindsay, 1917). Single-track roads also required outstanding traffic control to allow for up and down movement, a skill sadly lacking in II ANZAC Corps given the road congestion. Double-track roads on the other had took only marginally less time to build because they increased the efficiency of traffic flow, allowing construction materials to be moved forward quickly, and the use of the road by both lorries and mules.

6. The Battle begins

At 5.20 am on 9th October the attack was launched, but it was already in serious trouble before it began (Figure 3). Many of the assaulting troops who had begun their march on the evening of 8th October were either still crawling along the duckboards or floundering in the trackless morass where the boards ended, and failed to be in position on time. This left a 'rag, tag and bobtail' collection of exhausted support and reserve battalions to carry on the assault as best they could, with brigade commanders desperately trying to improvise some sort of disposition as the troops chased the barrage. However, the question on the lips of the soldiers was 'what barrage?' Normally an assault of this nature could expect support from a massive panoply of artillery numbered in the hundreds providing a dense protective barrage that acted as a guide for the assaulting troops. In support of the attack was a paltry 25 field guns with the majority of guns stuck uselessly on the blocked single-track roads further to the rear. By sheer exertion alone the troops struggled on before being stopped by machine gun fire and uncut wire at Bellevue Spur, just in front of Passchendaele village. Miraculously, a few remnants of the 3/5th Battalion, Lancashire Fusiliers actually made it into the village itself before being repulsed. By 6 PM the assaulting troops were occupying the same positions they started in, the furthest troops occupying shell-holes and pillboxes in and around the entrance to what is now Tyne Cot cemetery and very short of ammunition and rations. It would be at least three days before they could be rescued.

In 1917 information moved very slowly taking up to eight hours to move from Battalion to Brigade and the only messages that reached GHQ that day were the initial reports that indicated a successful capture of the first line and the 'capture' of Passchendaele by the Lancashire Fusiliers. By the time it was realised that the entire assault was a complete failure the order for the planned assault of the New Zealand and 4th and 5th Australian Divisions to take place on 12th October had already been given. With communications little improved the assault was as big a disaster as the assault of the 9th.

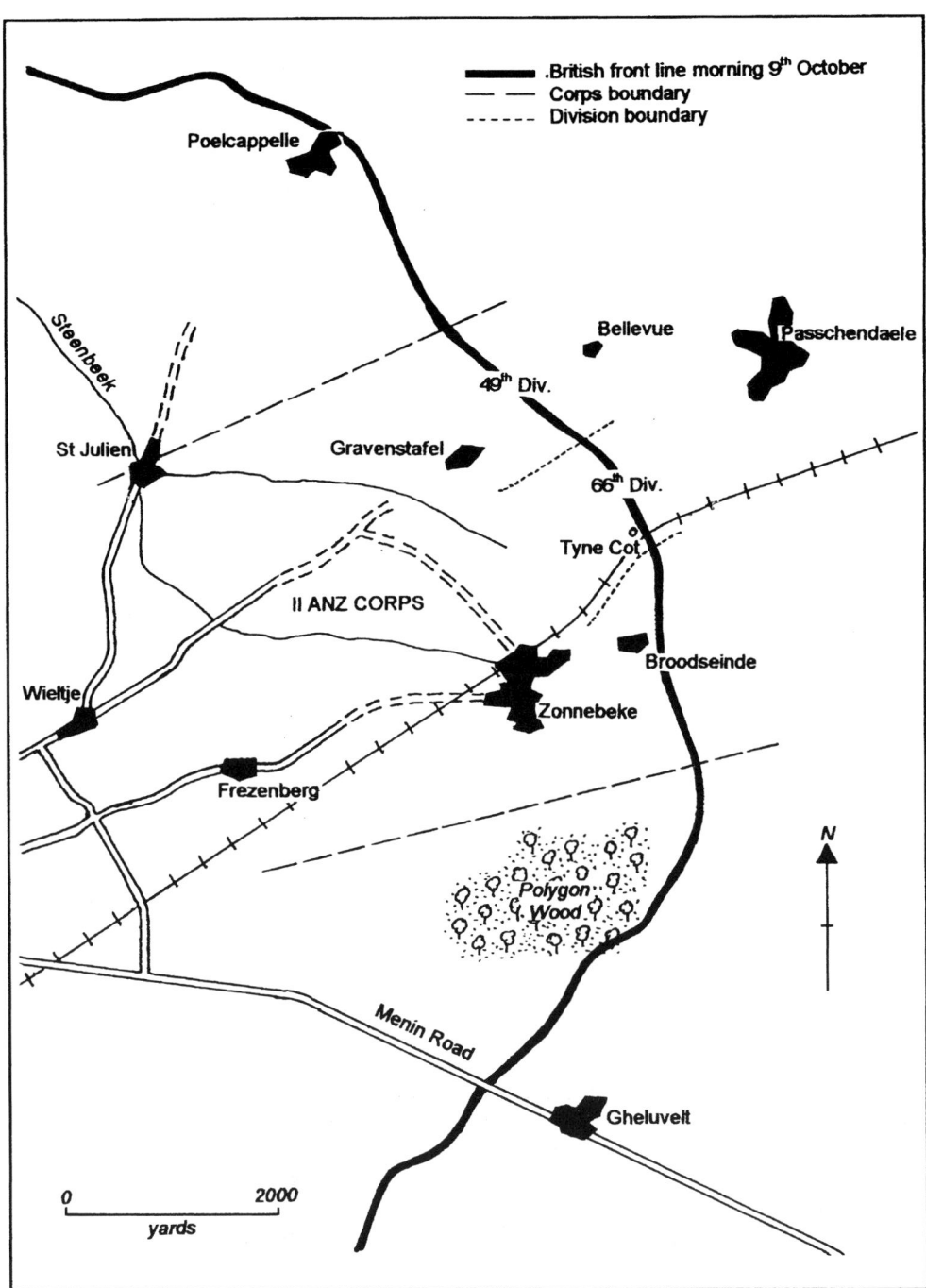

Figure 3. Specific attack frontage allotted to 66th and 49th Divisions on the morning of 9th October 1917

7. The arrival of the Canadians

The failure of the 9th and 12th October equated to the failure of the strategic element of the whole campaign and it would be spring before the tides were right for an amphibious assault. However, the BEF could not rest in such an exposed position and the tactical task of seizing the Passchendaele Ridge was given to the Canadian Corps under Lt.-Gen. Sir A.W. Currie. After the Battles of Vimy Ridge, Messines and Hill 70 Currie was acutely aware of the importance of preparation and the primary role of communications in defining operational advance, and planned his operation accordingly (Macfie, 1993). The Canadian Corps CE, Maj.-Gen. W.B. Lindsay (unlike his II ANZAC counterpart) inspected the entire road and track network and deemed it unworkable along with the Light Railway system. He ordered roads to be doubled and Light Railways abandoned in favour of push tramways that could be converted to Light Railway use at a later date. Given that road-building requirements alone demanded six trainloads of material per day, of which two trainloads could actually be supplied, he also demanded the time required to put the preparations in place. Finally Lindsay also demanded the manpower required to complete the job, which consisted not only of his own Canadian Corps assets, but also virtually the whole of 49th and 66th Divisions, as well as a myriad number of other formations (Lindsay, 1917)[2]. By using push tramways, properly constructed tracks, and doubled roads within an operational framework dominated by logistic and engineering factors, Lindsay increased the speed of the advance and the quality of communications. This ensured that the guns and ammunition required could be moved and supplied. Meanwhile Currie fully accepted these operational realities and limited advances to 500 m with long pauses in between. In a series or operations beginning on 26th October and ending on 10th November the Canadians finally took Passchendaele.

8. The aftermath

That Third Ypres was a terrible ordeal for the BEF is beyond all doubt, as is the fact that at certain places on the battlefield the mud was indeed impassable, but did the mud actually prevent forward movement? The British Official Historian, James Edmonds, blames the weather, whilst Bean, the Australian Official Historian, blames both the mud and the inexperience and poor staffwork of 66th Division. British Prime Minister David Lloyd George along with most military historians up until recently blame Haig for persistently throwing troops into an impassable bog. However, it is more enlightening to examine what the principal commanders and engineers thought, and these views are outlined below:

1. II ANZAC Corps — Godley and Panet. Unfortunately a great many of II ANZAC Corps documents relating to the period are missing from both the

Public Record Office in Britain and the Australian War Memorial in Canberra. A search of II ANZAC record in Canberra revealed a rather truncated and incomplete report compiled by Panet for Second Army. He describes the system of working as 'satisfactory' but is the only person to do so. Panet acknowledges difficulty with the roads but proffers no further explanation whilst tracks are dealt with as 'special orders', which are now missing. Regarding manpower, Panet's only recommendation was to make the temporary attachment of 100 infantry to each RE Field Company permanent, this at a time when each RE Field Company had between 330 and 1,000 men on attachment (Panet, 1917; Williams, 1917)[3].

2. New Zealand Division — Maj.-Gen. Sir A.H. Russell. Russell was the outstanding commander of the New Zealand Division. His position, eloquently and honestly stated in his personal diary was simple: II ANZAC staffwork was appalling; their responsibilities were left to Divisional staffs to solve, and the 66th and 49th Divisions were given no help or assistance and left to flounder along in the mud as best they could. The main problem was the co-ordination of the engineering effort expressed through poor road and track construction, and a total lack of appreciation of the time it takes to move guns and ammunition. Prior to the attack of the 9th, Russell attempted to warn Maj.-Gen. E.M. Perceval (commanding 49th Division) of the poor preparation by personally showing him the state of Panet Road (Pugsley, 1997).

3. 3rd Australian Division — Lt.-Col. H.O. Clogstoun. Maj.-Gen. J. Monash, a lawyer, civilian engineer, and future commander of the Australian Army Corps commanded the 3rd Australian Division. His Divisional CRE H.O. Clogstoun produced his own detailed report for Second Army that constituted a damning indictment of engineer administration, co-ordination, and preparation. He particularly criticised the lack and use of infantry manpower, advocating a large expansion of divisional engineer services to cope with the logistic and engineering demands of modern warfare. He further criticised infantry and mule track construction arguing that II ANZAC Corps dramatically underestimated the amount of transport and time it takes to move and lay duckboard tracks (Clogstoun, 1917).

4. 49th Division — Lt.-Col. D. Ogilvy. David Ogilvy was a Regular engineer officer and prior to his posting as 49th Division CRE had been a Royal Engineer Staff Officer. Ogilvy was far too careful to be caught expressing a position on paper and consequently his diary is brief, technical, and bereft of comment. However criticism was expressed by 148th Brigade Commander, Brig.-Gen. R.L. Adlercron who felt that the message runner relay system was inadequate due to a '... lack of proper system of transmission and supervision at relay posts' (Adlercron, 1917).

5. 66th Division — Lt.-Col. G.C. Williams. Guy Williams was a pre-war Regular engineer with much experience of the Western Front and his comments are copious and critical. On 10th October Williams inspected

Panet Road and warned 'that unless immediate steps are taken to maintain and widen it the roadway from Wieltje to Spree Fm [Farm]. Will give way altogether. CE so informed by telephone.' (Williams, 1917). Later that day at a meeting of CRE's and CE's Williams stated that unless the road was doubled no further materials for road construction could be brought forward, seriously dislocating operational movement and support. On 12th October Panet, failing to understand the nature and magnitude of the problem, offered to arrange two convoys of 40 lorries a day to move the materials. Williams politely informed Panet that even a surfeit of lorries still required a widened and maintained road down which to travel. By the 13th a carrying part of 2,000 infantry was required to move planking as no road widening had taken place. The tone of the War Diary entries makes it clear that Williams' relationship with Panet was becoming increasingly strained, and that the former regarded the latter as, at the very least, inefficient.

Two major themes are evident from these comments: they are all critical of logistic, engineering, manpower and administrative preparation and practice, and none of them identify mud as the principal operational factor. It is important to stress that the mud and weather represented major obstacles and under more tolerable conditions it is entirely probable that the system would have been sufficient. However, it is equally clear that these obstacles could and should have been overcome given better administration and practice. The failure to attend to these issues was indicative of a wider ignorance of the fundamental structural problems exposed by poor ground conditions and their implications for operational conduct.

The same problems that bedevilled Third Ypres were evident as early as 1916. During the Somme, reports from CRE's and Field Company Commanders criticised inadequate engineering practice and administration exposed by bad weather, especially in relation to numbers and quality of manpower (Irving, 1916), whilst at Vimy Ridge the central operational role of communications was clearly understood (Irving, 1917). Apart from Godley and Panet all Commanders recognised the problem, with Clogstoun, Russell and Williams being particularly scathing. However, it was Lindsay of the Canadian Corps who most succinctly summarised the problems, implications, and solutions that faced the BEF logistic and engineering system. Recognising that the leading edge of the battlefield was now some 5-8 km in advance of the Transport lines and RE dumps, itself a consequence of the development of 'deep battle' (Bailey, 1996), Lindsay knew that the assault moved forward according to the dictates of communications and not the wishes of High Command or the Artillery. He also recognised that engineering on the modern industrial battlefield consumed vast amounts of materials, manpower, and time, and that development in this area was handicapped by a failure to recognise the operational dominance of engineering and logistics. In a ground breaking report innocuously entitled *Method of Distribution of Ammunition Employed* Lindsay made several key suggestions (echoed by Clogstoun in his report to Second Army) which were profoundly to affect the organisation and efficiency of the Canadian Corps in 1918, and which had implications for the BEF as a whole (Lindsay, 1917). The five key points of the report are:

1. That the pace of the advance, operational planning, and method should be governed by logistic and engineering factors, primarily communication construction, and that this should take priority over the artillery.
2. That the Canadian Corps (and by implication all the BEF) should dramatically increase the size of their engineering assets by the creation of three engineering battalions per division, even if this meant reducing the infantry establishment. The importance of engineering can be gauged from the fact that this represents a call for a 500% increase in engineers when Lloyd George was already reducing divisional infantry establishments from 12 battalions to nine.
3. That the infantry working party system is unsound as the troops are not interested in the work and should concentrate on training and combat, leaving dedicated labour units to carry out the work. Not only are infantry uninterested, the constant demand made on them seriously interfered with vital rest periods and training programmes prior to the assault, thus reducing fighting efficiency.
4. That a civilian system of engineering organisation is far superior to a military one given the essentially civilian nature of the work. The increased depth and scope of operations required a centralisation of engineering command and assets at much higher levels in order that these assets could be deployed according to operational rather than unit dictates. This process had already taken place with theatre transportation in 1916 when Sir Eric Geddes radically overhauled the Directorate of Transportation utilising civilian techniques learnt as General Manager of the North Eastern Railway (Grieves, 1989). A similar process had occurred with the command and control of artillery and machine guns being centralised at ever-higher echelons in order to apply their firepower on an operational rather than unit basis.
5. That logistic and engineering resources should as far as possible be organic to the unit concerned. Field Companies (consisting of 200 men) were unable to do their work without infantry assistance, yet that assistance was inadequate, of poor quality, and quite often grudgingly given — if it was given at all. The expansion of companies to battalions would obviate this problem by providing the engineers with their own command chain, personnel, and logistic support. It would also release the infantry for training and combat, and improve infantry and engineering efficiency.

This report led to drastic changes in the establishment of the Canadian Corps that provided them with the logistic and engineering assets that were crucial to the Canadian victories of 1918, the value of which was quite clear to Currie: 'I am of the opinion that much of the success of the Canadian Corps in the final 100 days was due to the fact that they had sufficient engineers to do the engineering work and that in those closing battles we did not employ the infantry in that kind of work. We trained the infantry for fighting and used them only for fighting' (Rawling, 1992). It also constitutes a sweeping indictment of the Regular RE logistic and engineering system and the whole rationale underpinning operational methodology and its development. It put logistic and engineering factors at the heart of operations, placing the movement of artillery second.

It also highlighted a structural manpower problem that was not appreciated by either Lloyd George or Haig. The former was interested in cutting infantry establishments based upon the manipulation of manpower figures, whilst the latter still thought in terms of combat infantry. Both failed to understand the demands and nature of modern industrial war and it was units such as the 49th and 66th Divisions that paid the price for this mistake.

9. Conclusions

By 1917 the BEF had solved the problem of initial penetration by providing overwhelming artillery support for the infantry in the form of protective barrages. This required an acceptance of the dominance of artillery and the machine gun and both arms were expanded and re-organised at higher echelons in order to maximise efficiency as the scale of operations expanded. However, once beyond the protective barrage, the BEF became extremely vulnerable to counter-attack. Consequently operations tended to stall at this point, creating minor changes to the line at high cost. This caused a shift in operational thinking and an acceptance that the range of artillery support defined the limit of advance, whereupon the artillery would be moved forward and the whole procedure repeated in a series of set-piece operations. This development of operational thought is part of a wider change in attitudes to operational conduct described by modern revisionist historians as the 'Learning Curve' (Griffith, 1994). The timing of successive operations was crucial: too slow and the enemy recovers; too fast and the assault out-ranges its own artillery support. This is known as operational 'tempo' and its efficacy is entirely dependent upon the rapid forward movement of artillery, ammunition, and supplies in order to keep the enemy under pressure and off balance. In turn that depended upon battlefield conditions, geology, geography and the logistic and engineering services of the Royal Engineers and Pioneers who built and maintained the communications down which the materiel flowed. Despite the fact that this was indeed an 'engineers war', and despite the fact that the RE had expanded from a figure of 24,035 in 1914 to 359,064 in 1918 there was no attempt to redefine the role of the engineers to match that expansion (War Office, 1922). Consequently there was no commensurate logistic and engineering 'Learning Curve' to match that of the Directorate of Transportation, Machine Gun Corps, or Royal Artillery, so that whenever the BEF went into action it did so at a major disadvantage. The

Industrial Revolution and its technological products had altered warfare at the most fundamental level. War was now an industrial undertaking utilising civilian-industrial values of efficiency, specialisation, science, systems, and mass participation. Whilst there was a partial acceptance of the pre-eminence of civilian-industrial values demonstrated by the 'Learning Curve', there was still a Clauswitzian tendency to interpret success or failure through a martial 'prism'. This view defines battles as won or lost according to the quality of infantry training; accurate counter-battery fire, or the use of this or that tactical assault formation, but it does not recognise the operational centrality of engineering and logistics in modern warfare. The failure to understand and accept the operational pre-eminence of engineering and logistics was slowly rectified during 1918, beginning with the first Engineers Conferences in January 1918. However, the failure to act sooner despite the evidence of the Somme, Arras, Vimy Ridge, and Messines ultimately cost the BEF the Battle of Third Ypres.

References

Adlercron, Brig.-Gen. R.L. 1917. *War Diary of 148th Infantry Brigade, 49th Division. Narrative of Events 8th to 10th October 1917: Movement into assembly.* Public Record Office, WO 95/2768.

Adlercron, Brig.-Gen. R.L. 1917. *War Diary of 148th Infantry Brigade, 49th Division. Narrative of Events 8th to 10th October 1917: Communications.* Public Record Office, WO 95/2768.

Bailey, Col. J. 1996. The First World War and the Birth of the Modern Style of Warfare. *Strategic and Combat Studies Institute Occasional* Paper No. 22, Camberley.

Clogstoun, Lr.-Col H.O. 1917. *War Diary of 3rd Australian Division Commander, Royal Engineers, 19/11/17: Report on work carried out by Engineers during Ypres operations.* Public Record Office, WO 95/3366.

Edmonds, J.E. 1948. *History of the Great War, Military Operations France and Belgium 1917,* Volume I. HMSO, London.

Edmonds, J.E. 1948. *History of the Great War, Military Operations France and Belgium 1917,* Volume II. HMSO, London.

Fay, Sir S. 1937. *The War Office at War.* Hutchinson & Co., London. (Reprinted by EP Publishing, Wakefield. 1973).

Grieves, K. 1989. *Sir Eric Geddes: Business and Government in War and Peace.* Manchester University Press, Manchester.

Griffith, P. 1994. *Battle Tactics of the Western Front: The British Army's Art of Attack, 1916-1918.* Yale University Press, New Haven and London.

Guggisberg, Lt.-Gen. F.G. 1917. *War Diary of 66th Division Commander, Royal Engineers, 10/03/17.* Public Records Office, WO 95/3125.

Irving, Lt.-Col. T.C. 1916. *War Diary of 4th Canadian Division Engineers: Reports and opinions on recent Somme operations.* National Archives of Canada, RG9 III C5. Vol.4374, Folder 10, File 2.

Irving, Lt.-Col. T.C. 1917. *War Diary of 4th Canadian Division Engineers, 21/10/17: Report on mule tracks proposed for use in forthcoming operations.* National Archives of Canada, RG9 III C5. Vol.4373, Folder 9, File 15.

Irving, Lt.-Col. T.C. 1917. *War Diary of 4th Canadian Division Engineers: Organisation of Field Companies.* National Archives of Canada, RG9 III C5. Vol.4375, Folder 15, File 5.

Lewes, Brig.-Gen. C.G. 1917. *War Diary of 147th Infantry Brigade, 49th Division. General Narrative [of operations 8th-10th October] Appendix III:* Machine Guns, Public Record Office, WO 95/2768.

Lindsay, Maj.-Gen. W.B. 1917. *War Diary of Chief Engineer, Canadian Corps, October 1917, Appendix IV, Panet Road from Spree Farm to Seine: Observations re: building roads in similar subsoil — long roads* Public Record Office, WO 95/1063.

Lindsay, Maj.-Gen. W.B. 1917. *War Diary of Chief Engineer, Canadian Corps, October 1917, Appendix IV, Panet Road from Spree Farm to Seine: Rate of Progress.* Public Record Office, WO 95/1063.

Lindsay, Maj.-Gen. W.B. 1917. *War Diary of Chief Engineer, Canadian Corps, Appendix VI, Methods of Distribution of Ammunition Employed [in the forward area] 10.11.17.* Public Record Office, WO 95.1063.

Macfie, Corporal D.R. 1st Canadian Infantry Battalion, Transport Section. Testimony quoted in Macdonald, L. 1993. *They called It Passchendaele. The Story of the Third Battle of Ypres and of the men who fought in it.* Penguin Books, London, 220-1. (First Published by Michael Joseph Ltd., 1978).

Maxse, Sir I. 1917. *Notes relating to the proposed advance on Langemarck - St. Julien, August 1917.* Imperial War Museum, Box 69/53/10; File 34.

Panet, Brig.-Gen. E.A. 1917. *War Diary of Chief Engineer, II ANZAC Corps, Battle of Third Ypres 28/10-15/11/17. Report of Chief Engineer on the work carried out by technical troops of Divisions in the line during recent operations'.* Australian War Memorial, AWM 26 230/32.

Perceval, Maj.-Gen. E.M. 1917. *War Diary of 49th Division General Staff. Appendix XV: General narrative of events surrounding attack of 9th October 1917.* Public Record Office, WO 95/2768.

Prior, R. and Wilson, T. 1996. *Passchendaele: the untold story.* Yale University Press, New Haven and London.

Pugsley, C. 1997. The New Zealand Division at Passchendaele. In; Peter H. L. (Ed.), *Passchendaele in Perspective: The Third Battle of Ypres.* Leo Cooper, London, 279-82.

Rawling, B. 1992. *Surviving Trench Warfare: Technology and the Canadian Corps, 1914-1918.* University of Toronto Press, Toronto.

Shelton, Maj. R. 1917. *War Diary of Second Army Deputy Assistant Director, Transportation, 01/10/17 and 27/10/17.* Public Record Office, WO 95.292.

War Office, 1922. *Statistics of the Military Effort of the British Empire during the Great War 1914-1920.* War Office, London.

Williams, Lt.-Gen. G.C. 1917. *War Diary of 66th Division Commander, Royal Engineers, 06/10/17.* Public Record Office, WO 95/3125.

Williams, Lt.-Gen. G.C. 1917. *War Diary of 66th Division Commander, Royal Engineers 12/03/17 and 17/11/17.* Public Record Office, WO 95/3125.

Williams, Lt.-Gen. G.C. 1917. *War Diary of 66th Division Commander, Royal Engineers 10/10/17.* Public Record Office, WO 95/3125.

Notes

1. The importance of these two ports cannot be overstressed. By February 1917 Boulogne was handling 35% of total BEF ammunition requirements, and the Channel Ports combined handled 49% of total BEF requirements. Figures extrapolated from Henniker, Col. A.M. *Official History of the Great War, Transportation on the Western Front*, HMSO, London, (1937)
2. Definitive manpower figures are not available, though this document quotes a figure of six RE Field Companies and 1,800 infantry labour required to build just 250 yards of road in 24 hours. The War Diaries of the Second Army, II ANZAC Corps, Canadian Corps, and 49th, 66th, New Zealand, 4th Canadian, and 3rd Australian Divisions all indicate that the vast bulk of their manpower was engaged upon road and track construction during October and November 1917.
3. The former figure relates to March 1917, but is distorted by the absence of a Pioneer Battalion with 66th Division at that time. However it does demonstrate that the 'official' figure of 100 attached infantry was inadequate to the tasks expected of a RE Field Company.

Rob Thompson
Centre for Contemporary History and Politics
University of Salford
The Crescent
Salford, M5 4WT

Across the River: Interpreting the Battle of the Ebro *or* Battlefields as a Didactic Resource

Edmon Castell & Lluís Falcó

ABSTRACT: The battle of the Ebro (1938) is a singular military episode in the Spanish Civil War (1936-1939). It was an offensive initiated by the Republican forces on fording the river, and ended, after 116 days of intense combat, with the retreat of the Republican army from the Ebro. As a military manoeuvre, it was complex, but a military event is also the sum of its acts and meanings. The Battle of the Ebro transcends its chronological and geographical co-ordinates to become a cultural reference, a symbol of the Franco's victory and a legend of the Antifascist Resistance. In spite of this, sixty-one years later, the Ebro remains an incomprehensible, strange and invisible battlefield.

1. 'It's the closest we would like to be'

More than any other period of history, 'the 20th Century cannot be conceived without war, which was always present even when there was no sound of weapons nor bomb explosions' (Hosbawn, 1995, p. 30). The Great European War (1914-1918) initiated this era of slaughter and the spectre of total war came to stay.

From then on 'we got used to it' (Hosbawn, 1998, p. 253). War has become a commonplace, and has acquired the condition of ordinary experience that no longer surprises anybody. Beyond the actual conflict itself, wars have become instruments of massive consumption and they have invaded a great part of the daily life of our quiet western societies. As entertainment, information or shows, wars are shown in sophisticated strategy games, on the internet, in headlines and war reports, and in movies related to war events, such as *Saving Private Ryan* (1998). This film, with a budget of millions of dollars, is surely the movie that best approaches the Normandy landings of 1944, when the Allies invaded Hitler's Europe. As its star, the American actor Tom Hanks has remarked, 'It is the closest that we would like to be [to the landings]'.

But we can feel the past more immediately by visiting the very places where battles took place. The landing beaches of Sword, Omaha, Gold, Juno and Utah still exist and today, they have been set up as interpretative historical sites. Here aspects of the Normandy landings and of the recent history of the 20th century are explained.

2. The Battle of the Ebro remains unknown

In contrast to Normandy and other battlefields — Verdun, Somme, Gettysburg, etc. — the scenes of the Battle of the Ebro and of the Spanish Civil War of 1936-39 remain unknown and incomprehensible.

Sixty-one years after it ended, the Battle of the Ebro is still an enigma. Despite the publication of many memoirs and accounts of specific episodes of the battle, and the voluminous historical research done on the subject, the scenes of the Battle of the Ebro remain invisible, a battlefield in which the facts and events that are not explained through the landscape. People visiting the area of the Ebro Battle today are unable to understand the conflict, remaining invisible until people remember it, interprets the sites or erects explanatory panels. For example, a field of almond trees can seem an ordinary place until informed that a battalion or company advanced through it to suffer casualties.

Furthermore, battles are ephemeral events. Armies leave behind a multitude of artefacts and marks of their passage or of armed confrontation that 'are usually camouflaged with time and become difficult to locate', while military structures or fortifications 'are subject to a process of neglect and accelerated decay, a process of erosion carried out by natural or human agents that ends in the destruction of many of them' (Hernández Cardona, 1998, p. 3).

The Battle of the Ebro was a military episode of the Spanish Civil War. With the aim of stopping the advance of Franco's Nationalist troops towards Valencia, the Army of the Ebro developed a night offensive, during 24-25th July 1938, fording the river in an extensive stretch between Mequinensa and Amposta (about 100 km). Three and a half months later, the battle concluded with the retreat of the Republican troops to the left bank of the river. Between the initial offensive and the retreat, the battle had developed as trench warfare in the same way as the larger battles of the First World War would.

As a military operation, the fording of the Ebro involved the complex interaction of people, things (trucks, tanks, machine guns, boats, bridges, etc.), logistics, organisation, timing, and so on. But an event of war is also a sum of decisions, actions and meanings. And the Battle of the Ebro became a cultural reference with opposite meanings: on the one hand, a symbol of the Nationalist victory in the Spanish Civil War, and on the other hand, a legend of the international antifascist resistance movement.

Immediately following the combat, the victors fixed the memory of their victory on the landscape and they used the battle as an instrument to legitimate their power. Throughout the Dictatorship that followed the end of the Civil War, the Battle of the Ebro gained diverse meanings that would be expressed through literature, cinema, history textbooks, commemorative speeches, as well as through memorials on the scenes of the battle.

The victors filled several strategic locations of the counteroffensive in the battle with monuments and commemorative crosses, places like Coll del Moro

(Gandesa) or the Quatre Camins crossing (Vilalba dels Arcs). They also occupied central locations in towns involved in the conflict with more crosses, tombstones, plaques, monoliths and ostentatious monuments that commemorated their dead men, 'the combatants who attained glory in the Battle of the Ebro'[1].

All the commemorative monuments of the Ebro battle were inaugurated in massive ceremonies. Through this recreation of power, the victorious Franco past became highly visible on the old battlefields. These attempts to dominate public space became essential for maintaining social control during the Dictatorship. Over time, however, *franquismo* (the Franco doctrine) lost its significance where it was strong before, and the symbols that the victors of the battle erected in their day, such as the monument to the fallen of Tortosa, have become ghosts of a past that is being forgotten.

For the vanquished as well, the battle became a legend, although with a radically different meaning. The Republican offensive at the Ebro is a symbol of the international fight against Fascism.

The fording of the Ebro was something more than a military manoeuvre. It appeared in newspapers throughout the world on July of 1938. The fact that it has become a legend is demonstrated through popular culture throughout the second half of the century: in novels such as Malcolm Lowry's *Under the Volcano* (1941), in movies like Andrej Wajda's *Ash and Diamonds* (1958)[2], in songs such as *Across the Ebro* and *Ay, Carmela* that were made famous during the sixties by the North American singers Pete Seeger and Woody Guthrie. On the battlefield itself we find the vanquished relating to material that evokes the Battle of the Ebro[3].

The Battle of the Ebro therefore became a symbol of the international resistance against Fascism. On the battlefield itself we can find many examples of how popular culture considers it a legend. The Republican soldiers used marginal spaces to build monoliths dedicated to their comrades who fell in combat. In the Pandols Mountain Range, hidden among the undergrowth, a monolith erected in August 1938 honours the presence of international brigadiers that fought with the Republican troops. On it, the name of soldiers and officials from different countries are inscribed, still legible today despite the passage of time. In honour of their memory, some gardens have remained unplanted in the places where Republican soldiers who died during the battle were buried.

In 1988, Joan Brossa, a Catalan artist, installed a visual poem, *The Boot*, in memory of 'all the antifascist combatants who fought for the democratic and national freedom of Catalonia'. This poem occupies a central space of the former village of Corbera d'Ebre, completely destroyed during the conflict. In 1993, the Catalan Government declared the village of Corbera a site of cultural interest in the category of historical site. This is the only part of the battlefield protected today, and its preservation has allowed other action to be taken.

At present, close to the sixtieth anniversary of the battle, improvements on the battlefield continues, with for example, the restoration of the church square in the old village of Corbera. Or the erection of more monuments, such as L'alt dels Auts in Mequinensa, with the inscription: 'to all those who lost, which was everyone'.

Sixty years after the defeat of the Army of the Ebro, there are different interpretations of the battle. The debate has become more democratic, but there are often contradictory actions taken that do not help us appreciate our heritage. We must become aware that the beauty of the landscape of the Ebro battlefield is not just anyplace. It is a historical site, a milestone of resistance against Fascist totalitarianism.

Is no coincidence that a rock from the Pandols Mountains is exhibited in the Toronto Provincial Parliament in Canada as symbol of the international fight against Fascism. By the same token, many British municipalities have recently erected memorials in public spaces commemorating the presence of British citizens who fought on Spanish soil to defend democracy and freedom. In Canada, Britain, France, Italy, the United States, people celebrated the Allied victory in the Second World War half a century ago. Their history has been written from then on based on the defeat of fascism. These countries have transformed the battlefields into historical sites explaining this victory to the world.

The Ebro Battle was also a part of this fight. But only the former village of Corbera is protected today, while many archaeological remains subsist in a state of constant deterioration, or have simply disappeared. In Spain, local institutions, war veterans, etc. continue to diffuse contradictory messages. There are municipalities that, while they give their support to the memory of the battle through the erection of various commemorative monuments, plan to build windmill parks on some battlefield areas. In others, they have already destroyed significant sites of the conflict.

There is still no global view nor homogeneous treatment of the Ebro Battle sites that would allow us to understand the conflict. Numerous microscopic, contradictory and unclear visions are gaining hegemony and maintain the battlefield as an increasingly unknown and incomprehensible site.

3. A battlefield that hits you in the face

The Battle of the Ebro is over, but its 'memory is part of the landscape' (Winter, 1995, p. 1) It remains fixed in the landscape and is tangible through archaeological remains and some spatial reference points. These are the main elements that can be used to interpret the Battle of the Ebro on site. The principal types of remains or physical references that exist on the battlefield are fortifications, the scenes of action and the symbolic remains.

The fortifications are those structures built exclusively for military use. In contrast to other places, where the war front remained stable for a longer period, the fortifications of the Battle of the Ebro are of ephemeral character. The different types that we can find are trenches, bunkers and refuges, but also war roads, observatories, etc. Adapted to the relief and the geology of the area, they are often difficult to detect and reach, while at the same time, they are in an advanced stage of decay.

The scenes of action, consisting of geographical reference points and facilities or sites of common use, are the physical spaces where the events of the Battle of the Ebro occurred. They are the fords and bridges where the troops crossed the river (in Flix, Mora, Miravet, Benifallet, Amposta, etc.), the slopes of the mountains (606 m, 705 m on Pandols Mountain, etc.), crossroads (e.g. Camposines, Quatre Camins), streets, houses, chapels (such as La Fatarella, Corbera, La Fontcalda), tunnels, railroad tracks and stations (in Asco, Bot, etc.) that became command centres, parapets, field hospitals, and so on.

Symbolic remains are those spaces or physical elements that express an explicit message about the past. These are constructions that were generally erected by the dominant social classes to evoke specific events in the past and project them into the future. In the Battle of the Ebro, these consist of the set of crosses, tombstones, commemorative plaques and monuments that the victors erected, as well as the monoliths that the vanquished erected.

An analysis of the iconography and transformations — mutilations, graffiti, substitution of symbols, location changes, etc. — that these symbolic remains undergo, such as the monument erected at General Franco's command post at Coll del Moro, helps to measure the degree of consent to the Franco regime over the course of the years. At present, although most of the symbols of the battle erected by Franco remain in place and look nearly the same as ever, they are losing their intended meaning. In order for these remains to maintain part of their symbolic function, it is necessary to continually maintain them, exhibiting them, restoring them, adapting their surroundings Otherwise, such monuments lose their meaning, people pass by them with indifference, and they eventually become invisible.

But the elements of the landscape — scenes of action, fortifications and symbolic remains — cannot speak. 'The tangible past does not sustained itself. Relics are mute, they need to be interpreted to give voice to their role as reliquaries' (Lowenthal. 1998, p. 352-353). Considered separately, in their current state and out of their historical context, these spaces do not help us understand the battle. For this reason, it is necessary to give a general meaning to the terrain to transform the battlefield into an explicative space.

Research and selection of potential points of interest, their adaptation for visitors and organisation within a unifying historical view (through an interpretation centre, an historical itinerary, etc.), these are indispensable actions to make the Battle of the Ebro a comprehensible place, 'a landscape that acquires meaning through its association with a historical event' (Birnbaum, 1994, p. 1).

The interpretation of an historical landscape 'is the process of providing the visitor with the tools to experience the landscape just as it was in a specific period in the past...' (Birnbaum, 1994, p.10). It is a process of informal education that is based on sensorial and experimental contact with the elements of the terrain and is supported through secondary sources —

photographs, maps, accounts, memoirs, etc. — with the aim of understanding the past.

To interpret the Battle of the Ebro means acting on the battlefield, that is, taking action at the sites that played a major role in the battle. In our view, the Battle of the Ebro cannot be interpreted through books. Taking action involves, in the first place, the identification of places, material remains, and monuments that will allow us to recognise the historical event. 'If there were no explanatory signs, how many visitors to a former battlefield would actually know that it was a historical site?' In the second place, the material remains fixed on the battlefield must be protected and preserved to allow the development of the battle to be understood on site. In the case of the Battle of the Ebro, only the former village of Corbera is protected as an historical site, while numerous archaeological remains are in a state of continuous deterioration or have simply disappeared.

Preserving a battlefield implies a significant financial outlay. In times of economic crisis, the conservation and administration of these material remains may be considered a luxury. However, conservation is not just an expense, it creates the possession of a fixed asset. In any case, 'everything that survives the past has some value that gets lost unless it is conserved'. Local authorities, landowners, etc. should be convinced that is socially and economically positive to protect the sites of the battle. The past can also be a product for mass consumption, and the demand for history, cultural tourism and leisure convert it into a very potent stimulus for the local economy.

By the same token, the interpretation of the Battle of the Ebro must be open to the multidisciplinary points of view. A team of specialists in different sciences — archaeologists, museologists, historians, geographers, educators, forest engineers, etc. — these are the professionals that should design co-ordinate and manage the interpretation project. They should have the support of the majority of the local community, as much as possible, making agreements with institutions dedicated to the tourism, local administrations and politicians, associations, citizens, private companies, etc.

4. Mapping the human heart

In countries like France, Italy, Denmark, etc., the recovery and interpretation of Second World War historical sites is associated with 'the fact that the legitimacy of the post-war regimes and governments was fundamentally based on the participation in the resistance' (Hosbawm, 1994, p. 169) and in their victory against the different types of fascism in Europe. Arguably, the history of Europe since 1945 is based on the defeat of fascism.

In Spain, the Civil War placed the directing classes of Franco followers in power. The victors used the memory of the Battle of the Ebro and monumentalised the battlefield to legitimate their authority. After that, the transition to the democracy was built upon a tacit pact of silence about the war and the forty years of dictatorship. Today, the latest actions taken on the sites of the Battle of the Ebro continue to be an invitation to forget.

But the charm of the past is a potent stimulus. The democratisation of culture has caused a new social demand for history to arise, a demand that neither historians nor institutions of learning respond to adequately (Hernández Cardona, 1998). In this new context, battlefields have become a didactic resource that, beyond the textbook and the classroom, can provide honest answers to the new consumer demands for the past.

War events are factors that form part of historical processes. But in history it is not only the facts that count. The feelings, desires and actions of those who experienced these events are also a part of history. This dialogue between events and motives is what allows us to reconstruct and understand the Battle of the Ebro. The battlefield is a useful instrument to show this connection, to place us within the historical context of the past.

Interpreting the Battle of the Ebro through the ordering of its sites — establishing interpretation centres, historical itineraries, museums, etc. — will help us to develop a send for the place. As Tom Hanks, star of *Saving Private Ryan* was to comment in 1998, 'One of the definitive incidents in this personal experience took place when we filmed the cemetery scenes surrounded by some of the war victims. For me, that place was a necessary stop on the cultural itinerary of any American. Nevertheless, after working on the movie, it became a mysteriously religious place, like in an immense church. I felt that their souls were right there, in that exact place'.

Developing battlefields as a source of knowledge means allowing the senses, experimentation and intuition to take on a fundamental role in the learning process. It unstops a bottled emotional area associated with the sensation evoked by the place. In a similar way to *Saving Private Ryan*, visitors to the historical sites of the Battle of the Ebro will at least discover a small fraction of the feelings and yearnings of those who were involved in the conflict. A battlefield is a resource with infinite possibilities.

References

Birnbaum, C.A. 1994. *Protecting cultural landscapes*. ASAL, Washington.

Hernández Cardona, F.X. 1998. El patrimoni monumental militar con a bé cultural. Presentation, *European Symposium on Urban Military Spaces and Fortresses*.

Hernández Cardona, F.X. 1998. Métode històric I ensenyament. *Guix, Barcelona* 245, 4-8

Hobsbawn, E. 1995. *La época de la guerra total*. Historia del Siglo XX. Critica, Barcelona.

Hobsbawn, E. 1997. *La babarie: guia del usario*. Critica, Barcelona.

Lownthal, D. 1998. *La historia es un país estraño*, 3rd Edition. Akal Ediciones, Barcelona.

Winter, J. 1995. *Sites of Memory; sites of Mourning: The Great War in European Cultural History.* Cambridge University Press, Cambridge.

Notes
1. Inscription on the monument to the Battle of the Ebro in Tortosa. Inaugurated by the Head of State on 21st June, 1966.
2. On 8th May 1945, the Second World War in Europe had ended. It was a time for evaluating the half century of war that had preceded this momentous event. The film *Ashes and Diamonds* by Andrzej Wajda (1958) narrates the first 24 hours of this collective celebration in a Polish city. In a key scene, an Allied commander and a member of the secret services wonder about the uncertain future to come. A gramophone is playing *El paso del Ebro*. The melody evokes memories of the struggle that has brought them to this point, and of their fallen comrades in the long fight against fascism. 'Do you remember? Thats where everything started...' It was a moment of euphoria in which the hope of reconstructing the world was spreading through the continent of Europe. It was only fair to remember the Battle of the Ebro while Europe was drunk with victory.
3. This is the case of the garden l'Hort de Cardoneta in Mora d'Ebre, where numerous Republican soldiers remained buried during forty years of the Dictatorship. Elsewhere, in a carob field near Benifallet, a plaque commemorates the burial site of Marcel Savoïa, a French brigadier of the XIV Brigade, *La Marsellesa*, fallen during the battle.

Edmon Castell & Lluís Falcó
Departament de Didactica en les Ciencias Socials
Universidad de Barcelona
Edificio de Llevant
despatx 121
08035 Barcelona

Fortification of Island Terrain: Second World War German Military Engineering on the Channel Island of Jersey, a Classic Area of British Geology

Edward P.F. Rose, W. Michael Ginns & John T. Renouf

ABSTRACT: The 116 km^2 Channel island of Jersey was fortified by the British intermittently from the 13th to 19th centuries, and by the Germans more intensively during the Second World War — to form part of Hitler's 'Atlantic Wall'. British castles, forts and towers were sited to counter potential amphibious attack by the French. German forces (which at their peak totalled about 11,500 troops, supported by a construction workforce of a further 6,000 men) adapted many of these sites to counter amphibious attack by the Allies, augmenting them with batteries of heavier coastal artillery together with an impressive sequence of associated observation towers/posts, about 100 fortified infantry defence areas, a series of anti-tank walls and ditches, an extensive minefield complex, tunnels for storage and shelter, central and local command/control centres, and numerous anti-aircraft guns. Siting of fortress installations was influenced mainly by geomorphology, but the Germans made significant use of military geologist expertise to predict foundation characteristics, ensure adequate supplies of potable water, and locate sources of the considerable quantities of aggregate required for construction. Geologically the island forms part of the Armorican Massif more extensively exposed in the adjacent areas of Brittany and Normandy, a classic 'hard rock' terrain composed largely of metasediments and volcanics of 587-582 million years (late Proterozoic) age intruded by later Proterozoic and Ordovician granites and associated igneous rocks, modified overall by much later features of Quaternary erosion and sedimentary deposition. Jersey contrasts with the 6 km^2 Jurassic limestone Rock of Gibraltar, famous for over 200 years as a British fortress outpost, in geology as well as size and occupation history, but other aspects of their fortification were constrained by similar general principles.

1. Introduction

Jersey (116 km^2) is the largest of the Channel Islands — British territories which lie close to the Normandy coast of France (Figure 1). Between the 13th and 19th centuries the island was progressively fortified by the British to counter potential attack by the French. However, when France fell to German attack in the Second World War, the British garrison was deemed vulnerable and so quickly withdrawn. On 1st July 1940 German forces occupied Jersey unopposed, and from 1941 (and more intensively 1942) to 1944 they carried out major construction work to turn it into a 'German Gibraltar' — an impregnably fortified outpost — as part of Adolf Hitler's Atlantic Wall policy. In June 1944 the island was bypassed by the Allied invasion of northwest Europe (Figure 1), and at the close of the war in May 1945 surrendered intact — complete with key fortification records. Local fortifications therefore include some of the best surviving and well-documented examples of a line of massive German defensive works which had by 1945 stretched from northern Norway to the Spanish frontier.

Jersey is also one of the classic areas of British geology. The geological features which influence its terrain have been documented with increasing precision from the early 19th century to the present day. Relationships between geology and terrain, and consequently between geology and fortification, can therefore be demonstrated with particular clarity.

Plans of the fortifications discussed in this paper are illustrated in the site-specific leaflets produced by the Channel Islands Occupation Society, and many are also published in Partridge (1976). Calibre of weapons is given in SI units, to conform with current scientific and military practice in the citation of dimensions. German practice was to express calibre in centimetres rather than millimetres.

2. Terrain

Bedrock geology
Jersey is a small, hard-rock island. Its geology was first described in the early 19th century by MacCulloch (1811) on the opening pages of the *Transactions of the Geological Society of London*, and later by Nelson (1830). Both these authors by coincidence had British military associations. John MacCulloch lectured on chemistry at the Royal Military Academy, Woolwich, from 1804 to 1835, and on geology at the East India Company's military college, Addiscombe, from 1819 to 1835 (Rose, 1996, 1997). Richard John Nelson was a career officer in the Royal Engineers, who retired ultimately in the rank of major-general (Rose, in press a), but who published, amongst other things, on the geology of other islands — Bermuda and the Bahamas. These pioneer studies were followed by over a hundred publications by other authors (Prince, 1997), most notably a 1:25,000 scale solid and drift geological map (Bishop & Keen, 1982) and accompanying descriptive memoir (Bishop & Bisson, 1989).

Thus it has long been established that Jersey, like the other Channel Islands, is more closely related geologically to the adjacent parts of France (Normandy and Brittany) than to Great Britain. The Channel Islands and adjacent areas constitute part of the Armorican Massif, on the western margin of the Paris Basin (*cf.* Rose & Pareyn, 1998). Armorica was largely unaffected

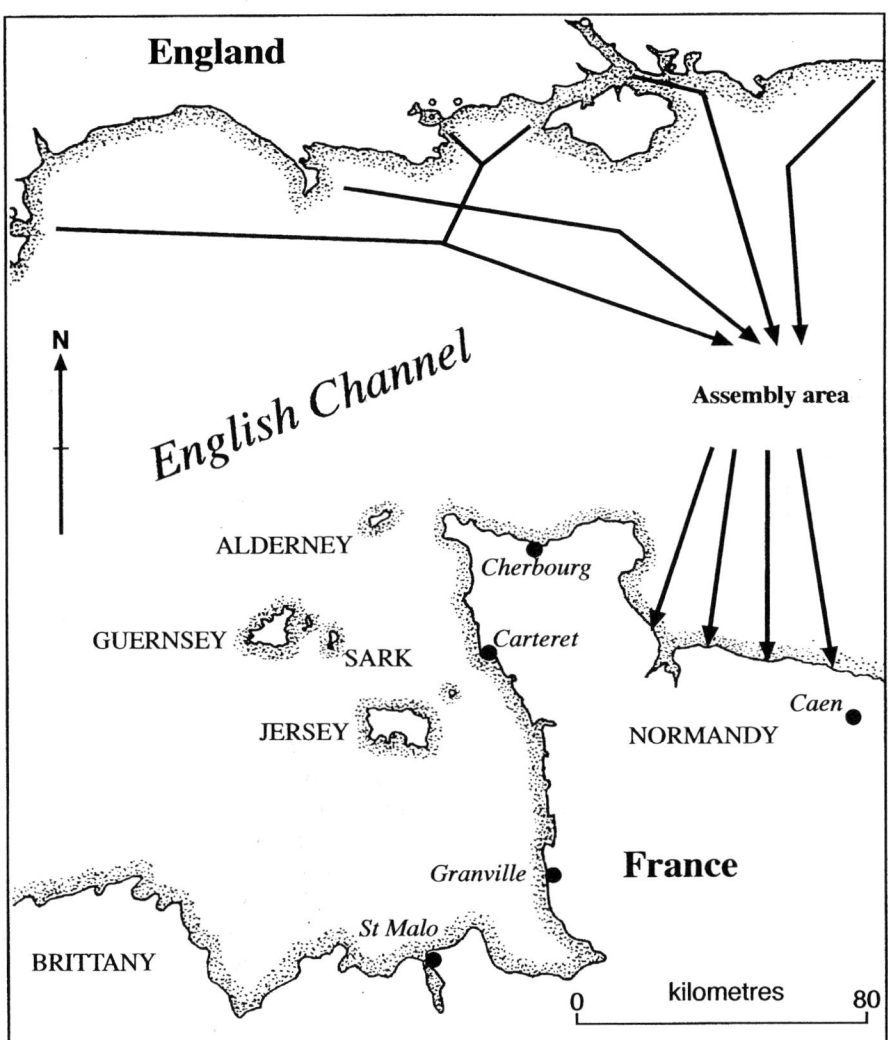

Figure 1. Map of the Channel Islands showing their closer proximity to France than England, and the main routes from England of ships assembled for the D-Day amphibious landings in Normandy of 6th June 1944, which all took place east of the Cherbourg Peninsula. Routes based on Young (1973)

by either Caledonian or Variscan earth movements (c. 400 & 290 million years, Ma, before present respectively), and so preserves a record of Precambrian to Early Palaeozoic sedimentary, tectonic, and igneous events different from those on the mainland of the British Isles.

Following Renouf (1985) and Bishop & Bisson (1989), the bedrock (solid) geology can be described in terms of five rock types (Figure 2):

1. *Late Proterozoic metasediments.* The oldest rocks exposed on Jersey are currently ascribed to the Jersey Shale Formation, which crops out in the western and central areas, and has an estimated minimum thickness of 2,500 m. The sequence comprises fine- to medium-grained sandstones with subordinate conglomerates, siltstones and mudstones, each with variable amounts of diagenetic calcite. On general petrological and stratigraphical grounds these rocks have been correlated with those of the Upper Brioverian widespread in Normandy and Brittany, which have been assigned a Late Proterozoic age. The Jersey rocks were largely deposited as thin (25 mm) to medium (250 mm) bedded units, turbidites within a submarine fan complex. Subsequently, they have been regionally metamorphosed to low greenschist facies, with the conversion of clay matrix to chlorite, and affected by tectonism and locally by contact metamorphism. Uranium-lead dating of zircons separated from the sediments now indicates 587.1±2.7 Ma Before Present as the maximum depositional age for the unit (Miller et al., 2001).

2. *Volcanics.* The Jersey Shale Formation is succeeded, apparently conformably, by the Jersey Volcanic Group, a sequence of lavas and pyroclastic rocks, cropping out in the northeast of the island, with probable total thickness of 2700 m. The St. Saviour's Andesite Formation (850 m) consists of lavas, tuffs and agglomerates of andesite and basaltic composition. This is succeeded, apparently unconformably, by the St. John's Rhyolite Formation (950 m). Much of the succession is ignimbrite, and dominantly porphyritic, but there are intraformational sediments, both mudstone and conglomerate. The overlying Bouley Rhyolite Formation (430 m) is dominated by aphyric rhyolitic lava flows, ignimbrites and air-fall or water-laid pyroclastics. Miller et al. (2001) have dated zircon crystals from near the base of the Bouley Rhyolite as about 582.7 Ma Before Present, and have inferred this to be the age of eruption. Overall the sequence appears to have originated on the flanks of volcanoes, and to have been metamorphosed with the Jersey Shale to greenschist facies.

3. *Plutonic complexes.* Three granitic complexes have intruded folded Jersey Shale sediments and the overlying volcanics. Contacts are sharp and steeply dipping, and the effects of thermal metamorphism can be detected in clean coastal exposures. The southwest complex (age about 580 Ma Before Present according to Miller et al., 2001) has an outer, major unit of coarse-grained granite surrounding a central distinctively pink aplogranite separated from the main mass by a 250 m wide belt of aplogranite with megacrysts of potassic feldspar and quartz. The northwest and southeast complexes include basic rocks and granite. The northwest complex includes layered gabbro and diorite intruded by a larger mass of coarse-grained granite, producing a suite of intermediate rocks. Miller et al. (2001) have now dated zircons from the granite at 482.7 Ma Before Present, indicating a renewed pulse of magmatism during the Ordovician. The southeast complex is the most complex of the three: a

diorite intruded by a distinctively megacrystic granite, producing a mass then invaded by basic dykes before intrusion of a younger granite.

4. *Cambro-Ordovician conglomerates.* The Rozel Conglomerate Formation is a post-orogenic molasse deposit which rests unconformably on the Bouley Rhyolite Formation in the northeast of the island, with two small outliers to the west of the main outcrop. It comprises coarse, poorly-sorted, polygenetic, clast-supported conglomerates, with subordinate sandstones and mudstones, deposited in beds mostly upwards of 1 m thick. Boulders within the conglomerate may reach 2 m in diameter, and the whole unit has an estimated thickness of 500 to 800 m. Deposition as continental alluvial fan deposits, and a post-Cadomian orogeny (*c.* 600 Ma) and therefore a Cambro-Ordovician age have been inferred, although without palaeontological evidence. Deposition in a minor extensional basin during mid Ordovician times has now been more confidently inferred by Went & Andrews (1990).

5. *Minor intrusions.* Most of the minor intrusions occur as dykes, but a few have sill-like form. Petrographically they are of three main types: basic, acid, and lamprophyric. The basic dykes range from true dolerites to epidiorites; acid dykes are predominantly porphyritic micro-granites; and mica-lamphrophyres are more common than hornblende-lamprophyres. Several periods of dyke emplacement can be recognised. Throughout the island, basic dykes with a roughly N-S alignment cut across dykes with more nearly E-W trends, although a few N-S dykes are earlier than the E-W ones. The dykes are particularly numerous in the south of the island, where the main swarm has a general E-W trend. These dykes, often composite intrusions, are commonly 4 to 10 m wide, but other dykes are generally < 3 m wide.

The present configuration of the Jersey Shale Formation and overlying volcanics has been determined largely by fold movements attributed to the late Proterozoic to early Palaeozoic Cadomian orogeny which occurred about 600 Ma Before Present — an age of deformation now more tightly constrained to 583-580 Ma Before Present within Jersey itself (Miller *et al.*, 2001). Early E-W compression gave rise to N-S folds, and subsequent N-S compression refolded these about E-W axes. After the Cadomian folding and regional metamorphism, the plutonic complexes were intruded. Uplift and erosion followed, the products of erosion accumulating in hollows to form the Rozel Conglomerate. Arguably, two further phases of folding followed, and further erosion, so that the bedrock as presently exposed comprises a broad synclinorium with NE-SW axis in the northeast of the island, separated by an approximately E-W trending anticline from a SSW plunging syncline to the east of St. Aubin's Bay on the central south coast. Principal faults strike WNW-ESE, but many other trends are represented.

Quaternary geology
Deposition of the Rozel Conglomerate was followed by a long (c. 400 Ma) period which has left no sedimentary record in Jersey. The youngest rocks on the island are unconsolidated, superficial sedimentary deposits ('drift') of Quaternary age, both Pleistocene (Figure 3) and Holocene (Figure 4), which reflect the changing climates and sea levels from the Middle Pleistocene to the present day (Keen, 1978, 1981, 1993; Jones et al., 1990) — a timespan of several hundred thousand years. Largely deposits of interglacial or periglacial origin, they comprise:

1. *Raised beaches.* Gravels associated with Pleistocene shoreline landforms occur locally around the present coast of Jersey at heights of 30 m (two localities), 18 m (in caves and on some fragmentary benches), and 8 m (widespread gravels, notch or platform) (Figure 3).
2. *Blown sands.* An older (Pleistocene) blown sand near Jersey Airport, < 5 m in thickness, can be distinguished from a younger (Holocene) blown sand which occurs in a coastal strip < 500 m wide in the west, north, and southern bays of Jersey (Figure 4). This sand is typically a structureless, quartz-feldspar sand, up to 27 m in thickness.
3. *Head.* Pleistocene head — the unsorted reworked weathered remnants of the local solid rocks — is a breccia which occurs at the foot of cliffs along the southwest, north, and northeast coasts of Jersey; as undissected cones beneath blown sand in the major bays of the island; and along valley sides inland (Figure 3). Locally, against high cliffs in the north, maximum thickness may reach 30 m.
4. *Loess.* Pleistocene loess is an orange-yellow to pale brown sediment in which 80% of the particles are of silt grade. It is composed largely of quartz and feldspar, but is calcareous in patches. The well-sorted nature and lack of sedimentary structure of the loess are consistent with an aeolian origin, and the sediment was perhaps derived from the floor of the English Channel exposed during a period of low sea level. The main areas of loess in Jersey are in the centre and east of the island, with greatest thickness about 5 m (Figure 3).
5. *Peat and alluvium.* Holocene peat and alluvium occur as narrow deposits along most of the river valleys (Figure 4), but the main occurrences are at the mouths of the major valleys which discharge along the south and southeast coasts. Here up to 8 m are present, but coastal peats elsewhere are < 1 m thick.

Except for the loess which covers the upland plateau of the island and the valley-side head deposits, these drift formations occur largely in coastal areas.

Topography, climate and soils
The island is a plateau < 150 m high, divided by a series of incised valleys which drain mainly from the higher ground in the north towards the south

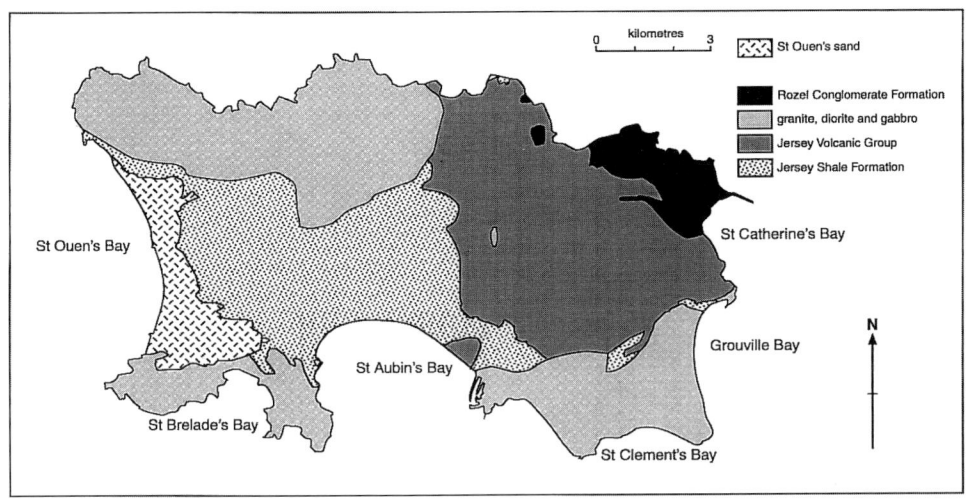

Figure 2. Bedrock (solid) geological map of Jersey, after British Geological Survey (1992), simplified from Bishop & Keen (1982), Bishop & Bisson (1989). Thick deposits of Holocene sand obscure the bedrock in St. Ouen's Bay. Minor intrusions and faults cannot conveniently be illustrated at this scale. (Reproduced by permission of the British Geological Survey. Crown copyright reserved)

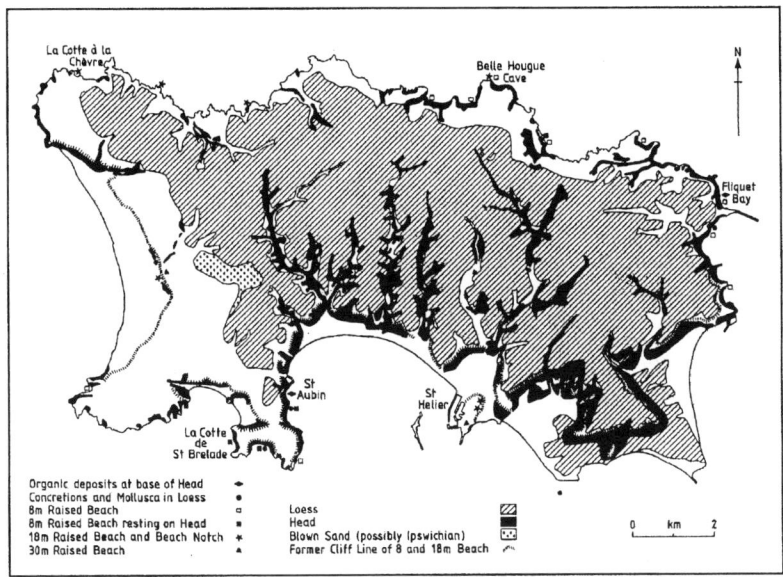

Figure 3. Map of Jersey showing the main Pleistocene deposits (Modified from: Jones et al., 1990; Keen, 1978. Reproduced by permission of the British Geological Survey. Crown copyright reserved)

coast (Figure 5). Contemporary beaches with adjacent areas of blown sand and therefore low relief are well developed adjoining St. Ouen's Bay in the west, St. Aubin to St. Helier in the south, and the Royal Bay of Grouville, to the south of Gorey, in the east (Figure 4). Smaller beaches are developed elsewhere, but much of the coast is rocky and, especially in the north, fringed by steep cliffs to the sea.

As discussed by Jones et al. (1990), the principal controls on the island's topography are the major elements of its solid geology, especially lithology and structure. These authors suggest that the southerly tilt of the island is almost certainly due to Neogene (Late Tertiary and Quaternary) movement of the old rocks which make up its current land area. Superimposed on the southerly tilt is a series of flatter, bench-like features, which reflect periods of marine erosion during the recent tectonic history of the island. They also suggest that, tectonics notwithstanding, lithology has had a significant impact on topography. Major topographic differences are caused by the resistance to erosion offered by the relatively homogeneous igneous rocks, and the lightly-metamorphosed, fractured and cleaved shales. The efficacy with which the latter can be removed has led to the excavation of the deep bays of Saint Ouen and Saint Aubin between granite headlands (cf. Figures 2, 5).

Jersey's climate is influenced by the relatively warm ocean water that flows around the island; exposure to westerly air masses; and shelter afforded by the French mainland. Together these ensure that Jersey experiences comparatively genial weather and climate, in which marked variations or extremes are lacking (Jones et al., 1990). Average mean daily air temperature is 11.5 °C; on average groundfrosts occur on only 60 days per year, and snow or sleet falls on only 12 days per year.

The average annual rainfall, derived from 131 years of data (1865-1995), is 849 mm and is well-distributed through the year (McCartney & Houghton-Carr, 1998). Average annual potential evaporation lies in the range 648 to 754 mm (Blackie & Jones, 1993). Rainfall is highest on the high ground to the north of the island and generally diminishes southwards with topography (Figure 5).

No complete survey of the island's soils has been accomplished, but Jones et al. (1990) note that a complex pattern exists, in which major soil types relate mainly to underlying rock type and to drainage characteristics:

1. *Trinité Series*. Fine sandy and silty loams, usually about one metre deep, developed over loess, loessic head, and fine-textured blown sand; well to moderately-well drained.
2. *Colombier Series*. Very fine sandy to silt loams, shallow (<0.35 m) or deep (>0.35 m), developed on Jersey Shale or Pleistocene deposits derived from it; well drained.
3. *Noirmont Series*. A sandy loam, also of shallow and moderately deep phases, developed on granite; well drained.

4. *Samarès Series*. Fine sandy to silt loams, on average 0.30-0.45 m deep, developed below *c.* 12 m above sea level on loess or blown sand; imperfectly or poorly drained.
5. *Rozel Series*. Sandy to silty loams, generally 0.30-0.50 m deep, developed on sloping ground over head or igneous rock; well drained.
6. *Radier Series*. Fine sandy to silty clay-loams, usually not exceeding 0.10 m depth, developed over alluvium or other drift in valleys; badly drained.
7. *Saint Ouen Series*. Loamy sands to sandy loams containing little organic matter, with shallow and deep phases, occurring over blown sand; well drained.

The soils are largely fertile, and Jersey is predominantly rural, consisting mainly of intensively worked arable land and grassland used for dairy farming. Some 60% of the island area is currently in agricultural use — a landscape which has resulted from cultural activity over several millenia (Jones *et al.*, 1990). The current resident population is about 84,000, most of whom live in the south-coast towns of St. Helier, St. Aubin and the residential area of St. Brelade, although there are also numerous small villages scattered throughout the area.

3. British fortification

From the Norman Conquest of England in 1066 through to 1204, Normandy, as well as other areas of France, owed allegiance to the English monarch. In 1200 the English King John separated the administration of the Channel Islands from that of Normandy and the King of France, Philip Augustus, seized the opportunity to strengthen his own position and regain control of the Normandy peninsula. From that time the Channel Islands faced potential French attack, and were progressively fortified by the English (Figure 6).

The earliest stone-built fortification on the island is Mont Orgueil Castle (Figure 7), built on a granite promontory overlooking the wide sweep of the Royal Bay of Grouville (Figure 4) and facing the coastline of Normandy from the northern tip of the Cherbourg perninsula to Granville in the south (Figure 1). The granite, bounded to the west by Jersey Shales and to the north by Jersey Volcanics, is the northernmost outcrop of the southeastern granitic complex. Erosion to produce steep slopes and high cliffs on three sides offered a naturally defended site. The castle was founded in the early 13th century and progressively strengthened thereafter to counter developments in weapon technology, and perceived threats from France. Its role as a fortress lapsed from the late 16th century, but between 1942 and 1944 German occupying forces developed it as a self-contained strongpoint. It was adapted to provide fire control and observation towers, dugouts, trenches, and positions for automatic and other small arms. Twenty-four 'roll bombs' were placed on the outer

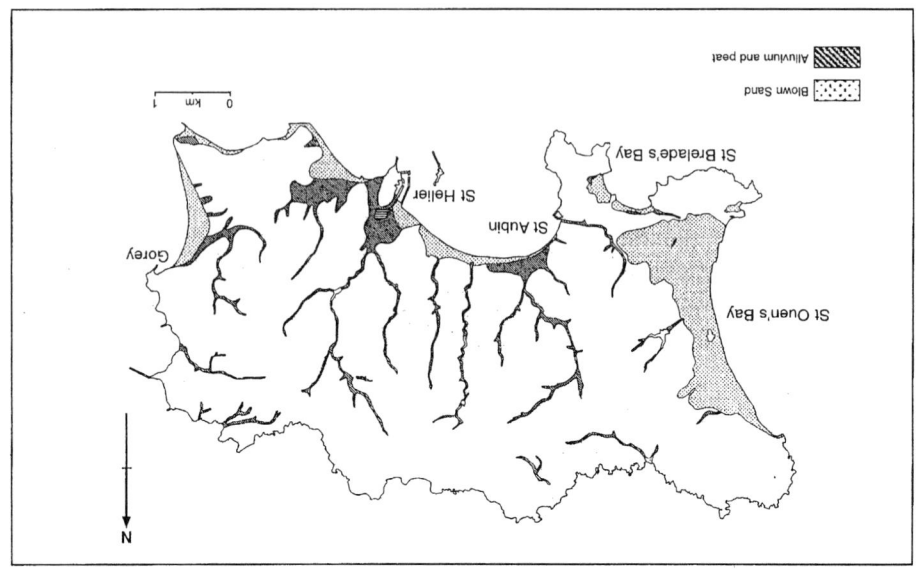

Figure 4. Map of Jersey showing distribution of Holocene deposits (sediments deposited within the last 10,000 years. (Modified from: Jones et al., 1990; Keen, 1981. Reproduced by permission of the British Geological Survey. Crown copyright reserved)

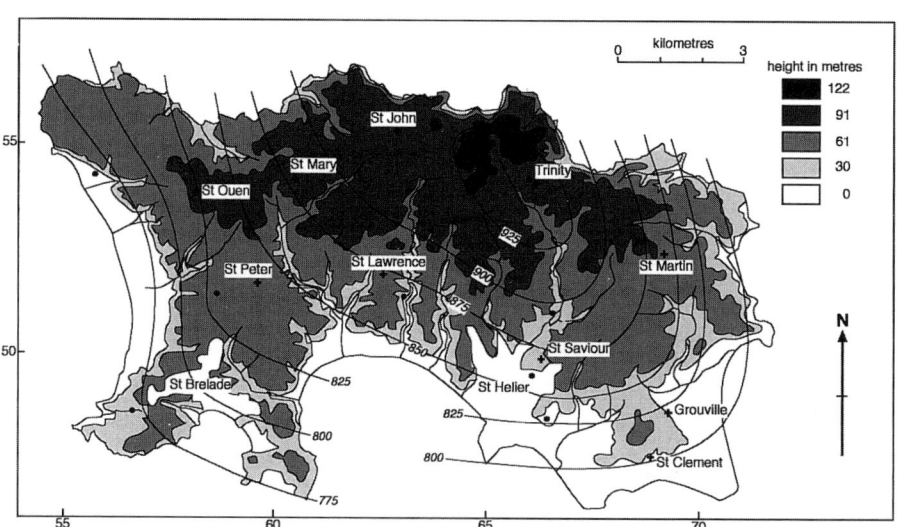

Figure 5. Map of Jersey showing the relief (contoured at 100 foot, c. 30 m intervals), together with average rainfall (indicated by isohyets at 25 mm intervals), and Universal Transverse Mercator grid eastings and northings indicated at 5 km intervals. (Modified from: British Geological Survey, 1992. Reproduced by permission of the British Geological Survey. Crown copyright reserved)

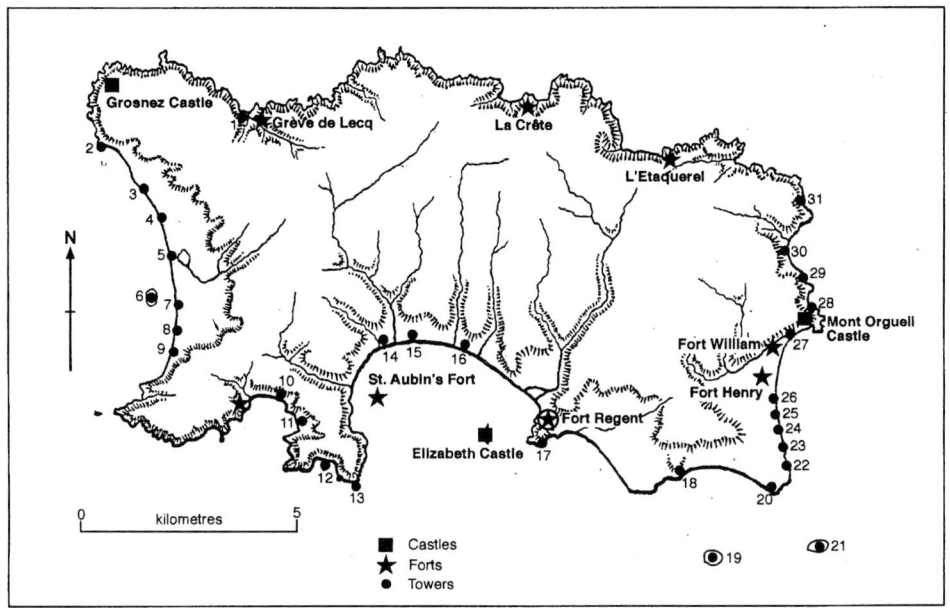

Figure 6. Map of Jersey indicating sites of British-built 13th to 19th century fortifications. (Compiled from various sources, notably The Wayfarer Map of Jersey, *& Grimsley, 1988)*

walls, and a flame thrower at the top of the steps leading to the keep. The northeast outworks were heavily reinforced to house a battle headquarters. Electric apparatus was installed and the water supply improved. Large rooms in the keep were converted into fully-furnished barracks (Rybot, 1978; Brown, undated).

Grosnez Castle (Figure 6) was built in the early 14th century upon granite at the northwest point of Jersey. Before the invention of firearms it provided a place of refuge for the western islanders, but had no secure water supply to enable it to withstand siege, and no garrison. It was easily (if briefly) captured by the French in 1373, and had fallen into ruin by the 16th century (Balleine, 1950).

From the late 16th century Mont Orgueil was superseded as the principal island fortress by Elizabeth Castle (Figure 8), constructed on a granophyre-diorite island facing Jersey's principal port, St Helier (Figure 2). With the development of cannon, Mont Orgueil was no longer a safe refuge, despite remodelling of its defences. The first part of a new castle (the keep, known as the Upper Ward), adapted to contemporary fighting methods, was largely constructed between 1594 and 1601, during the reign of Queen Elizabeth the

First. By 1668 fortification had progressively extended to form the Lower and Outer Wards. With some re-modelling, Elizabeth Castle remained a British garrison until 1923, when it was sold to the Jersey government for use as an historic monument. The Germans returned it to its former use. In the Outer Ward, three light automatic guns were installed, together with a 105 mm calibre gun protected by a bomb-proof casemate, plus other bomb-proof shelters and searchlight positions. Similar installations were constructed in the Lower Ward. The keep was in part adapted for modern weapons, modified to provide increased barrack room accommodation, and surmounted by a massive concrete artillery fire control tower (Partridge, 1976, p. 136; Rybot, 1986).

In the 18th and 19th centuries redoubts and batteries were constructed to defend those embayments deemed vulnerable to sea-borne invasion, and although long disused, several were adapted and re-activated by the Germans. Examples are the two redoubts of Grouville Common on Jersey's east coast: Forts William and Henry (Figure 6). Fort William (built originally in 1760) was strengthened with concrete emplacements and heavy machine guns. Fort Henry (an 18th century infantry base; Figure 9) was similarly converted to a 'resistance nest' — with two 105 mm guns in concrete emplacements, two 50 mm mortar emplacements connected by trenches to the fort, two light and three heavy machine guns, ten flame throwers,

Figure 7. Mont Orgueil Castle, viewed from the southwest, sited on a granite promontory and overlooking the village of Gorey at the northern end of Grouville Bay. The flat tops on the three turrets of the medieval castle are German concrete additions to convert the turrets to artillery fire control/observation towers (M7 on Figure 14)

Figure 8. Elizabeth Castle, sited on a granophyre-diorite island. View of the Upper Ward from the Lower Ward, showing beneath the Jersey flag the massive concrete fire control tower constructed by the Germans on the castle's keep

one anti-aircraft gun, and four searchlights, manned by an officer and 35 troops of other ranks (Ginns, 1973).

In 1778, when the French became the allies of the Americans in their War for Independence, the governor of Jersey was certain that the French would seize the opportunity to attack the Channel Islands. He therefore proposed the construction of 30 coastal towers to resist enemy landings (Pocock, 1971; Grimsley, 1988) — round towers about 10 m high and 500 m apart, the walls pierced with loop holes for musketry in two stages (Figure 10). They were designed to defend the beaches, and for musket fire because of the shortage of cannon. Twenty-two round towers and one square tower were eventually built by 1798; a further three round (Martello) towers between 1807 and 1814; and five more between 1834 and 1838 (e.g. Figure 11) — 31 towers in total. Jersey contains both the earliest and latest examples of such British-built fortifications in Europe — which by 1838 fringed all but the steep northern coast of the island (Figure 6). All but four of these granite-built towers were still standing at the time of the German occupation. Three were destroyed by the Germans to make way for other works, but many were modified for contemporary use (see Partridge, 1976, p. 134).

The last British fortress to be constructed on Jersey, Fort Regent, was founded in 1806, when defensive works throughout the island were in hand to counter the threat from Napoleonic France, and completed in 1814 as the threat came to an end (Davies, 1971). By this time Elizabeth Castle was considered too isolated to be of much value, except as a refuge of last resort, and the extensive new fort was constructed on the igneous promontory overlooking St. Helier (Figures 4 & 6). It was garrisoned by the British regular army until 1932, but from that date onward no longer regularly occupied. Ironically, the only shots ever fired in anger from the Fort were by German occupying forces against Allied aircraft during the Second World War.

4. German occupation and the Atlantic Wall

During the Second World War, following evacuation of the British Expeditionary Force from Dunkirk and the fall of France to German attack, the islands were demilitarised and soon occupied by German forces — an occupation already much documented (e.g. Wood & Wood, 1955; Ramsey, 1981).

Figure 9. *Fort Henry, viewed from inland. In 1781 known as Fort Conway and the barracks for the 93rd Regiment of Foot, the site was most recently re-fortified and manned by the Germans*

Figure 10. Le Hoque tower (number 18 on Figure 6), a typical example of the Jersey round towers completed between 1780 and 1798 during the threat of invasion from revolutionary France, built to defend potential landing areas (at a time when there was a shortage of artillery) by musket fire from loopholes and the roof. The 'machicoules' which jut out at the top of the tower enabled the defenders to fire straight down the walls whilst protected from enemy fire, and so prevent the enemy from undermining the tower

Figure 11. Kempt tower (number 4 on Figure 6), one of the five 'Martello' towers constructed on Jersey between 1834 and 1838. Elliptical rather than round in plan, with its thickest wall towards the sea, this contained a trifoil gun platform — and was designed for defence by artillery rather than by musketry. The door inserted at ground level was part of later adaptation for German use

The early years were relatively uneventful, but from early in 1942 the style of occupation changed. Hitler decreed in December 1941 that the Channel Islands were to be permanently fortified outposts of the Third Reich — as Gibraltar was for the British Empire (*cf.* Rose, 2000, in press b). Fortifications were begun that were to represent some 8% of the entire German Atlantic Wall.

The 'Atlantic Wall' was the name given by Adolf Hitler to the line of coastal fortifications, stretching from Norway to the Spanish frontier, which was constructed between 1940 and 1944 as a defensive rampart to protect the western limit of Nazi Germany's European expansion (Figure 12). In little more than four years the Germans carried out what was probably the most impressive building programme since the time of the Romans (Partridge, 1976). Moreover, the strength of the 'Wall' was exaggerated by a deliberate programme of propaganda, aimed at fostering the myth of 'Fortress Europe', and whilst it failed to deter the Allied invasion of Normandy in June 1944, it was measurably successful in determining Allied plans for this Second Front.

The building programme for the Atlantic Wall as a whole provided for the construction of 15,000 heavy bunkers along the Dutch, Belgian and French coasts. The German army, navy and air force cooperated in the enterprise, which was facilitated by the adoption of standardised designs developed from earlier (1938-1940) experience of fortification on Germany's western frontier — the West Wall (Rolf, 1988). Labour for the construction work was provided largely by the 'Organisation Todt': a workforce founded in 1933 by Dr Fritz Todt, a civil engineer of great competence and expertise, when he was charged by Hitler with the construction of Germany's pre-war motorway system. By 1943 this had expanded to a uniformed organisation auxiliary to the armed forces of over 1.5 million workers, mostly foreign volunteers and forced labour under German planners and overseers (Ginns, 1994).

5. German fortification

German fortifications largely comprised artillery batteries and associated observation posts to prevent Allied ships from approaching the coast; infantry defence areas to oppose beach landings from ships that had survived artillery opposition; anti-tank walls and ditches to prevent armour moving inland from potential landing beach areas; minefields to form a similar barrier to troop movements; tunnels for storage and shelter; command/control bunkers safe from aerial or naval bombardment; and prolific anti-aircraft batteries (since as the largest of the Channel Islands, Jersey was deemed to be the most vulnerable to Allied airborne assault) (Ginns & Bryans, 1978; Ginns, 1999).

Artillery batteries
For their size, there was a greater concentration of artillery fire power in the Channel Islands than in any other part of the Atlantic Wall proper. The

largest calibre guns were sited (Figure 13) as coastal batteries with a threefold task: to protect the Islands and the entire Gulf of St. Malo from Allied invasion; to protect German shipping passing between the Islands and the adjacent French coast; and to protect units of the German navy passing down a 'core route' between Cherbourg, Brest and the French Atlantic ports. Three batteries arrived on Jersey in March 1941, and these were progressively upgraded by 1944 to nine (Figure 14 & Table 1). Lothringen was under naval command, other batteries were under army command. Sites were initially concentrated in the west of the island; Endrass was a harbour-blocking battery with weapons in casemates having restricted fields of fire, and the Schlieffen and Haeseler batteries were moved from Guernsey to Jersey only after American capture of the Cherbourg peninsula and consequent threat from the east in July 1944. 'Batterie Moltke' at Les Landes is currently illustrative of the armament (Figure 15) and bunker complex (Figure 16) associated with a coastal artillery site.

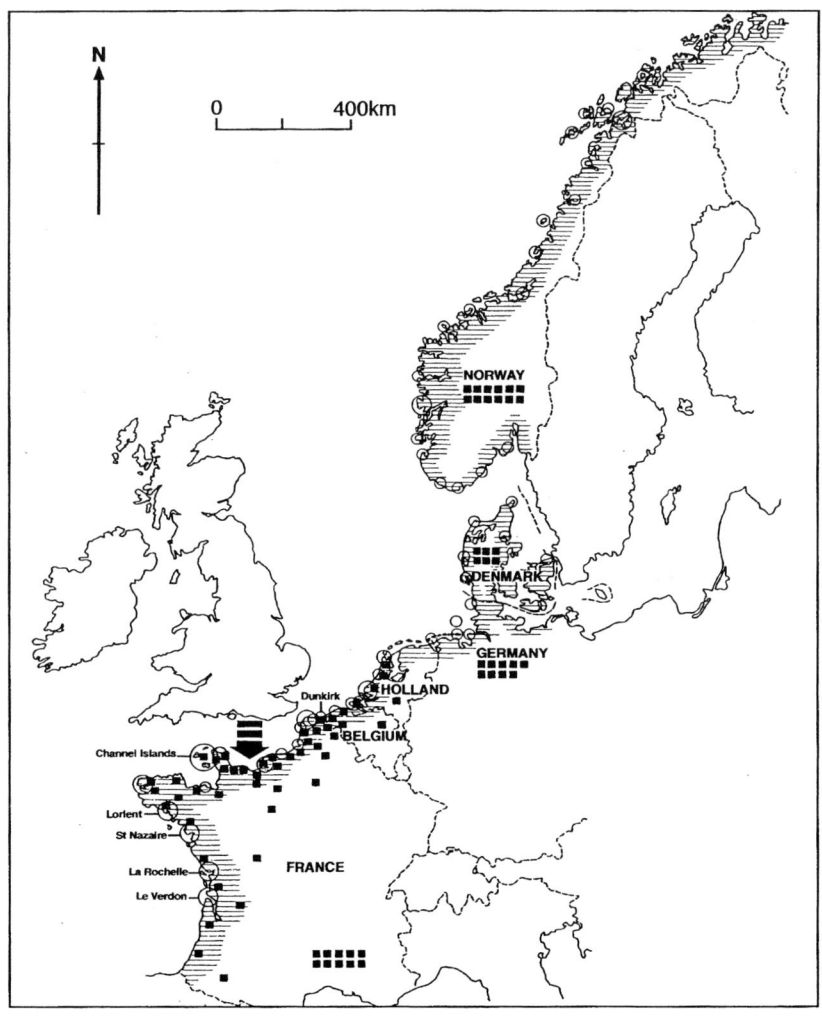

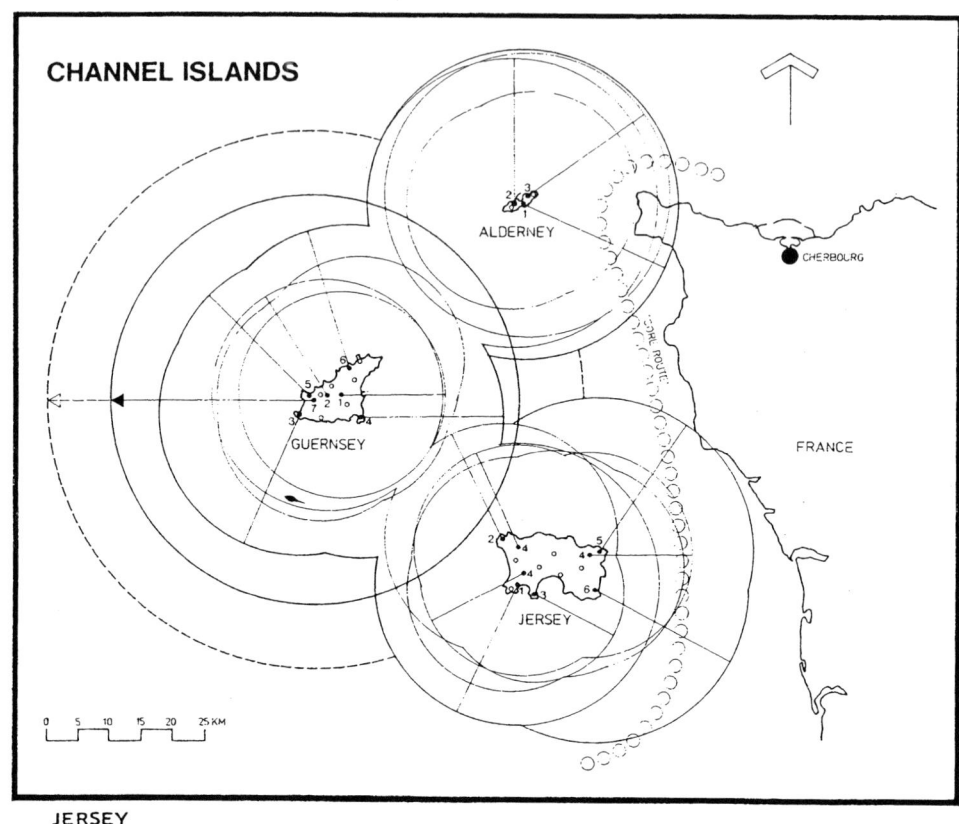

JERSEY				
1.	4 x 22cm	Army	22,000m	
2.	4 x 15.5cm	Army	18,000m	
3.	4 x 15cm	Navy	14,000m	
4.	3 x 21cm	Army	18,000m	
5.	4 x 15cm	Army	24,000m	
6.	4 x 15cm	Army	24,000m	
ALDERNEY				
1.	4 x 15cm	Army	24,000m	
2.	4 x 15cm	Navy	22,000m	

3.	3 x 17cm	Navy	22,000m
GUERNSEY			
1.	9 x 21cm	Army	16,000m
2.	4 x 15.5cm	Army	18,000m
3.	4 x 22cm	Army	22,000m
4.	4 x 22cm	Navy	22,000m
5.	4 x 22cm	Army	19,500m
6.	4 x 15cm	Navy	22,000m
7.	4 x 30.5cm	Navy	38,000m

Figure 13. Map of the Channel Islands showing co-ordinated fire plan for the major German coastal batteries established on Jersey, Guernsey and Alderney in the Second World War, together with the 'core route' for German shipping between the Channel Islands and the French coast. (Modified from: Partridge, 1976; Partridge & Wallbridge, 1983; courtesy of Colin Partridge)

Figure 12. Map showing fortified areas (circled) which together formed the Atlantic Wall, stretching some 2,685 km from the North Cape of Norway to the Spanish frontier (from Partridge, 1976, courtesy of the author). Names of coastal sites indicate fortresses which remained intact until the end of the war. Black squares indicate distribution of German army divisions in the west as at June 1944

Table 1. *German coastal artillery deployed on Jersey during the Second World War (Modified from: Ginns & Bryans, 1978)*

Battery name	Weapons
Lothringen	Four 150 mm SK L/45
Hindenburg	Three 210 mm Mrs 18
Ludendorff	Three 210 mm Mrs 18
Mackensen	Three 210 mm Mrs 18
Endrass	Four 105 mm K331(f)
Roon	Four 220 mm K532(f)
Moltke	Four 155 mm K418(f)
Schlieffen	Four 150 mm K18
Haeseler	Four 150 mm K18

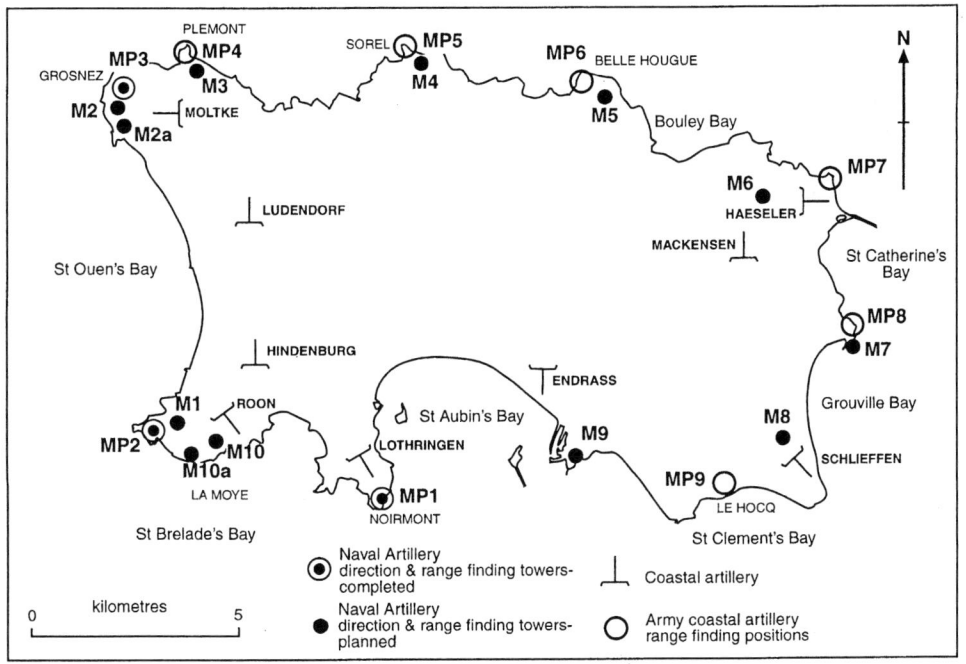

Figure 14. *Map of Jersey showing sites of German coastal artillery batteries (cf. Table 1), and both navy and army direction/range-finding positions, as at 1945. (Modified from: Ginns, 1989; Baker, undated)*

Figure 15. Partially restored section of coastal artillery battery 'Moltke' at Les Landes, Jersey: one of originally four 155 mm guns, with a range of 18 km, which were all dismounted soon after the war

Batteries on Jersey enjoyed all-round fields of fire. Direction and range to potential rargets out of direct sight were communicated from observation posts sited as a continuous chain around the coast (Figure 14). The most impressive of these are the naval observation towers (Figure 17), of which nine were planned but only three actually constructed (Ginns, 1989). These are unique to the Channel Islands and not found elsewhere in the Atlantic Wall. Whilst the building programme for the naval towers was in hand (each tower reportedly consumed 5,000 bags of cement during construction), the army adapted a variety of existing sites (including Mont Orgeuil castle) as range-finding positions, and where no suitable building was available, constructed split-level field positions (i.e. positions only lightly reinforced, with concrete < 1 m thick).

The coastal batteries were supported in a local defence role by an artillery regiment comprising six fully mobile light batteries.

Infantry defence areas
Infantry defence areas were of three types:

1. *'Action posts'*. Lightly defended and often minimally protected sites manned normally during battle alerts only.
2. *'Resistance nests'*. Extensive defence works including open field positions linked to self-contained fortified emplacements.
3. *'Strongpoints'*. Large, heavily fortified and elaborate defence areas. Several resistance nests covering a specific area were frequently grouped as a collectively-named strongpoint.

Together these comprise the most numerous of German fortifications, and account for most of the bunkers concentrated along the beaches which formed the most likely areas for amphibious attack (Figure 18). Many were equipped with 105 mm calibre coastal defence guns (Figure 19), commonly sited within massive casemates of reinforced concrete (Figure 20; e.g. Resistance Point La Carrière), although some were in more lightly fortified 'field' positions. Fortifications built to 'fortress' as contrasted with 'field' standard had concrete > 2 m rather than > 1 m thick. A typical casemate during construction required excavation of 815 m^3 of spoil and consumed 730 m^3 of concrete, 40 tonnes of reinforcing steel, and 6.2 tonnes of steel girders. Other sites were equipped with 47 mm anti-tank guns (Figure 21), also usually within a standard casemate (e.g. Resistance Points La Carrière and Millbrook). By 1944 Jersey's armament included about 125 large-calibre weapons deployed to support infantry in a coastal defence role (Table 2). Structures associated with infantry defence areas commonly composed not only casemates for weapons but also personnel and ammunition shelters. A second line of infantry defence installations was planned to support the coastal defences with fire from their main weapons, contain any Allied assaults that might have broken through, and repel them with counter attacks. However, building of this second line of fortification was given a lower priority than that of the coastal defences, and construction developed only in the hinterland of St. Ouen's Bay — the area deemed to be the most vulnerable to amphibious attack.

Anti-tank walls and ditches
Sea walls constructed from granite blocks during the latter part of the 19th century and the first two decades of the 20th century to protect the coast from marine erosion were so massively constructed that they needed little if any modification to form a barrier to armoured assault (Ginns, 1974). Gaps between the existing walls were, however, filled by nine lengths of concrete wall to form a barrier to tank movement landward of the beach areas lying to the west, south and east of the island (Figure 22). Altogether 8,200 m of anti-tank wall were planned, of which 7,397 m were completed. Seldom was any length built to a common plan, but these reinforced concrete structures remain impressively massive — rising in places to 6 m height, with foundations 2 m thick (Figure 23). Inland from St. Ouen's Bay (Figure 22), a nearly 2 km series of anti-tank ditches was constructed to form an additional barrier.

Minefields
Over 100 minefields were laid in Jersey (Gander, 1991), away from the populated area of St. Helier on the south coast (Figure 24). They supplemented the coastal defences of the western and eastern beach areas of the island, and provided an almost continuous obstacle along its less intensively fortified and garrisoned north coast. Most fields were of anti-personnel mines: over 20,000 *Shrapnellminen* 35 were emplaced, over 2000 of them tripwired, and over 20,000 *Schützenminen* 42, many of them improvised (Table 3). Anti-tank mines were much less numerous: only some 2,000

Tellerminen of various types, plus an additional 1,000 *Tellerminen* 43 *Pilz* employed as beach obstacles (probably attached to anti-tank tetrahedra). 'Roll bombs' (made to roll down slopes towards an enemy) and other charges improvised from captured or obsolete ammunition were also deployed in a defensive role.

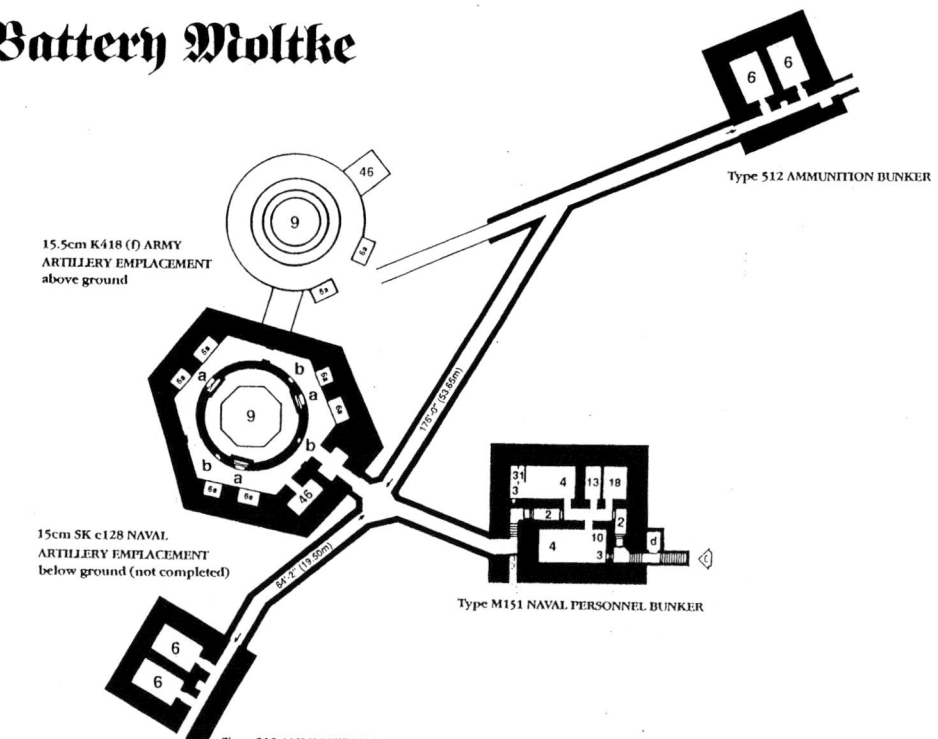

Figure 16. Plan of part of German battery 'Moltke', showing ammunition and personnel bunkers adjacent to artillery emplacements. The roof and outer wall of each installation is of concrete > 2 m thick. Key: 2, gas lock; 3, entrance defence; 4, crew rooms; 6, ammunition room; 6a, ammunition storage; 9, gun emplacement; 10, ventilation; 13, communications room; 18, NCO's room; 31, central heating; 46, crew shelter; a, ammunition hoist; b, access ladder to gun emplacement; c, entrance; d, generator. (Reproduced from: Channel Islands Occupation Society leaflet — 'Batterie Moltke', courtesy of the Society)

Table 2. German coastal infantry defence weapons deployed on Jersey during the Second World War. (Modified from: Ginns & Bryans, 1978)

Type		Number deployed
105 mm	K331(f)	30
37 mm	KwK 144(f)	30
47 mm	Pak 36(t)	23
75 mm	Pak 40	12
37 mm	Pak 35/36	14
50 mm	Pak 38	8
75 mm	FK 231(f)	4
80 mm	FK 30(t)	4

Figure 17. Naval observation tower MP3 (cf. Figure 14) at Les Landes, Jersey, viewed from the south, sited at the edge of a granite cliff and with the island of Sark visible in the distance. Each observation slit in the tower face was intended to permit control of fire from a separate artillery battery. Tower height 16 m

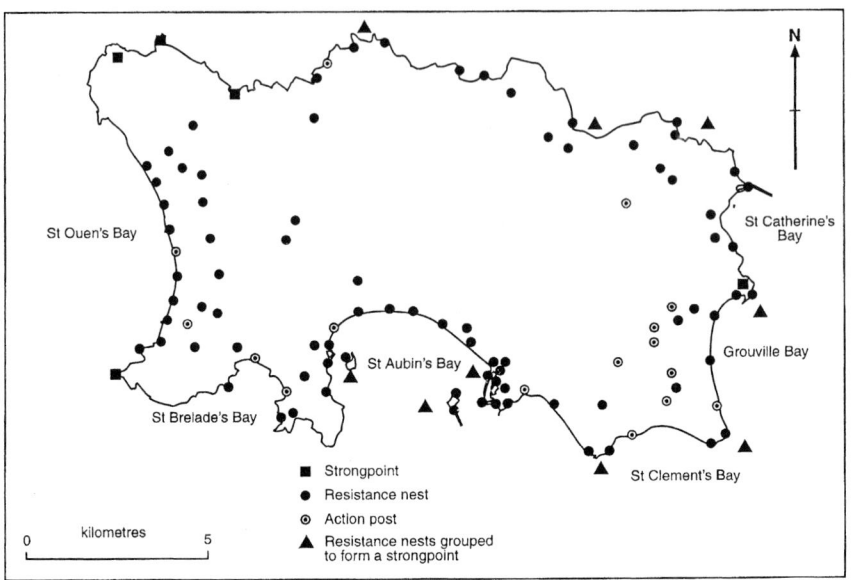

Figure 18. Map of Jersey showing main German infantry defence areas as at 1945. Symbols indicating resistance nests grouped to form a strongpoint relate to resistance nests shown on land nearby: they are not additional defence areas. (Modified from: Ginns & Bryans, 1978; Baker, undated)

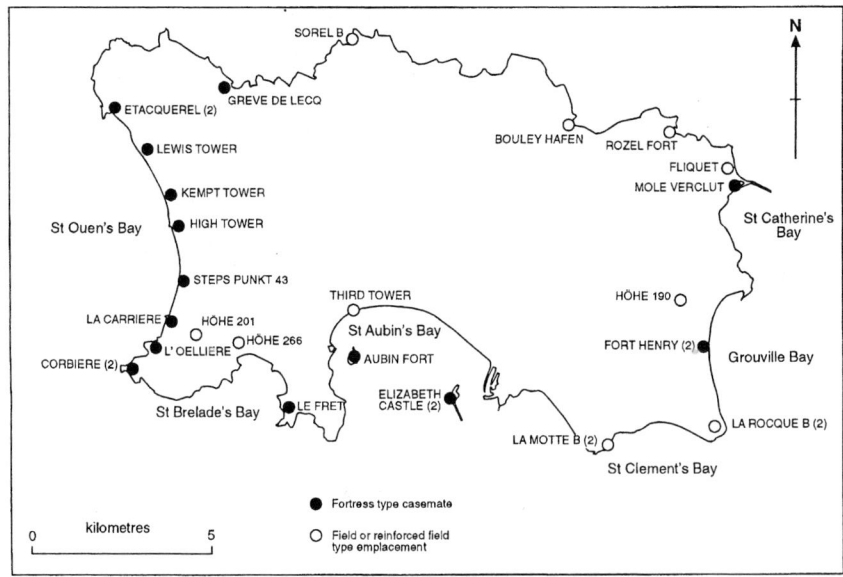

Figure 19. Map of Jersey showing deployment of 105 mm K331(f) coastal defence guns, and casemate type, as at 1945. (Modified from: Channel Islands Occupation Society Jersey leaflet 'Resistance Point La Carrière')

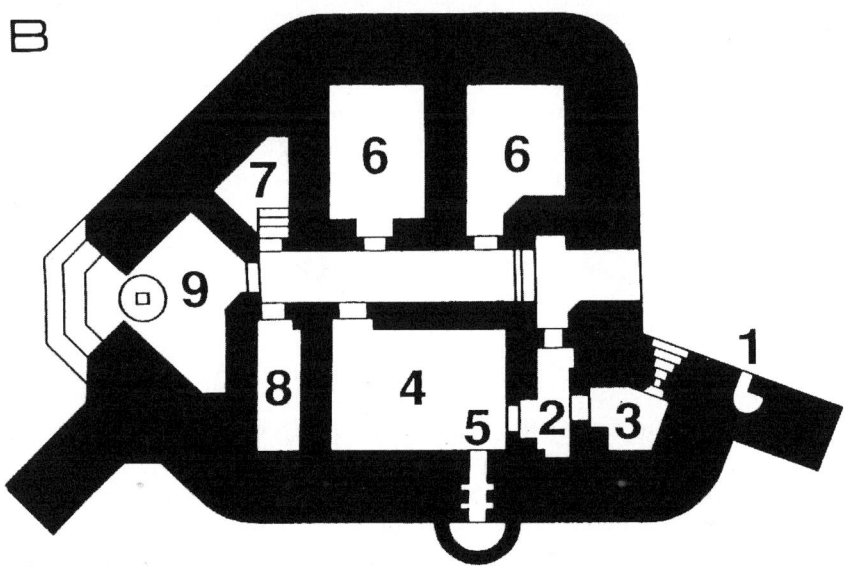

Figure 20. **A.** *105 mm K331(f) coastal defence gun and casemate at Strongpoint La Corbière.* **B.** *plan of 'Channel Islands' standard casemate for 105 mm K331(f) coastal defence gun The roof and outer walls are of reinforced concrete > 2 m thick. Key: 1, Tobruk/observer's open position; 2, gas lock; 3, entrance defence; 4, standby room; 5, escape hatch; 6, ammunition room; 7, empty shellcase room; 8, foul air extraction plant; 9, gun room. (From Channel Islands Occupation Society Jersey leaflet 'Resistance Point La Carrière', courtesy of the Society)*

Tunnels
At least 16 tunnels (*Hohlgangsanlagen*, cave passage installations) were planned to provide accommodation stores for rations, fuel and ammunition, as well as troop shelters and electricity works (Figure 25 & Table 4; Ginns, 1993). Tunnel excavation was facilitated by the island's steep-sided narrow valleys which permitted vehicle access to sites with almost immediate potential protective rock cover of up to 36 m — proof against the most prolonged shelling or aerial bombardment. Only two tunnels seem to have been completed (Ho 5 and the defensive tunnel for Strongpoint Etaquerel), but four were in a sufficiently advanced state for partial use (Ho 1, 4, 8, 19) and four more actively under construction (Ho 2, 10, 13, 15) by the close of hostilities. Ho 8 (Figure 26) was converted for use as an underground hospital, and is currently preserved as such as a tourist attraction and museum.

Command and control centres
The Kernwerk, a complex of six bunkers dispersed to minimise vulnerability to aerial attack, was situated in the approximate geographical centre of Jersey — and close to the airport (Ginns & Bryans, 1978; Figure 27). There were three command bunkers (housing fortress, artillery and infantry headquarters), two communications bunkers, and a bunker for pumping and storing potable water. The three command bunkers were identical in design: built on two levels they incorporated working and living quarters for the staff with wash rooms, flush toilets and a central heating system. Bunker walls and ceilings were of reinforced concrete > 2 m thick, and command/communication bunkers were disguised externally as dwelling houses with dummy windows, shutters and chimney pots (Ginns, 1999).

Outside the Kernwerk, command of the Naval coastal artillery was exercised from the Noirmont Point Command Bunker (Figure 28), and of the north, east, south and west defence sectors by regional headquarters situated inland.

Anti-aircraft batteries
Once the decision to fortify the Channel Islands had been taken, anti-aircraft gun batteries multiplied prodigiously. By September 1944 Jersey had at least 165 guns in position — thirty-six 88 mm Flak 36 and 37 guns, fifteen 37 mm Flak 41 guns, about 104 light 20 mm Flak 30, Flak 38 or Flak *Oerlikon* guns, and ten of the four-barrelled 20 mm Flak *vierling* weapons — excluding mobile weapons and those mounted on ships in the harbour (Ginns & Bryans, 1978).

6. German resource use

The German forces adapted many of the earlier British fortifications, even those dating from medieval times, to contemporary use — for they were well sited to dominate areas of terrain vulnerable to amphibious attack, and were sufficiently massive in stone-built construction to withstand fire from small-calibre weapons. Construction of the massive new fortifications, however, demanded a large labour force, and large supplies of aggregate and water.

Table 3. Mine types and total numbers deployed by the German army in the Channel Islands during the Second World War. (Modified from: Gander, 1991)

	Alderney	Guernsey	Jersey	Sark
Tellerminen	1657	5163(a)	2253	29
Shrapnellminen	6536(b)	23 912	21,127(c)	3232
Schützenminen	10,611	16,352	20,597(d)	6872
'Improvised', locally manufactured	8310	13 851	17,276	2622
Stockminen	201	800	1781(e)	351
Panzerabwehrminen 407(f)	2206	-	998	-
Panzerabwehrminen 408(b)	554	-	1987	-
270 mm shells,	43	78	-	-
Concealed charges (unspecified)	-	-	62	-
Beach Obstacles:				
Tellerminen 43 Pilz	-	-	1075	-
'Rollbomben'	-	-	248	-
Air Landing Obstacles:				
Tellerminen 43 Pilz	51	-	-	-
270 mm shells	176	-	19	-
Unspecified charges	-	-	460	-

Notes: (a) of which 360 were tripwired; (b) of which 2204 were fitted with anti-handling devices; (c) of which 2319 were tripwired; (d) of which 15922 were improvised; (e) of which 1026 were improvised; and (f) total includes all captured types and may include 408(b).

The labour force
The first strengthening of German defences in the Channel Islands began in March 1941, by construction units of the German army and navy. Fortress Engineer Staffs arrived in July 1941 to plan conversion of the Channel Islands into impregnable fortresses, and roles were defined for Fortress Construction Battalions (Figure 29) and associated units. Fortress Engineer Staff 14 was established on Jersey (Figure 30), and the strength of the Engineers under its command enhanced following Hitler's order of 15th December 1941 for the construction of the Atlantic Wall (Ginns, 1994).

The major workforce, however, on Jersey as elsewhere along the Atlantic Wall, was provided by the Organisation Todt. This had its origins in the workforce gathered together in 1933 by Dr Fritz Todt to construct Germany's motorway system. In 1938 it was deployed to assist Fortress Engineers in

Table 4. Tunnels (Hohlgangsanlagen) scheduled for construction on Jersey by German forces during the Second World War, and their purpose. (Modified from: Ginns, 1993)

Number	Purpose
1	Munition store I
2	Ration store I
3	Munition store II
4	Munition store III
5	Fuel store I
6	Personnel shelter I
7	Artillery reserves
8	Artillery quarters
9	Electricity works I
10	Ration store II
11	Personnel shelter II
12	Fuel store II
13	Munition store IV
14	Fuel store III
15	unknown
16-18	nothing known
19	Electricity works II

construction of the West Wall along the German border with France, and during the early years of the war it became a uniformed organisation auxiliary to the German army. Todt himself visited Jersey in 1941, and by mid 1942 his organisation had an established strength of 1,621 German and 4,265 foreign workers on the island. The 'foreign' workers included volunteers from occupied countries such as Belgium and Holland, large numbers of forced labourers (French North Africans, Spaniards who had fled to France after the Spanish Civil War, and especially Russians), plus some local recruits and citizens of the (neutral) Republic of Ireland. Numbers increased slightly to a peak in 1943, but decreased thereafter as with change in the fortunes of war construction effort was diverted to Germany and to fortification of the Franco-Italian border.

To accommodate Organisation Todt workers, some use was made of existing buildings but 14 hutted camps were developed conveniently close to fortification construction sites.

Aggregates
Construction of the Atlantic Wall required very large quantities of concrete (Figure 31). On Jersey cement had to be imported from France, since there were

FORTIFICATION OF ISLAND TERRAIN 293

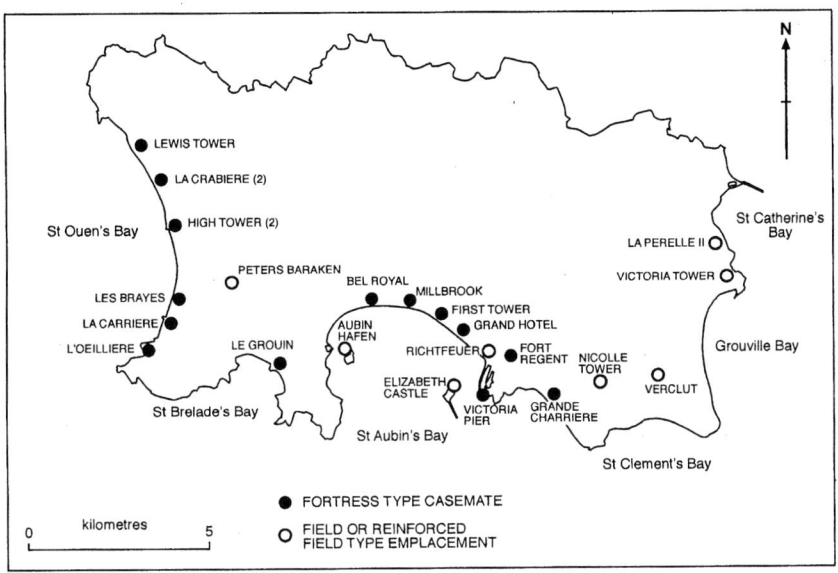

Figure 21. Map of Jersey showing deployment of 47 mm Pak K36(t) anti-tank guns, and casemate type, as at 1945. (Modified from: Channel Islands Occupation Society Jersey leaflet 'Resistance Point Millbrook')

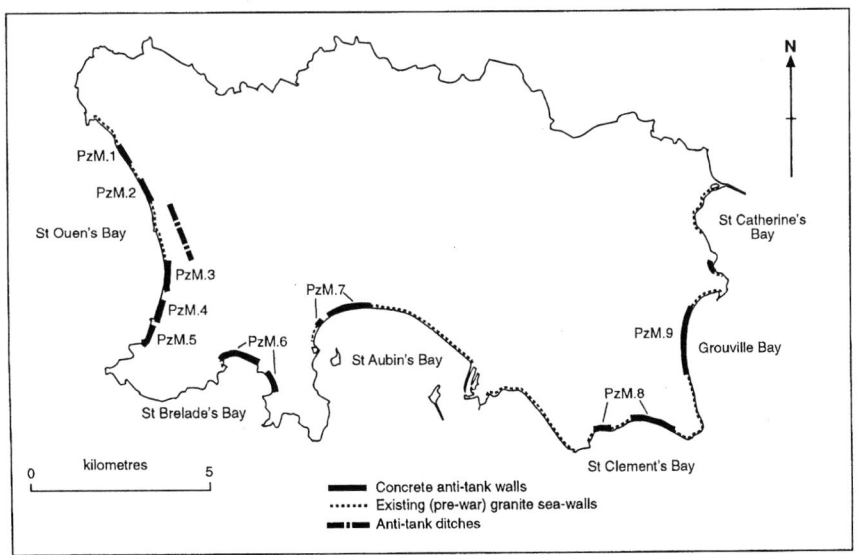

Figure 22. Map of Jersey showing sites of pre-war granite-block sea walls and of concrete anti-tank walls of German construction (after Ginns, 1974), plus anti-tank ditches inland from St. Ouen's Bay. (Modified from: Baker, undated)

Figure 23. Concrete anti-tank wall constructed by the Germans at the south end of St. Ouen's Bay, viewed from the south, with part of the stone-built sea wall in the foreground. Note the wide flat beach, shown here at low tide, amenable to amphibious landing. Jersey beaches all have a low-high tidal range of many metres

no limestones in the Channel Islands to permit its local production. However, the other essential ingredient, aggregate, was readily available from two sources:

1. *Sand and gravel from local beach areas.* A coarse sand deemed particularly suitable for use in concrete was excavated from the 2 km beach area in Grouville Bay (Ginns, 1973, 1978). About a million tonnes of sand were removed, initially by road but from early 1942 by narrow gauge railway laid specifically for the purpose (Figure 32). Smaller quantities of beach sand were extracted at St. Brelade's Bay and also at First Tower near St. Helier, for local use and onward transportation respectively. Beach gravels from St. Catherine's Bay were used for field fortifications in the spring of 1941, before implementation of the Atlantic Wall policy provided the manpower and equipment necessary for quarrying. Local beach gravels were also used to construct anti-tank walls in parts of St. Ouen's and St. Brelade's bays — but the smooth sea-worn pebbles have not keyed themselves into the concrete as satisfactorily as crushed stone, and these areas of wall have deteriorated more quickly than those built using crushed stone aggregate.
2. *Crushed stone.* This was obtained from eight quarry sites (Figure 31; Ginns, 1978). The existing quarry at La Crête, working rhyolite from the Jersey Volcanic Group, was one of the first to be taken over for German use but closed in August 1942 following mechanical breakdown and lack of spare parts for its operating equipment. The former quarry at Ronez, yielding

diorite and gabbro from the northern complex, was reactivated from early 1942 to late 1943 (and again after the war (Figure 33). The adjacent quarries of L'Etacq (producing white granite) and La Thiebaud (in Brioverian shale; Figure 34) were taken over in 1942. La Carrière (producing red granite), dormant for many years, was reactivated in 1941 to facilitate fortification in the St. Ouen's Bay area. Two quarries near Gigoulande both produced granite. The old quarry at Les Maltières, producing micro-granite as well as some diorite and dolerite, was reactivated early in the summer of 1942 and remained open until May 1945, partly because its products could be moved onward by sea from Gorey Pier as well as overland by rail, and partly because fortification nearby in the east of the island was delayed relative to that in the west.

When quarrying ceased in the west of the island, some use was made of spoil from tunnel excavation as a source of aggregate. About 14,000 tonnes of Brioverian shale were available from Ho 8 alone, but this material seems to have been dumped as waste largely on the adjacent valley floor. In contrast, spoil from Ho 1 and Ho 2 was crushed on site, and used as an aggregate both during the war and afterwards.

Water
During the German occupation, demand for water was increased by its use in concrete production, and by the need to supply a construction workforce of up to 6,000 men and a garrison of up to 11,500 troops in addition to the civilian population.

Currently, a third of total water requirements are obtained from groundwater sources, the rest from surface water streams and reservoirs (McCartney & Houghton-Carr, 1998). Jersey's hydrogeology and hydrochemistry have been subject to particularly detailed postwar study (e.g. Robbins & Smedley, 1991, 1994) and are depicted on a recent 1:25,000 scale map (British Geological Survey, 1992). Such detailed information was not available at the time of the German occupation. In consequence, concrete was often made with seawater rather than freshwater despite risk to long-term stability from included salt, and a geologist was deployed to monitor and develop groundwater resources.

7. Geology and terrain use

Siting of British fortifications was determined largely by topographic features and the range and effectiveness of contemporary weapons. These factors too were to influence the much more intensive German fortification of the island, but to ensure best use of ground the Germans deployed a new resource — a geologist.

The Allies made use of military geologists in both world wars, but only in small numbers. The British army, despite a long-term association with

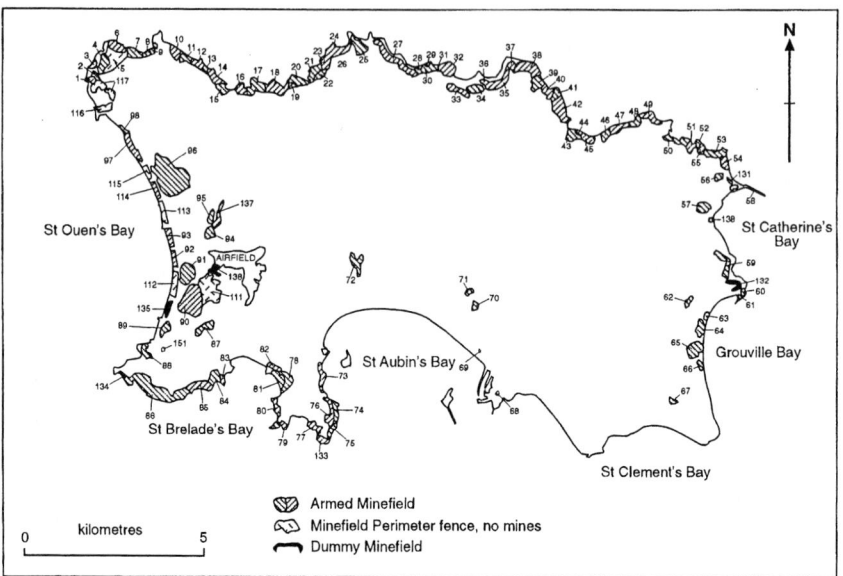

Figure 24. Map of Jersey showing German minefields as at 1945, numbered from Grosnez in clockwise sequence as in Royal Engineer records. (Modified from: Gander, 1991)

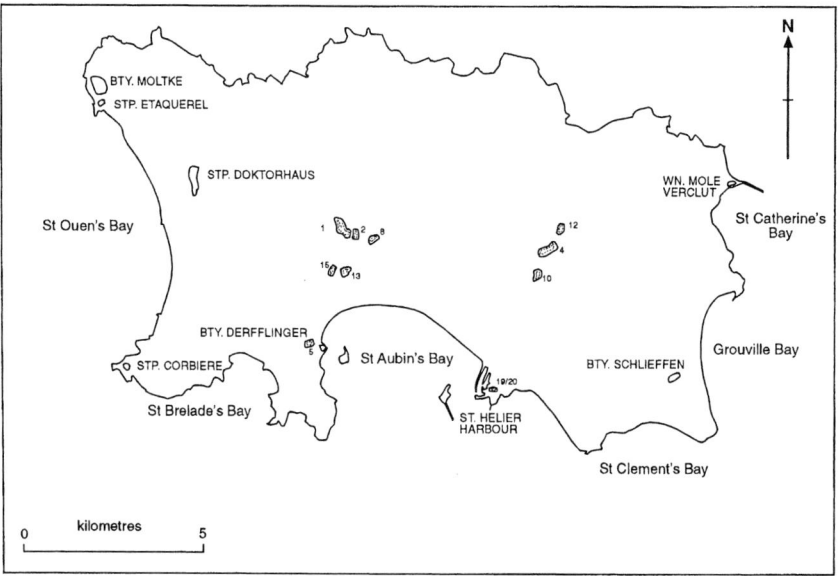

Figure 25. Map of Jersey showing positions of storage tunnels (numbered as for Table 4) and defensive tunnels (named) wholly or partly completed by 1945. (Modified from: Ginns, 1993)

geology through the 19th century (Rose, 1996, 1997) made significant use of only two military geologists during the First World War (Rose & Rosenbaum, 1993a); of only a few more during the European, African and Far Eastern campaigns of the Second World War (Rose & Rosenbaum, 1993b); and of no more than seven reserve army (and therefore part-time) military geologists in any year of the following Cold War (Rose & Hughes, 1993a, 1993b, 1993c). The American army deployed up to nine military geologists in the First World War, but too late to be of much operational use (Brooks, 1920); supported operations in the Second World War with a Military Geology Unit staffed by less than 90 geologists, largely based in the United States (Hunt, 1950; Terman, 1998a); and during the early years of the Cold War (1945 to 1972) made use of only about 150 geologists in total through a Military Geology Branch of the United States Geological Survey (Terman, 1998b).

In contrast, the German army made far more extensive use of military geologists (Rose et al., 2000). In the First World War, a military geological organisation was progressively developed from 1915 to comprise some 250 geologists by the close of hostilities in 1918. Germany re-founded a military geological organisation in 1937, and had five military geological textbooks in use by 1938 (Wilser, 1921; Wasmund, 1937; von Bülow et al., 1938; Mordziol, 1938; Kranz, 1938). In the Second World War, the German army had developed 32 military geology teams (Wehrgeologenstellen) by 1941, and 40 such teams by 1943 (Rose, 1980; Häusler, 1995; 2000; Häusler & Willig, 2000). Jersey provides a case-history of how such geotechnical expertise was put to military use — and on British terrain.

Although a military geologist (Dr Dieter Hoenes) from the German military geological organisation as such was deployed only briefly in the Channel Islands (in 1942; Häusler, 1995), and only on Guernsey and Alderney (Bishop & Launert, 1977), Jersey had a resident military geologist on its Fortress Engineer Staff from at least August 1941 until 1944, seemingly until the time of the Allied invasion of Normandy in June of that year — Lieutenant (later Captain) Walther Klüpfel (Figure 35). Klüpfel's personal notebooks are now preserved in the archives of the British Geological Survey, Keyworth, and much of his work on Jersey has been documented by Bishop & Launert (1977, 1979), providing a uniquely detailed record of the work of a German military geologist over a three-year timespan.

Like many of the German geologists put to military use in the Second World War, Klüpfel had considerable practical experience. Born on 28th May 1888, he was already 53 years of age at the time of his call-up for military service on 31st March 1941. He had graduated with a doctorate in geology from the University of Strasbourg in 1914, prior to wartime service as a military geologist during which he received the Iron Cross, Second Class, for work in water supply to a section of the front line. After the war he held several industrial posts in Germany and abroad before obtaining a non-established lectureship at the University of Giessen, at which time he supported himself largely by consulting work for the quarrying industry.

Figure 26. Part of an unfinished tunnel in the German Underground Hospital, excavated through late Precambrian (Brioverian) turbidites, with original timber roof supports. Dummy figures indicate working conditions. (Reproduced from: Rosenbaum & Rose, 1992, courtesy of Blackwell Publications; © German Underground Hospital)

Called-up as a second lieutenant, he was soon posted to Normandy, arriving there on 2nd April to head a small cartographic unit. In June 1941 Hitler decreed that the Channel Island garrison was to be increased to a full division, and issued two programmes for the Islands' fortification, one to last four months and the other seven years. In consequence, 319 Infantry Division, to which Klüpfel was attached, was ordered to relieve units of 216 Infantry Division which had occupied the islands for most of the time since the invasion on 1st July 1940. Klüpfel's movement order to Jersey was dated 27th July 1941, and his notebooks record that his fieldwork had begun by 22nd August.

Klüpfel's first tasks were concerned with the preparation of a defence report which was to include maps illustrating the geology of the island, its building material, mineral resources, and water supply. Initial priorities were the selection of sites for gun emplacements, observation towers, strongpoints, headquarter positions, and storage areas. Secondary priorities were then to determine sources of construction materials (aggregate, sand, cement, and water) and select sites for accommodation of the workforce.

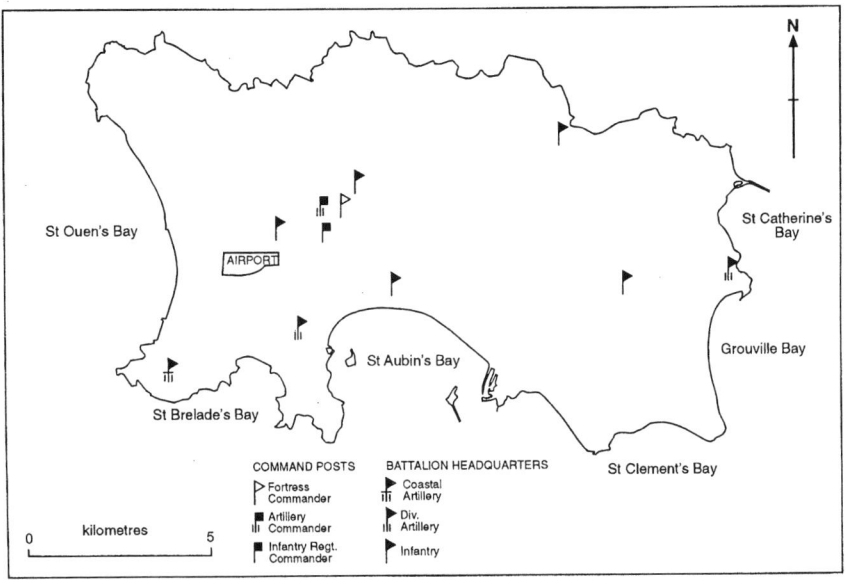

Figure 27. Map of Jersey showing German command/control centres as at 1943 (Modified from: Ginns & Bryans, 1978; Baker, undated). The fortress commander was sited adjacent to artillery and infantry commanders within a command area known as the Kernwerk, northeast of the airport

Figure 28. The Noirmont Point Command Bunker, largely concealed beneath ground, viewed from the west. The camouflaged range-finder periscope is visible, and the twin armoured observation turrets. A two-storey building with seventeen rooms in total lies below ground, its roof and outer walls of reinforced concrete >2 m thick

Klüpfel was able to compile a geological database from published maps and literature available in France, supplemented by his own field studies and by discussions with a local geological enthusiast, Micia Casimir, whose delight in finding a kindred academic enthusiasm in the war-isolated island outweighed the convention not to associate with an enemy. Their geological correspondence was to continue postwar, until the then Professor Klüpfel died on 16th September 1964.

Siting of surface installations on Jersey posed few problems: the 'hard rock' geology gave strong foundations except where concealed beneath a significant thickness of Quaternary cover, for soils were generally less than a metre in thickness. Tunnelling posed more of a problem, and Klüpfel was to visit virtually every tunnel site, although Ho 1 and Ho 8 were to occupy most of his time in this respect. Ho 1 was being constructed through three different rock types, separated by faults with displacements calculated as 100 m or more, and Klüpfel was influential in routing part of the tunnel system into sound rather than fractured rock. Ho 8 had similarly encountered faulted, disturbed rock in part of the tunnel area, and Klüpfel recommended development in sounder rock to the south rather than faulted rock to the west.

Field fortification had begun before Klüpfel's arrival on Jersey, and a shipload of sand and gravel arrived from France on 21st March 1941 to facilitate construction. However, it was soon realised that such imports were unnecessary. By October and November 1941 Klüpfel was making careful surveys of the sand and gravel resources of the beaches (Figure 32). He recorded profiles for the larger beaches; particle sizes for both pebbles and sand; and rock types present among the pebbles. His estimated reserves for St. Ouen's beach were 10,000 m^3 of gravel, 20,000 m^3 of sand; Grouville 15 000 m^3 gravel, 30 000 m^3 sand; Anne Port 2,000 m^3 gravel, 5,000 m^3 sand; St. Catherine's 3,000 m^3 gravel, 5,000 m^3 sand. A geological reason for the development of Grouville beach as a major resource area is therefore evident.

Klüpfel also visited the quarries on the island, although the exact reasons for his visits have not been recorded. The 'hard rock' geology of the island provided stone suitable for crushing as aggregate almost everywhere. Problems were not those of locating suitable stone, merely of installing crushing equipment and providing transport adequate to cope with demand.

Potable water was much more of a problem on a small 'hard rock' island. Provision of adequate water supply was one of the major factors taken into consideration in siting camps for the Organisation Todt workforce, and Klüpfel was involved in related field studies and planning discussions through much of 1942. He was also involved in the provision of secure water supplies to the many military sites on the island. Some had significant numbers of permanently-based troops. For example, a record of 5th May 1942 shows that 40 m^3 of water per day were required for the 160 men at the La Moye Battery. Generally, some supplies were extracted from existing wells on the island (whose water levels and extraction rates were regularly monitored); some were surface supplies provided by streams and reservoirs;

FORTIFICATION OF ISLAND TERRAIN 301

Figure 29. Construction Battalion at work in St. Ouen's Bay (Modified from: Ginns, 1994). The plateau typical of the Jersey terrain is clearly visible in the background, and the steep cliffs which fringe much of the coastline

Figure 30. Officers of Fortress Engineer Staff 14, on Jersey in 1942. The geologist Walther Klüpfel is standing far left. (Reproduced from: Bishop & Launert, 1977, 1979; photo courtesy of the British Geological Survey)

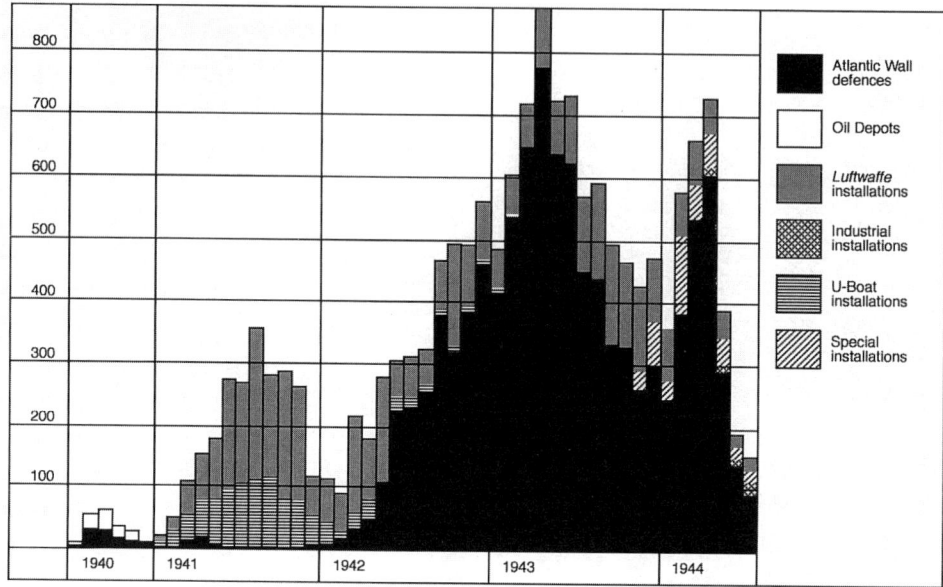

Figure 31. Monthly concrete output (per thousand m^3) for Organisation Todt, Einsatzgruppe West, from July 1940 to July 1944, to indicate large quantities required for construction of the Atlantic Wall. (Modified from: Partridge, 1976)

others were site-specific supplies for which new boreholes had to be drilled, on geological advice.

8. Conclusions

The adjacent Channel Island of Guernsey (65 km²) was large enough to have merited similar British fortification since medieval times, major construction work during the German occupation, and the deployment of German military geologist expertise (Dr D. Hoenes). Even the smaller island of Alderney (8 km²) had a chain of British 19th century forts flanking its northern and eastern coasts which were strengthened by the Germans. Naval and army batteries of coastal artillery (Figure 13), numerous infantry strongpoints and resistance nests, and anti-aircraft batteries were emplaced as on the larger islands (for map, see Partridge, 1976, p. 136), and Dieter Hoenes provided a geological appraisal. Each of the three islands could be supplied by sea via at least one major harbour, and by air through an airfield. The other Channel Islands (Sark 5 km², and the smaller islands) were too small to support either a major harbour or an airfield, or to merit significant fortification by either the British or the Germans.

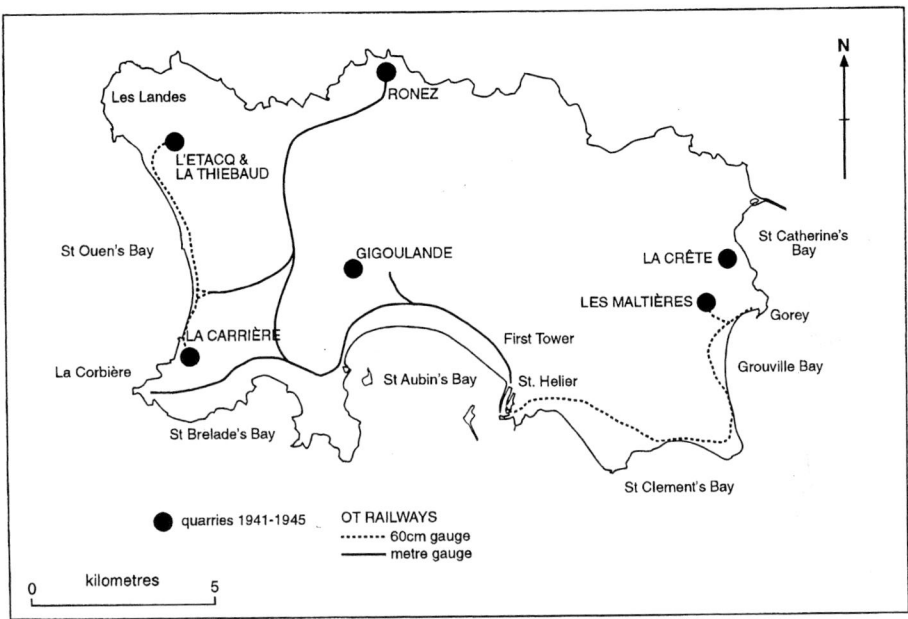

Figure 32. Map of Jersey showing sites of beaches and quarries used by the Germans as sources of aggregate, and narrow gauge railway systems developed to transport these materials. (Modified from: Ginns, 1978)

There are striking differences between Jersey and its supposed British equivalent as a fortress outpost, Gibraltar. The most obvious differences are in size (Gibraltar is barely 6 km^2 in land area relative to Jersey's 116 km^2), topography (the Rock of Gibraltar forms a sharply-ridged crest with peaks over 400 m high, contrasting with the plateau < 150 m high of Jersey), and degree of isolation (Gibraltar is a peninsula rather than an island, connected to the adjacent mainland by a low, sandy isthmus). There are, however, major differences in geology. The bedrock of Gibraltar is primarily dolomitic limestone (Rose, 1998), of relatively young (Jurassic) age, whereas the bedrock of Jersey contains no limestone at all, but comprises a complex of much older and stronger rocks, primarily igneous rocks and metamorphosed clastic sediments. Moreover, whereas the overlying Quaternary (drift) deposits are volumetrically significant on Gibraltar but there are only small amounts of contemporary beach sands and gravels, the converse is true of Jersey.

That said, both Jersey and Gibraltar have been progressively fortified from medieval times, the fortifications culminating in a period of massive construction works during the Second World War. In both areas steep coastal cliffs formed a major but incomplete topographic barrier to amphibious attack, and these natural defences were enhanced by numerous batteries of coastal artillery. In both areas tunnelling was developed as a means of locating stores and personnel safe from aerial attack — although tunnelling

Figure 33. Ronez Quarry (cf. Figure 32) in 1998, still in operation, viewed from the west

Figure 34. La Thiebaud quarry (cf. Figure 32), in operation during 1943, showing the diesel locomotive and narrow-gauge railway used to transport aggregate. (Courtesy of the Channel Islands Occupation Society)

was far more extensive on Gibraltar because its surface area was smaller, much of it steeply sloping, and the limestone bedrock easier to excavate (*cf.* Rosenbaum & Rose, 1992). Both areas were supplied by sea via a major harbour and by air via a militarily-controlled airport. Both the British in Gibraltar and the Germans in Jersey during the Second World War made use of military geologist expertise to facilitate their fortification of the terrain. The fortification of Gibraltar has already been documented in considerable detail (e.g. by Hughes & Migos, 1995), as has the role of Royal Engineer geologists on the Rock (Rose & Rosenbaum, 1990; Rose, 2000, in press b). This paper provides a comparative summary of fortification and military geologist activities on Jersey, hitherto documented mainly in a variety of publications local to the island (e.g. those by Ginns, between 1973 and 1998).

Acknowledgements

We are grateful to G. McKenna, Chief Librarian and Archivist of the British Geological Survey, for providing access to Walther Klüpfel's notebooks and associated papers preserved in the Survey's archives at Keyworth, Nottingham; to M.M. Billot, Librarian of the Société Jersiaise for access to the 'Green Books', 14 volumes of plans of German fortifications and island defences captured by British forces when the island was liberated in May 1945, plus a volume compiled by the liberating force, and presented to the Société by the War Office in October 1947; to N. Wilson of the Department of Geology, Royal Holloway, for all artwork in this paper; to copyright owners as acknowledged in individual captions for permission to reproduce specified illustrations; and to T.E. Eastler for kindness in refereeing the submitted manuscript. Photographs are by the senior author except where indicated, printed by K. D'Souza at Royal Holloway. EPFR also acknowledges helpful provision of background data by Pam Nisbet.

References

Baker, H.B. Undated. *The German occupation of Jersey 1940/1045. Reference maps with supporting text and comprehensive history.* Jersey. 1 sheet (2 sides).

Balleine, G.R. 1950. *A history of the island of Jersey.* Staples, London.

Bishop, A.C. & Bisson, G. 1989. *Classical Areas of British Geology: Jersey: description of 1:25,000 Channel Islands Sheet 2.* HMSO for British Geological Survey, London.

Bishop, A.C. & Keen, D.H. 1982. *Jersey: Channel Islands Sheet 2.* 1:25,000 Map Series, Institute of Geological Sciences, London.

Bishop, A.C. & Launert, E. 1977. Geology in Jersey during the Occupation. *Annual Bulletin of the Société Jersiaise* 22, 51-60.

Bishop, A.C. & Launert, E. 1979. A Wehrmacht geologist in Jersey. *Channel Islands Occupation Review* 7, 30-37.

Blackie, J.R. & Jones, T.K. 1993. *Estimates of open water evaporation and of potential transpiration in Jersey.* Institute of Hydrology Technical Report, Wallingford.

British Geological Survey. 1992. *Hydrogeological Map of Jersey.* 1:25,000. British Geological Survey, Keyworth.

Brooks, A.H. 1920. The use of geology on the Western Front: *United States Geological Survey Professional Paper* 128-D, 85-124.

Brown, M. Undated. *Guide to Mont Orgueil Castle.* The Guernsey Press, Guernsey.

Bülow, K. von, Kranz, W. & Sonne, E. 1938. *Wehrgeologie.* Quelle & Meyer, Leipzig.

Davies, W. 1971. *Fort Regent: a History.* Private publication, St. Helier, Jersey.

Gander, T. 1991. Land mine warfare in the Channel Islands. *Channel Islands Occupation Review* 19, 13-31.

Ginns, [W.] M. 1973. Grouville Common during the German occupation. *Annual Bulletin of the Société Jersiaise* 21, 194-199.

Ginns, [W.] M. 1974. Anti-tank walls in Jersey. *Channel Islands Occupation Review* 2, 57-66.

Ginns, [W.] M. 1978. Wartime quarries in Jersey. *Channel Islands Occupation Review* 6, 24-34.

Ginns, [W.] M. 1989. Coastal artillery observation posts. *Channel Islands Occupation Review* 17, 12-37.

Ginns, [W.] M. 1993. *German Tunnels in the Channel Islands. Archive Book No. 7.* Channel Islands Occupation Society, Jersey.

Ginns, [W.] M. 1994. *The Organisation Todt and the Fortress Engineers in the Channel Islands. Archive Book No. 8.* Channel Islands Occupation Society, Jersey.

Ginns, [W.] M. 1999. *Jersey's German bunkers. Archive Book No. 9.* Channel Islands Occupation Society, Jersey.

Ginns, [W.] M. & Bryans, P. 1978. *German Fortifications in Jersey.* Meadowbank, Jersey.

Grimsley, E.J. 1988. *The Historical Development of the Martello Tower in the Channel Islands.* Sarnian Publications, Guernsey.

Häusler, H. 1995. Die Wehrgeologie im Rahmen der Deutschen Wehrmacht und Kriegswirtschaft. *Informationen des Militärischen Geo-Dienstes, Bundesministerium für Landesverteidigung,* Wien, 47, 1-155; 48, 1-119.

Häusler, H. 2000. Deployment and role of military geology teams in the German army 1941-45. *In:* Rose, E.P.F. & Nathanail, C.P. (Eds), *Geology and Warfare: Examples of the Influence of Terrain and Geologists on Military Operations.* Geological Society, London, 159-175.

Häusler, H. & Willig, D. 2000 Development of military geology in the German Wehrmacht 1939-45. *In:* Rose, E.P.F. & Nathanail, C.P. (Eds),

Geology and Warfare: Examples of the Influence of Terrain and Geologists on Military Operations. Geological Society, London, 141-158.

Hughes, Q. & Migos, A. 1995. *Strong as the Rock of Gibraltar.* Exchange Publications, Gibraltar.

Hunt, C.B. 1950. Military geology. *In:* Paige, S. (Ed.), *Application of Geology to Engineering Practice. Berkey Volume.* Geological Society of America, Boulder, Colorado, 295-327.

Jones, R.L., Keen, D.H., Birnie, J.F. & Waton, P.V. 1990. *Past Landscapes of Jersey: Environmental Changes During the Last Ten Thousand Years.* Société Jersiaise, Jersey.

Keen, D.H. 1978. The Pleistocene deposits of the Channel Islands. *Report of the Institute of Geological Sciences* 78/26, 1-14.

Keen, D.H. 1981. The Holocene deposits of the Channel Islands. *Report of the Institute of Geological Sciences* 81/10, 1-13.

Keen, D.H. (Ed.) 1993. *The Quaternary of Jersey: Field Guide.* Quaternary Research Association, London.

Kranz, W. 1938. *Technische Wehrgeologie. Wegweiser für Soldaten, Geologen, Techniker, Ärzte, Chemiker und andere Fachleute.* Jänecke, Leipzig.

MacCulloch, J. 1811. Account of Guernsey and the other Channel Islands. *Transactions of the Geological Society of London* 1, 1-22.

McCartney, M.P. & Houghton-Carr, H.A. 1998. An assessment of groundwater recharge on the Channel Island of Jersey. *Journal of the Chartered Institute of Water and Environmental Management* 12, 445-451.

Miller, B.V., Samson, S.D. & D'Lemos, R.S. 2001. U-Pb geochronological constraints on the timing of plutonism, volcanism and sedimentation, Jersey, Channel Islands, UK. *Journal of the Geological Society* 158, 243-252.

Mordziol, C. 1938. *Einführung in die Wehrgeologie.* Salle, Frankfurt am Main.

Nelson, R.J. 1830. Geological survey of the island of Jersey. *Quarterly Journal of Science, Literature and Arts,* new series 6, 359-378.

Partridge, C. [W.] 1976. *Hitler's Atlantic Wall.* D.I. Publications, Guernsey.

Partridge, C.[W.] & Wallbridge, J. 1983. *'Mirus' The Making of a Battery.* The Ampersand Press, Alderney.

Pocock, H.R.S. 1971. Jersey's Martello Towers. *Annual Bulletin of the Société Jersiaise* 20, 289-298.

Prince, S.J. 1997. The geology of the Channel Islands - an introductory resource. *British Geological Survey Technical Report* WO/97/2.

Ramsey, W.G. 1981. *The war in the Channel Islands: Then and Now.* Battle of Britain Prints International Limited, London

Renouf, J. [T.] 1985. Geological excursion guide 1. Jersey and Guernsey, Channel Islands. *Geology Today* 1, 90-93.

Robbins, N.S. & Smedley, P.L. 1991. Hydrogeological and hydrochemical survey of Jersey. *British Geological Survey, Hydrology Series, Technical Report,* WD/91/15.

Robbins, N.S. & Smedley, P.L. 1994. Hydrogeology and hydrochemistry of a small, hard-rock island - the heavily stressed aquifer of Jersey. *Journal of Hydrology* 163, 249-269.

Rolf, R. 1988. *Atlantic Wall typology*. AMA, Beetserwaag, Holland.

Rose, E.P.F. 1980. German military geologists in the Second World War. *Royal Engineers Journal* 94, 14-16.

Rose, E.P.F. 1996. Geologists and the army in nineteenth century Britain: a scientific and educational symbiosis? *Proceedings of the Geologists' Association* 107, 129-141.

Rose, E.P.F. 1997. Geological training for British army officers: a long-lost cause? *Royal Engineers Journal* 111, 23-29.

Rose, E.P.F. 1998. Environmental geology of Gibraltar: living with limited resources. *In:* Bennett, M.R. & Doyle, P. (Eds), *Issues in Environmental Geology: a British Perspective*. Geological Society, London, 81-121.

Rose, E.P.F. 2000. Geology and the fortress of Gibraltar. *In:* Rose, E.P.F. & Nathanail, C.P. (Eds), *Geology and Warfare: Examples of the Influence of Terrain and Geologists on Military Operations*. Geological Society, London, 236-274.

Rose, E.P.F. in press a. Nelson, John Richard (1803-1877). *In:* Matthew, H.G.C. & Harrison, B. (Eds) *New Dictionary of National Biography*. Oxford University Press, Oxford.

Rose, E.P.F. in press b. Military engineering on the Rock of Gibraltar and its geoenvironmental legacy. *In:* Ehlen, J. & Harmon, R. S. (Eds), *The Environmental Legacy of Military Operations*. Reviews in Engineering Geology XIV. Geological Society of America, Boulder, Colorado

Rose E.P.F., Häusler, H. & Willig, D. 2000. A comparison of British and German military applications of geology in World War. *In:* Rose, E.P.F. & Nathanail, C.P. (Eds), *Geology and Warfare: Examples of the Influence of Terrain and Geologists on Military Operations*. Geological Society, London, 107-140.

Rose, E.P.F. & Hughes, N.F., 1993a. Sapper Geology: Part 1. Lessons learnt from world war. *Royal Engineers Journal* 107, 27-33.

Rose, E.P.F. & Hughes, N.F., 1993b. Sapper Geology: Part 2. Geologist pools in the reserve army. *Royal Engineers Journal* 107, 173-181.

Rose, E.P.F. & Hughes, N.F., 1993c. Sapper Geology: Part 3. Engineer Specialist Pool geologists. *Royal Engineers Journal* 107, 306-316.

Rose, E.P.F. & Pareyn, C. 1998. British applications of military geology for 'Operation Overlord' and the battle in Normandy, France, 1944. *In:* Underwood, J.R., Jr., & Guth, P.L. (Eds), *Military Geology in War and Peace. Reviews in Engineering Geology XIII*. Geological Society of America, Boulder, Colorado, 55-66.

Rose, E.P.F. & Rosenbaum, M.S. 1990. *Royal Engineer Geologists and the Geology of Gibraltar*. Gibraltar Museum, Gibraltar.

Rose, E.P.F. & Rosenbaum, M.S. 1993a. British military geologists: the formative years to the end of the First World War. *Proceedings of the Geologists' Association* 104, 41-49.

Rose, E.P.F. & Rosenbaum, M.S. 1993b. British military geologists: through the Second World War to the end of the Cold War. *Proceedings of the Geologists' Association* 104, 95-108.

Rosenbaum, M.S. & Rose, E.P.F. 1992. Geology and military tunnels. *Geology Today* 8, 92-98.

Rybot, N.L.V. 1978. *Gorey Castle*, 2nd Edition. States of Jersey, Jersey.

Rybot, N.L.V. 1986. *The Islet of St. Helier and Elizabeth Castle.* 9th Edition. Société Jersiaise, Jersey.

Terman, M. 1998a. Military Geology Unit of the U.S. Geological Survey during World War II. *In:* Underwood, J.R., Jr., & Guth, P.L. (Eds), *Military Geology in War and Peace. Reviews in Engineering Geology XIII.* Geological Society of America, Boulder, Colorado, 49-54.

Terman, M. 1998b. Military Geology Branch of the U.S. Geological Survey from 1945-1972. *In:* Underwood, J.R., Jr., & Guth, P.L. (Eds), *Military Geology in War and Peace. Reviews in Engineering Geology XIII.* Geological Society of America, Boulder, Colorado, 75-81.

Wasmund, E. 1937. *Wehrgeologie in ihrer Bedeutung für die Landesverteidigung.* Mittler & Sohn, Berlin.

Went, D. & Andrews, M. 1990. Post-Cadomian erosion, deposition and basin development in the Channel Islands and northern Brittany. *In:* D'Lemos, R.S., Strachan, R.S. & Topley, C.G. (Eds), The Cadomian Orogeny. *Special Publications of the Geological Society, London,* 51: 293-304.

Wilser, G. 1921. *Grundriss der angewandten Geologie unter Berücksichtigung der Kriegserfahrungen für Geologen und Techniker.* Borntraeger, Berlin.

Wood, A. & Wood, M.S. 1955. (New edition 1976). *Islands in Danger.* New English Library, Sevenoaks.

Young, P. 1973. *Atlas of the Second World War.* Weidendenfeld & Nicholson, London.

Edward P.F. Rose
Department of Geology
Royal Holloway, University of London
Egham, Surrey, TW20 0EX

W. Michael Ginns
Channel Islands Occupation Society (Jersey)
'Les Geonnais de Bas', Rue des Geonnais
St. Ouen
Jersey, JE3 2BS

John T. Renouf
Le Côtill des Pelles
Petit Port
St. Brelade
Jersey, JE3 4TB

Piracy on the High Desert: the Long-Range Desert Group 1940-1943

James R. Underwood, Jr. & Robert F. Giegengack

ABSTRACT: From June 1940, when General Sir Archibald Wavell, Commander in Chief of British Forces, Middle East, authorised then Major Ralph Bagnold to form the Long Range Desert Group (LRDG), until near the end of the North African campaign in April 1943, the highly mobile, self-contained desert patrols of the LRDG maintained close watch on the movement of enemy troops and vehicles and attacked enemy military installations hundreds of kilometres behind their lines. This spirited 'piracy on the high desert' not only cost the Axis forces dearly in men and equipment, but tied up great numbers of troops and large quantities of arms and supplies to protect Axis positions from the threat of slashing hit-and-run attacks of LRDG patrols that seemed to materialise out of the desert suddenly and without warning. Key to the success of the LRDG was its skilful and imaginative utilisation of the terrain of the Libyan Desert, based on the experience and insight gained by Bagnold and a few close associates during their private expeditions in the region during pre-war years.

1. Introduction

The Libyan Desert (Figure 1) includes the Western Desert of Egypt, that part of Libya south of the coastal ranges, northeast Chad, and northwest Sudan, and is comparable in size to the peninsula of India (Bagnold, 1935). Far south, such as at Aswan, daily temperature differences of as much as 33°C are common (Tignor, 2000), and rainfall at any one locality in the deep desert may occur only at intervals of decades. Conditions are so severe in most of the region that they preclude the existence even of wandering Bedouin and their herds of camels and sheep; vegetation on which to graze is virtually nonexistent. Inhabitants of the scattered oases travel between them only on well-known desert tracks.

The topographic relief of the region in which most LRDG operations took place is not great, perhaps averaging only 300 m; the common geomorphical features are sand sheets or plains, sand dunes, and bedrock buttes, mesas and plateaux, all landforms common to arid regions with nearly flat-lying stratified rocks. Although rainfall is rare, surface runoff in the geological past has been responsible for carving the major elements of the landscape. In

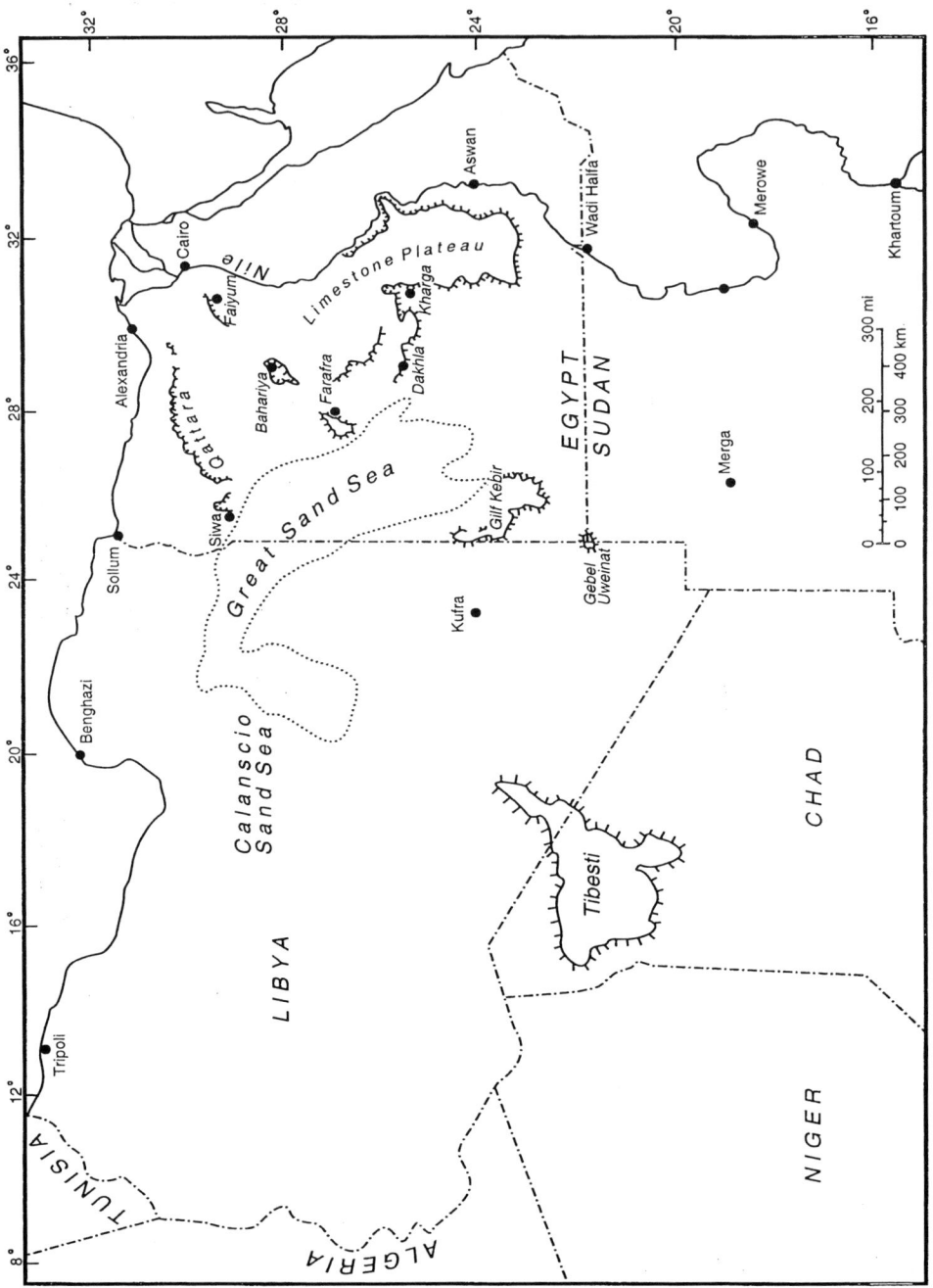

Figure 1. Map of the northeast Sahara showing the Libyan Desert, which includes western Egypt, eastern Libya, northeastern Chad, and northwestern Sudan

the current climatic regime, stream valleys carry runoff sporadically and only close to the Mediterranean during the brief season of winter rainfall.

In Libya, this formidable desert restricted most vehicular traffic to the Via Balbia, the principal east-west route connecting Tripoli and Benghazi and extending eastward to the Egyptian-Libyan border. The Via Balbia, one of the few paved roads in Libya at the time, lay within the green coastal strip, a narrow belt of vegetation and cultivation a few kilometres to a few tens of kilomtres wide. This strip was created by Italian colonialists during their occupation of Libya from 1911 until their eviction by the Allied forces in March 1943. South of the coastal green belt lies the emptiness of the northeast Sahara, where widely scattered oases are separated by vast areas of rock and sand, and are connected only by tracks across the intervening desert. Few Europeans, including the occupying Italians, knew much of the deep desert or felt at home in it, and it was this barren relatively unknown emptiness that the Axis powers counted on to provide them a secure southern flank. Just how wrong they were was demonstrated time and again by patrols of the Long Range Desert Group (LRDG) as they attacked widely separated targets, at times on the same day (Constable, 1971). This relatively small group of men was organised and led in its first year by Major (later Brigadier) Ralph Bagnold, a Royal Engineers signals officer, who had lived and travelled extensively in the desert in the years between the two world wars.

2. Ralph Alger Bagnold

Born in Devon in 1896, graduate of the Royal Military Academy and the Military Engineering School, Ralph Alger Bagnold survived three years of trench warfare in France and Belgium as a signals officer in the Royal Engineers. Following the close of the First World War in 1918, Bagnold took an educational leave from the military, and in 1921 completed a second-class honours degree in engineering at Cambridge. Back on active duty, he had assignments in Ireland (1921-1923); Signals Training Centre, Sussex (1923-1925); Egypt (1926-1928); India (1928-1931); School of Signals, Catterick (1931-1933); and Hong Kong (1933-1934). Ill health forced his return to England in 1934 and retirement from the army in 1935 (Bagnold, 1990).

During his stay in Egypt, Bagnold, and a number of associates who had become fascinated by the challenge of desert exploration, made increasingly longer trips into the dry wastes both east and west of the Nile (Bagnold, 1935). These trips were conducted on personal time using personal equipment, supplies, and vehicles (at first Ford Model Ts and later, Ford Model As), The most extensive of those journeys covered about 10,000 km. It was during these forays into the desert that Bagnold and his friends developed their skill in desert navigation and perfected various items of equipment, including: (1) a sun compass for use during daylight hours; (2) a closed system for the radiator cooling water of their vehicles; and (3) sand mats and tracks to place beneath the tyres when stuck in, or traversing, soft sand (Bagnold, 1990).

It was during those many miles of desert travel that Bagnold developed a keen interest in the transport of sand by wind and intense curiosity about the origin and evolution of the numerous and varied large and small landforms, composed of sand and created by the wind. Fortunately, his health was not as severely impaired as doctors had feared initially, and in the years between his retirement in 1934 and his recall to active duty when war was declared between Germany and Britain in 1939, he had the time and the opportunity to undertake a detailed study of the transport of sand by wind. Designing and building his own wind tunnel and the instruments required to measure the controlling parameters, and applying the basic laws of physics and his observations from his desert travels, Bagnold described in detail, both quantitatively and qualitatively, the processes and principles involved in the transport of sand by wind and the evolution of the astonishing variety of aeolian landforms he had observed in the desert (Bagnold, 1990). His classic *Physics of Blown Sand and Desert Dunes*, published in 1941 while he was on active military duty, is still in print and remains a basic reference.

3. Collision at sea

Bagnold was recalled to active duty in August 1939 and sent by troop transport to a post in East Africa. His ship was involved in a collision in the Mediterranean that required repairs in Port Said. Bagnold took the opportunity to visit friends in Cairo, and while there his presence came to the attention of General Archibald Wavell, who knew something of Bagnold's background. Following a brief meeting with him, Wavell arranged for Bagnold to be reassigned to duty with the signals unit of an armoured division stationed along the coast west of Alexandria (Bagnold, 1990). Thus did fate place Major Ralph Bagnold, the one person in the British army who not only knew the Libyan Desert well but who had mastered the techniques of living and travelling in it, in a position to contribute his unique skills and insights in the months ahead to containing the large enemy force to the west, first that of Italian Marshall Rodolfo Graziani and later that of German Field Marshall Irwin Rommel.

At the time of declaration of war between Britain and Germany, British forces in Egypt consisted only of one experimental, under-strength armoured division (Bagnold, 1990). Hoping not to provoke the Italians into a declaration of war, overt preparations in Egypt for hostilities were kept to a minimum. Upon Italy's entry into the war in June 1940, immediate concern arose about three possible routes of Italian troop movement: (1) eastward from Libya along the coast road toward Alexandria and Cairo; (2) eastward from southeast Libya to the Nile, thereby isolating the Sudan to the south; and (3) northward from Abyssinia into the Sudan (Bagnold, 1990).

4. The Long Range Desert Group

At the time of Italy's declaration of war, little was known about the Italian military preparations in Libya for action against British troops in Egypt or the Sudan. The lack of information about possible Italian military activity, especially in southern Libya, led Major Bagnold on 23rd June 1940 to propose to General Wavell, now Commander in Chief, British Troops, Middle East, that a unit similar to the Light Car Patrol of the First World War be organised (Lloyd Owen, 1981). Based on his extensive travels in the Libyan Desert, Bagnold suggested that small, self-contained motorised patrols could move westward undetected into Libya. By studying vehicle tracks connecting the Libyan oases, the patrols could determine if preparations were under way by the 15-division Italian Army for operations against Egypt or the Sudan. In answer to Wavell's question, 'If no such preparations are underway, what then do you propose to do?' Bagnold replied, 'How about a little piracy on the high desert?' That response brought a broad smile to Wavell's face as he saw immediately the possibility of creating great uncertainty and anxiety on the part of the enemy about the number, size, and location of British units operating behind the Italian lines. General Wavell requested that Bagnold's proposed desert-patrol group be ready to begin operations in six weeks, and provided him with a remarkable memorandum giving him carte blanche to obtain the necessary men and equipment to do so (Bagnold, 1990).

Leaving the meeting astounded that his modest proposal had been received by Wavell with such enthusiasm, Bagnold then remembered that Wavell '...was the leading exponent of strategic deception. Faced with the probability of being attacked in overwhelming numbers, he was going to delay the enemy by bluff until he could get reinforcements...' (Bagnold, 1990, p. 125).

At the outset, Bagnold had the good judgement to insist that the LRDG headquarters representative in Cairo be directly under the highest command directing operations in the desert. 'This was essential, because only at that level could the Long Range Desert Group be fitted into the overall strategic picture, and correct priority between offensive action and intelligence work be weighed up' (Lloyd Owen, 1981, p. 11). Further, it was necessary to co-ordinate the activities of '...deeply penetrating patrols with the long-range attacks of the Royal Air Force' (Lloyd Owen, 1981, p. 11).

It has been said that Ralph Bagnold was the only person who had the vision, the experience, and the drive to have organised, equipped, trained, and readied the LRDG for operations in six weeks. 'Some had the necessary knowledge of the army, others the necessary experience of the desert, none had both' (Kennedy Shaw, 1945, p. 22). Bagnold was an exceptional person and officer. 'He has been described by those who worked with him during this period...as tough, wiry, reticent, and undemonstrative, with frugal habits, inexhaustible stamina, and a natural modesty. He was a perfectionist and set for himself and others high standards of physical fitness, attention to detail, and sense of duty' (Underwood, 1996, p. 40).

5. Selection of personnel

Critical to the success of the proposed desert patrols would be the selection of men who were energetic, innovative, self-reliant, physically and mentally tough, and attracted to the prospect of living and fighting in small, self-contained units in the broad expanse of the Libyan Desert. Bagnold felt that New Zealand sheep farmers would be most likely to possess most of these attributes, and he obtained permission to 'borrow' selected men and officers, volunteers all, from the New Zealand division. En route to Egypt, their weapons and other equipment had fallen prey to enemy torpedoes and thus they were in a lull of activity (Bagnold, 1990). Half of the division volunteered for some 150 available billets. In addition, Bagnold arranged for a number of his pre-war companions in desert exploration to join him as key personnel in the LRDG.

The size of the group was increased from three to six patrols, then to 12. Initially, they were almost exclusively New Zealanders. They were joined later by volunteers from different military units: the Brigade of Guards, Rhodesians, Indians, and the Yeomanry regiments (Bagnold, 1990).

Figure 2. Long Range Desert Group patrol crossing stony ground in Libya, January 1941. The trucks are 1939 Chevrolets modified for operations in the desert. (Reprinted from: Bagnold, R.A. 1990. Sand, Wind, and War: Memoirs of a Desert Explorer, *courtesy of The University of Arizona Press and with the permission of Stephen Bagnold)*

6. Outfitting the patrols

The demands on the patrols and their vehicles would be great. Carrying up to a metric tonne of fuel, food, water, spare parts, tyres, weapons, and ammunition, they were to be self-contained for trips of up to three weeks over distances of 2,400-3,200 km. Each of the three initial patrols of 2 officers and 28 enlisted men was provided with one 15 cwt truck for the patrol commander, driver, and gunner and, for the other personnel, ten 30-cwt trucks, initially Chevrolets and later Fords (Figure 2). Patrols eventually were pared down to 5-6 vehicles and 15-18 men. Each of the trucks was stripped of all but the basic requirements for desert travel and subsistence and specially fitted with extra springs, map containers, sand tracks, and condensing radiator systems. The trucks were two-wheel drive vehicles, chosen because they were lighter and used less fuel than four-wheel-drive vehicles then available. Vital also to navigating in the desert were sun compasses specially designed by Bagnold, and each patrol also carried a theodolite with which to determine their location each evening by astronomical fix. Each patrol also had a designated navigator, radio operator, and medical orderly, in addition to a complement of maintenance specialists (Lloyd Owen, 1981).

The vehicles were not armoured. Armament for each patrol initially consisted of eleven Lewis machine guns, four Boys antitank rifles, and one 37-mm Bofors gun distributed among the 11 vehicles. Later experience led to some modification in the choice of the heavier weapons, and patrols employed land mines from time to time. In addition, each person was equipped with a pistol or a rifle and with hand grenades (Lloyd Owen, 1981).

Each patrol also carried a radio with which they communicated with the base camp, which, in turn, relayed information to, and received instructions from, LRDG headquarters. The radio sets proved to be very dependable, despite the severe conditions under which they operated. Signalling was done entirely by key, and great effort — frequent change of frequency, altering of call signs, and encoding — was made to maintain secure communications. Insofar as is known, no LRDG communication was ever compromised (Lloyd Owen, 1981).

Perhaps even more vital than equipment when operating in the desert was the supply of food and water rationed to the troops. Appropriate items and daily amounts had been determined and tested by Bagnold and his companions during their numerous pre-Second World War desert journeys. Their well-tested daily rations of food and water were immediately adaptable to the LRDG upon its formation and proved to be highly satisfactory. Considering the trying circumstances of living in the open for week after week and in extremes of winter and summer temperatures, Bagnold added a daily ration of rum, which had long been abolished in the British military services (Bagnold, 1990).

7. LRDG and the Libyan Desert

The immense size of the Libyan Desert, the small local population, and the lack of major landforms meant that travel in the desert was much like moving over the ocean. The patrols of the LRDG, using the sun compass, navigated during the day by dead reckoning, and each evening, when weather permitted, determined their location with an astronomical fix.

Along much of the border with Libya, travel westward from Egypt was blocked by the Great Sand Sea (Figure 1) and especially by a field of nearly north-south linear dunes. During pre-war exploration, Bagnold and his associates had identified and used a passage through this formidable barrier. The route led more or less due west from an artesian spring, Ain Dalla, that lies 70 km west northwest of Farafra Oasis (Figure 1). Although pre-war maps available in Cairo prior to the war showed this route, Bagnold counted on the Italians being unaware of the maps and the potential military significance of the route through the Sand Sea (Constable, 1971). He was correct, and the LRDG patrols made repeated use of the passage during their operations in the Libyan Desert.

Bagnold and his small cadre of experienced pre-war desert explorers taught the men of the LRDG to identify and avoid most areas of dry quicksand, and when they did become stuck, to use sand tracks and sand mats to extract their trucks and move on. With practice and using the optimum combination speed and gears, the mean learned to drive up and over the 100 m high dunes of the Great Sand Sea (Bagnold, 1990). Bagnold's men also became adept at adjusting tyre pressure on the vehicles so that relatively soft sand or rugged rock fields could be traversed with minimum chance of sticking and blowouts. These insights and skills gave the LRDG a significant advantage over enemy troops, many of whom never adjusted to living and operating in the desert.

From its inception, the primary focus of the LRDG was to obtain information about the enemy behind his lines (Lloyd Owen, 1981). Bagnold's men quickly learned to live in harmony with the desert, and they utilised movement patterns of the native Arabs and of the Italians and the Germans to the advantage of the LRDG. Of great importance was the tracking of movements of troops, vehicles, and equipment, especially along the coast road between Tripoli and Benghazi and points farther east. 'Perhaps the LRDG's most astonishing achievement was the undetected maintenance for eighteen months of a well-camouflaged observation post beside a lonely stretch of the coastal road far inside enemy territory. That post was able to record and report by radio every troop and transport movement to and from the enemy's front' (Bagnold, 1990). In many places along that route, the placement of LRDG observation posts was enhanced by the presence of low hills and scattered vegetation just south of the bordering coastal lowlands and the Via Balbia. The vehicles in which LRDG patrols travelled to, or near, their observation points were hidden during the day beneath camouflage netting and branches and leaves of desert shrubs; in some places, the presence of gullies cutting into the low hills provided additional cover for the vehicles.

At times, LRDG patrols provided 'taxi service' for such specialised units as the SAS (Special Air Service or 'parashots') and for individuals separated from their patrols or those seriously in need of medical attention. It was a matter of pride to the LRDG that they could deliver a person or persons, or equipment or supplies, to a specified locality virtually anywhere in the Libyan Desert on a desired date and at a specified time. The LRDG assisted David Sterling's SAS on a number of occasions by transporting them close to their targets and by rendezvous with them following their attacks in order to provide return transport to their bases (Lloyd Owen, 1981).

In addition to direct observation of enemy troop and vehicle movements, LRDG personnel were skilled in 'reading tracks' left by recent vehicle traffic. The types of vehicles and direction of travel could be determined and an estimate made of their number (Constable, 1971). Such observations, together with information from the road watch, were radioed to LRDG base camp and relayed to LRDG headquarters. Reports arriving from a number of different localities of troop and vehicle movements provided an overview of current enemy activity and insight to possible future operations.

The relatively firm surface and subdued topography of much of the Libyan Desert meant that tracked vehicles were not necessary and that wheeled vehicles could move about with relative ease. The vehicles were fitted with large, wide-tread sand tyres that, at low pressure, could traverse loose sand without sticking and, at high pressure, could traverse much rougher, even rocky, surfaces safely. These tyres proved far superior to the earlier narrow-tread, high-pressure tyres that Bagnold had used before the war.

A typical attack on an isolated installation would involve hundreds of kilometres of desert travel and, upon reaching the vicinity of the target, a period spent lying unseen in nearby hills or sand dunes until nightfall. Under cover of darkness, the patrol would attack with all of the weapons at its disposal, then, in the midst of the resulting bedlam, steal quietly away into the desert and put as much distance as possible between the patrol and its victims before sunrise. Morning light almost inevitably would bring enemy aircraft flown by pilots eager to retaliate. In many areas, the relatively smooth terrain meant that vehicles could move at speeds up to 100 $kmhr^{-1}$ and, when pursued by enemy aircraft, manoeuvre freely to avoid them. If the terrain were rough and irregular, the vehicles would scatter and present only small targets difficult to hit during strafing runs. A common procedure during such runs was for each patrol member not manning a machine gun to grab food or water and move quickly away from the vehicles, repeating the routine with each strafing run. One or two vehicles might be destroyed, but patrol members together with salvaged food, water, weapons, and ammunition would pile on the remaining vehicles and resume their journey. The limited range of the attacking planes and the vastness of the desert worked much in favour of the LRDG patrols. Furthermore, vehicle tracks are notoriously difficult to see and especially to follow from a fast-moving aeroplane. In some areas, in the daylight hours following a typical night attack on an isolated post, fort, or airfield, an LRDG patrol could escape detection by driving along slowly in the shadow of an overlying cloud or by hiding in the shadows of steep cliffs. Unfortunately, clouds and steep cliffs are not common in the Libyan Desert. Occasional intense dust storms or 'ghiblis' also provided cover for patrols but were extremely hard on men and equipment.

The general uniformity of colour of the Libyan Desert meant that in most places the beige and brown camouflage paint, and the irregular patterns with which it was applied to the vehicles, was effective in reducing the possibility of detection from the ground and from the air. This allowed much greater freedom of movement of vehicles than otherwise would have been possible.

The limited range of reconnaissance and attack aircraft and the sparse human population of the desert, together with its great size, meant that the relatively small LRDG patrols could travel great distances without being detected. This freedom of movement was enhanced by the reluctance of the enemy, even of the Arabs in the oases, to venture off the well-marked desert tracks. It was a great advantage to the LRDG to be able to predict with some certainty just where the enemy and local natives might be encountered, and over which routes the enemy would move personnel and supplies.

The relatively low topographic relief also enhanced radio communication between the LRDG patrols and their base camp and between the base camp and army headquarters. On one occasion, radio communications were achieved over a distance of 2,250 km (Lloyd Owen, 1981).

The size and emptiness of the Libyan Desert provided a distinct advantage to LRDG because of the experience of their leaders in moving about in it. Distances were so great that patrols were little bothered by enemy overflights except those following a raid. Useful maps were virtually non-existent, but patrols were issued 1:500,000 scale maps, almost blank in many areas, on which to record terrain characteristics and landmarks. The location of rock cairns and distinctive natural landforms, together with the surface roughness or 'going', were carefully noted until, eventually, useful terrain maps of many areas were compiled. Eventually, the LRDG developed its own survey section to oversee acquisition, compilation, and distribution of topographic data (Bagnold, 1990).

The efficiency of LRDG operations was significantly enhanced by the independent purchase from an Egyptian pasha of two single-engine WACO (Weaver Aircraft Company) biplanes (Figure 3). One cruised at 185 kmhr^{-1}, and the other at 240 kmhr^{-1}; when modified for desert operations, they had a range of some 480 km. The widespread distribution of sand plains and other relatively smooth surfaces meant that the small aeroplanes were seldom far from a suitable landing ground. Normally two-seater aircraft, they could carry a third person in an emergency. The aircraft were used primarily to bring wounded in from the desert and to deliver spare parts and mail to LRDG patrols. Their effectiveness was enhanced by placing caches of fuel, water, and food along the routes usually flown by the aircraft (Kennedy Shaw, 1945).

Figure 3. Long Range Desert Group 'air force' consisting of two privately purchased and maintained WACO biplanes. Major Ralph A. Bagnold, left, confers with one of the pilots near the Libyan-Egyptian border, 1941. (Reprinted from: Bagnold, R.A. 1990. Sand, Wind, and War: Memoirs of a Desert Explorer, *courtesy of The University of Arizona Press and with the permission of Stephen Bagnold)*

The Royal Air Force was displeased that LRDG had its own 'air arm', and provided no assistance in acquiring, operating, or maintaining the aeroplanes. Of the three pilots, two were from the LRDG and one was on short-term loan from the Chad Army. Maintenance and minor repairs were done by the pilots; major repairs were contracted to Egyptian mechanics at the Cairo civil airport (Bagnold, 1960).

Finally, the men of the LRDG must have benefited from the Libyan Desert in still another way, perhaps even unconsciously. The desert, especially at night with the stars overhead, can provide a soothing calmness, and sound of the desert silence, especially after a turbulent day, can heal and rejuvenate the spirit (Lloyd Owen, 1981).

8. Conclusions

The advantages that accrued to Allied forces through the deception visualised by Bagnold and Wavell and carried out brilliantly by the LRDG prompted Wavell to write, in October 1941: 'I would like to take this opportunity to bring to notice a small body of men who have for a year past done inconspicuous but invaluable service, the Long Range Desert Group. It was formed under Major (now Colonel) R.A. Bagnold in July 1940 to reconnoitre the great Libyan Desert on the western border of Egypt and the Sudan. Operating in small independent columns, the group has penetrated into nearly every part of the desert Libya, an area comparable to that of India. Not only have the patrols brought back much information, but they have attacked enemy forts, captured personnel [and] transport and grounded aircraft as far as 800 miles inside hostile territory. They have protected Egypt and the Sudan from any possibility of raids, and have caused the enemy, in lively apprehension of their activities, to tie up considerable forces in the defence of distant outposts. Their journeys across vast regions of unexplored desert have entailed the crossing of physical obstacles and the endurance of extreme summer temperatures, both of which would, a year ago, have been deemed impossible. Their exploits have been achieved only by careful organisation and a very high standard of enterprise, discipline, mechanical maintenance and desert navigation. The personnel of these patrols was originally drawn almost entirely from the New Zealand forces; later officers and men from British units and from Southern Rhodesia joined the group. A special word of praise must be added for the RAOC fitters whose work contributed so much to the mechanical endurance of the vehicles in such unprecedented conditions' (Bagnold, 1990, p. 137-138). In April 1943, a similar letter of appreciation was written to the then commanding officer of the LRDG by General Bernard Montgomery, Commanding General of the British Eighth Army, after British forces, following a route earlier identified by units of the LRDG and guided by one of those units, swung south through rough Tunisian terrain to outflank Rommel's forces and hasten the end of the war in North Africa (Kennedy Shaw, 1945).

Although Bagnold left the LRDG on 1st August 1941, when he was transferred to Cairo as Inspector of Desert Troops (Lloyd Owen, 1981), the LRDG continued to live up to his high expectations of performance and achievement throughout its existence, until April 1943 in North Africa, and May 1945 in the Aegean, Italy, and the Balkans (Lloyd Owen, 1981). Meanwhile, in Cairo, Bagnold eventually became deputy chief signal officer with the rank of brigadier; late in 1941, he received the Order of the British Empire (Military).

Bagnold had opted for a military career because of 'family tradition,' but he had said even as a young officer that he would rather be a Fellow of the Royal Society than a Major General (Bagnold, 1990). He was, in fact, elected to the Royal Society in 1944 while serving overseas, unaware that he had even been nominated. In the official Royal Society biographical memoir, the soldier/scientist who solved some of the mysteries of windblown sand and who established the LRDG was described as possessed of '...innate determination, fortitude, imagination, thoroughness, acute observation, perceptive intellect and quiet modesty' (Kenn, 1991, p. 57).

In reviewing the many contributions of the LRDG to Allied victory in North Africa, certainly the many and varied successful tactical operations undertaken by the LRDG in Libya come to mind, but the LRDG also made a profound strategic contribution. It demonstrated dramatically, first to the Italians and later to the Germans following their appearance in North Africa early in 1941, that a great desert need not provide a secure defensive flank (Bagnold, 1990). To the contrary, 'Bagnold's original concept, his detailed development of it, and his far-seeing organisation had transformed the inner desert from a text-book 'defensive flank' into a serious liability to the enemy' (Constable, 1971, p. 141).

Acknowledgements

Many people and institutions were helpful in our having made numerous trips into various part of the Libyan Desert, thus experiencing first-hand the challenges of travelling and working in that barren but intriguing region. The list is long and will not be provided here, but you know who you are. Please accept our heartfelt thanks. One of us (JRU) wishes to acknowledge the fascinating description of LRDG operations provided by a former member of the group who, in 1970, was a radio operator at the Kufra Agricultural Project; his name, unfortunately was not recorded and cannot be recalled. Jennifer Smith drafted the map of Figure 1.

References

Bagnold, R.A. 1935. *Libyan Sands: Travel in a Dead World*. The Bath Press, Bath. Republished in 1987 with 4 page Epilogue, Michael Haag, London.
Bagnold, R.A. 1990. *Sand, Wind, and War: Memoirs of a Desert Explorer*. The University of Arizona Press, Tuscon.
Constable, T. 1971. *Hidden heroes*. Arthur Barker Ltd., London.
Kenn, M.J. 1991. Ralph Alger Bagnold, 3 April 1896-28 May 1990. *Biographical Memoirs of Fellows of the Royal Society* 37, 56-68.
Kennedy Shaw W.B. 1945. *Long Range Desert Group*. Collins, London.
Lloyd Owen D. 1981. *Providence their Guide: The Long Range Desert Group, 1940-1945*. The Battery Press, Nashville.
Tignor, R.L. 2000. Egypt. *In:* Janus, R. J., (Ed.), *World Book 2000, Volume 6*, World Book, Inc., Chicago, 127-128.

Underwood, J.R., Jr. 1996. *Bagnold, Ralph Alger. In*: Dash, E. J., (Ed.), *Macmillan Encyclopedia of Earth Sciences*, Macmillan Reference USA, New York, 39-41.

James R. Underwood, Jr.
Department of Geology
Kansas State University
Manhattan, KF 66506-3201

Robert F. Giegengack
Department of Geology
University of Pennsylvania
Philadelphia, PA 19104-6316

The Geology of the Battle of Monte Cassino, 1944[1]

John A. Ciciarelli

ABSTRACT: In 1943/44 German troops were retreating northwards out of Southern Italy. To prevent Allied forces from taking the city, Adolf Hitler ordered a defensive stand to made to the south of Rome. The Gustav Line, a system of defensive fortifications stretching across the Italian Peninsula, was constructed. The line utilised the geology and terrain as obstacles/barriers and Monte Cassino was its strong-point. Approaching the Gustav Line, the Allies first encountered lacustrine muds from Pleistocene Lake Lirino. This lake was formed behind a dam of volcanic ejecta from the Roccamonfina volcano, but was drained in the Late Pleistocene when the dam was breached. The Allies' heavy equipment quickly became mired in the muds, forcing the soldiers to move on foot. Crossing the Rapido River developed into a major defeat for the Americans. The defenders utilised the steep banks, inundated flood plains, and fired on the Allies from fortifications in travertine outcrops. Monte Cassino presented steep slopes veneered with colluvium, hence slides were common. Bedrock, exposed nearly everywhere, prevented troops from taking cover from enemy fire and from the elements. Exploding shells propelled shards of brittle rock through the air causing many head and facial injuries.

1. Introduction

It has been said that during the Second World War the battles for Cassino and the fighting in Italy were 'tactically the most absurd and strategically the most senseless campaign in the whole war' (Smith, 1989, p.10). The 350,000 casualties suffered by both sides during those engagements of 1944, to a large extent, resulted from the siege-like fighting that raged over a period of four months (Piekalkiewicz, 1980). This was the type of warfare that neither side wanted nor was prepared to fight. But because the only practical road to Rome wove its way past Monte Cassino from Naples, the allied armies, like many armies, of centuries past, had to pass Cassino (Figure 1). It is easy to understand why Monte Cassino became such an important element in political contests and military struggles, for its possession meant control of the way to Rome. As the poet of the 11th century sang, *'Hinc est iter ad urbem apostolicam'* (Leccisotti, 1987, p. 13). This was obvious to the German planners as well, who made Cassino the strongest point of their defences.

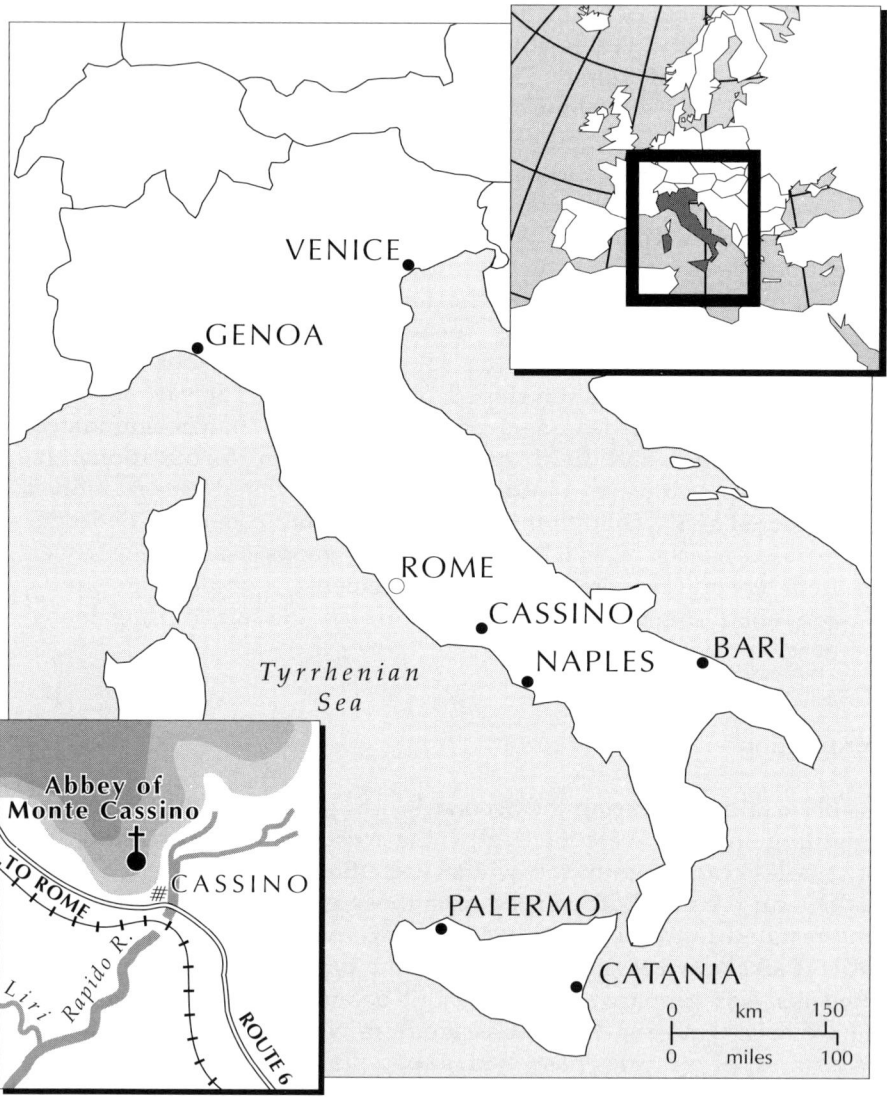

Figure 1. The highway distance between Naples and Rome is 200 km, with Cassino situated approximately half way between them. (Modified from: *Carta Fisico-Politica d'Italia*, 1973)

The crucial moment in the conception of the Italian campaign occurred on 6th October 1943 when Hitler decided to make a stand south of Rome (Bennett, 1989). As the Germans retreated northward from Sicily and Southern Italy, Hitler ordered his armies to make a defensive stand to keep the Allies from taking Tome, a city which the Allies greatly prized. In compliance, the field commanders decided to construct their defensive line (the Gustav Line) at the narrowest point of the Italian Peninsula (Figures 2 & 3). In addition to their military defences, the Germans very skilfully incorporated the natural features of the area as barriers and obstacles, and Monte Cassino, which for centuries had been cited as being a model of impregnable terrain, lay at the southeastern end of the Gustav Line. It was situated squarely along the route the Allies intended to follow to Rome. So from that day in October onward, a battle at Cassino became inevitable.

It is particularly appropriate at this time, when attention is being focused on the anniversaries of many of the battles of the Second World War, that this crucial battle be remembered. It is also appropriate that the geology of the area be highlighted. For probably in no other battle of that war was the topography, geology, climate and other physical features interwoven so intimately as to profoundly influence the activities of the combatants and so to determine the course which the battle followed.

2. Scope of this investigation

It is very difficult, and perhaps misleading, to discuss out of context just a few battles in a large and lengthy campaign such as the Italian Campaign of 1944. For the way any battle commences, is fought and is finally resolved, is determined by the outcome of previous battles, by countless tactical decisions, as well as by the influence of the terrain in the areas where the conflict took place.

Moreover, it is well beyond the scope of this paper and the expertise of the author to discuss the regional and global political strategies that led to the carrying out of the Italian Campaign. What is being attempted here is to explain how the physical features of the Cassino area came to be so important in the fighting.

3. General physical setting

About 80% of the Italian peninsula consists of rugged mountains and hilly topography. The Apennines, which run almost the entire length of the peninsula, are situated roughly along its centre line. Flanking the Apennines on the east and west are gently rolling hills and plains, which extend to both coasts (Ahlgren, 1950).

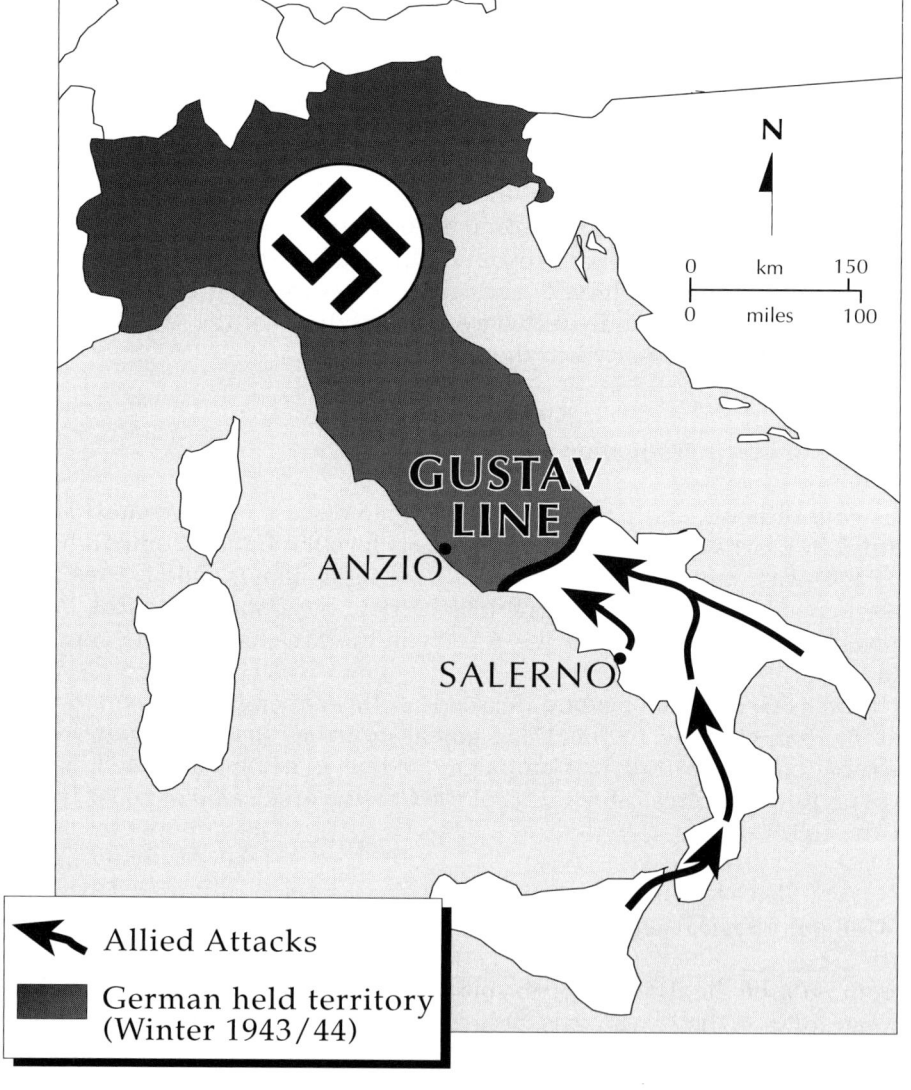

Figure 2. The Gustav Line was constructed at the narrowest point of the Italian Peninsula in order to halt Allied advances out of southern Italy

East of the Apennine range, on the Adriatic side of the peninsula, the gentle hills and coastal plains extend from the southern tip of the heel of Italy northward about 350 km. North of the Gargano promontory, however, the flatlands are gradually pinched out against the sea by the impinging mountainous terrain. So, from the point of view of the military attempting to move northward, these flatlands lead nowhere except into the mountains.

On the western side of the peninsula between Naples and Rome, except for a kilometre or two of flat land and gentle hills located here and there along the coast, a wall of mountains rises steeply from the Tyrrhenian Sea. There is a highway following this coastal route, the famous *Via Appia* (Appian Way), today known as Route 7 (Figure 4).

However, a few kilometres eastward from the sea, in the interior, there are a few fl-bottomed river valleys varying in width from 3 to 10 km and, extending from just north of Naples, virtually the entire distance of about 200 km to Rome. The most notable and most relevant to this study is the valley of the Liri River and one of its tributaries, the Rapido River (Figure 4). Through this interior valley stretches another major highway between Naples and Rome, Route 6, *Via Casalina*. Because of the valley's sheer size and gentle terrain, the Allies decided that it was the only place where they could hope to develop an assault in strength (Smith, 1989). Thus, the Liri Valley, over which the Abbey on top of Monte Cassino has a commanding view, was chosen as the route for the Allied push to Rome.

4. Geology of the Cassino area

The bedrock in the Cassino area consists of highly fractured Mesozoic carbonates. Most of the strata are limestones, which are thickly bedded. There are a few interbeds of hard calcareous shales containing layers of white and grey calcareous nodules. Both the limestones and dolomites are quite varied in their lithology, ranging from thick crystalline and micro-granular strata, where bedding is poorly defined, to a coarse calcareous breccia (Servicio Geologico d'Italia, 1960; Devoto, 1965).

Structurally, the carbonate sequences are cut by steeply dipping normal faults that roughly parallel the general trend of the Apennines. The Liri Valley, in fact, is a graben bounded by faults that trend northwest-southeast. Also playing a significant role in the geomorphic evolution of the area, is a second series of faults that cut perpendicularly across the main structural and morphological trend of the Liri Valley. These faults dismember the surrounding carbonate mountain blocks into isolated 'horsts' (Martini & Wightman, 1987). The Garigliano River, into which the Liri, Rapido and other tributaries flow, occupies one of these cross-cutting faults (Figure 4).

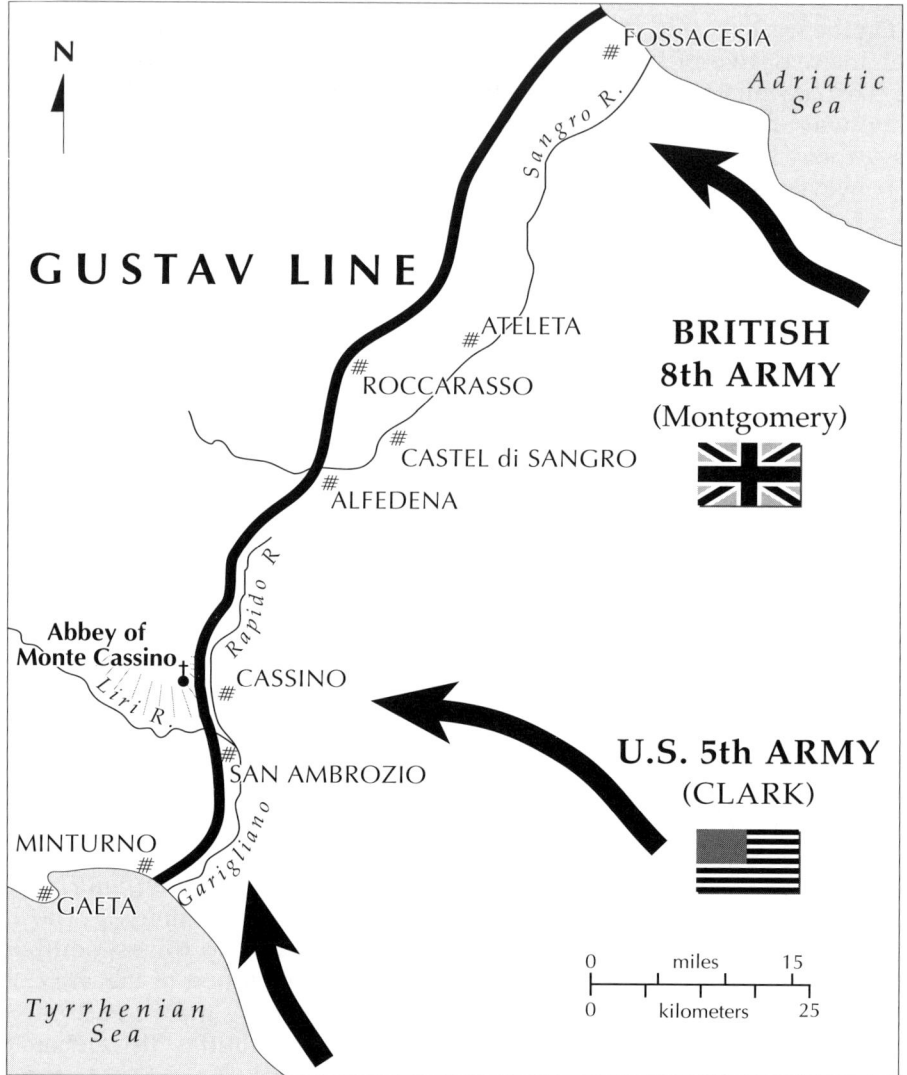

Figure 3. Along the entire length of the Gustav Line the Germans incorporated the natural features into their fortifications. In the Cassino area, they skilfully used the Rapido and Garigliano rivers. In other areas, it was the Sangro River and the rugged Abruzzi Mountains to the northwest that were used. (Modified from: Goodenough, 1982)

According to other researchers in the area, these faults have also localised volcanoes and paravolcanic activity. The Roccamonfina volcano is one of the volcanoes that evolved very recently, from mid to late Pleistocene. Its ejecta profoundly affected the modern landscape. Drainage out of the lower Liri Valley was blocked by this ejecta creating a large (10 km x 35 km) lake, Lake Lirino, which persisted through most of the late Pleistocene (Devoto, 1965; Figure 5). The Lake Lirino lacustrine sediments filled the intratectonic basin to a depth of about 100 metres and rest unconformably on Mesozoic carbonates. The lake sediments consist of a complex interlayered sequence of calcareous muds, volcanic ash and scoria. There are some interbeds of calcareous sands near the mountain bases but grain size diminishes rapidly towards the distal areas (Martini & Wightman, 1987).

According to Devoto, during the late Pleistocene, the dam of volcanic debris was breached and eroded, allowing unrestricted drainage out of the entire Liri Valley. Since then, the lacustrine sequence has been dissected by the Liri, Garigliano and Rapido Rivers. Typical alluvial features and deposits, such as flood plains and terraces as wide as 1.6 km, have developed in the lacustrine sediments. Travertine outcrops are also numerous in the valley and are intermixed with the lacustrine sediments. The outcrops have a northeast to southwest alignment and can be traced to fractures that extend down to the Mesozoic carbonate substratum. The fractures apparently localised $CaCO_3$-rich spring waters that were the source of the travertine (Devoto, 1965). Several villages are built atop these travertine deposits and hence have slightly higher elevations than the surrounding lake sediments. The village of San Angelo, which is situated on the Rapido River, 4 or 5 km south of the town of Cassino, rests on one of these deposits (Figures 6 & 7).

5. Fighting in the valley

The battle of Cassino began on the night of Monday, 17th January, 1944 when the American 36th Infantry Division slogged, under heavy enemy fire, across the flat valley meadows and marshes. The soldiers were moving toward the Rapido River for a crossing at sites just north and south of the village of San Angelo (Figure 6). Because San Angelo occupies a high area underlain by the travertine deposits, the main thrust of the crossing element was forced to split into smaller units. Approaching the travertine bluff of San Angelo, they faced the waiting defenders concealed in and around the Village. Crossing the lacustrine plain to this point was very difficult, for the mud made vehicular travel virtually impossible. Trucks and even tracked vehicles bogged down and became useless to the advancing soldiers. This meant that the troops, making their way to the crossing sites, had to carry all of their equipment a distance of 3 or 4 km from the staging areas (Blumenson, 1970; Figure 6).

As the troops moved closer to the river, they eventually encountered the flood plain of the Rapido River itself. The flood plain, normally soft and

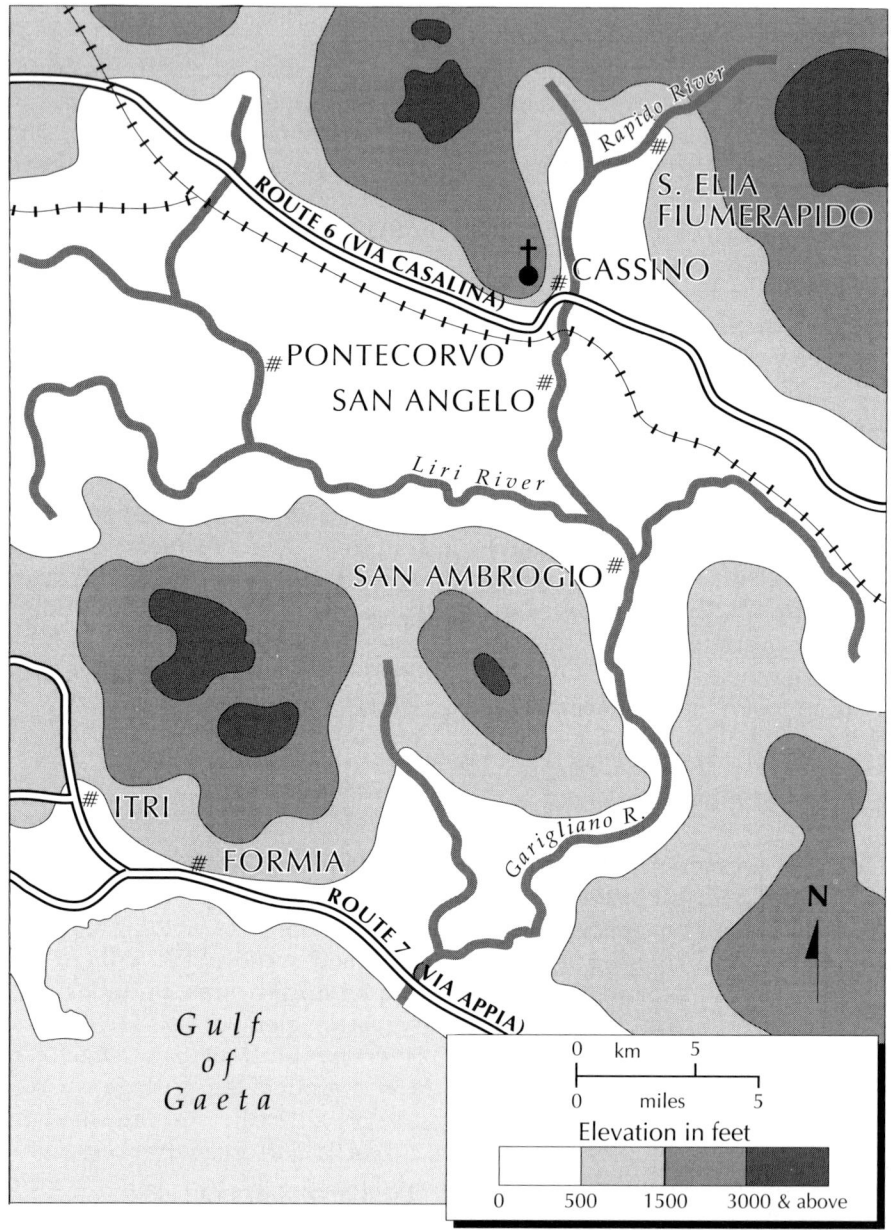

Figure 4. Generalised topographic map of the Cassino area. It is apparent why the Allies chose the broad Liri Valley as their invasion route over the narrower coastal route. (Modified from: Goodenough, 1982)

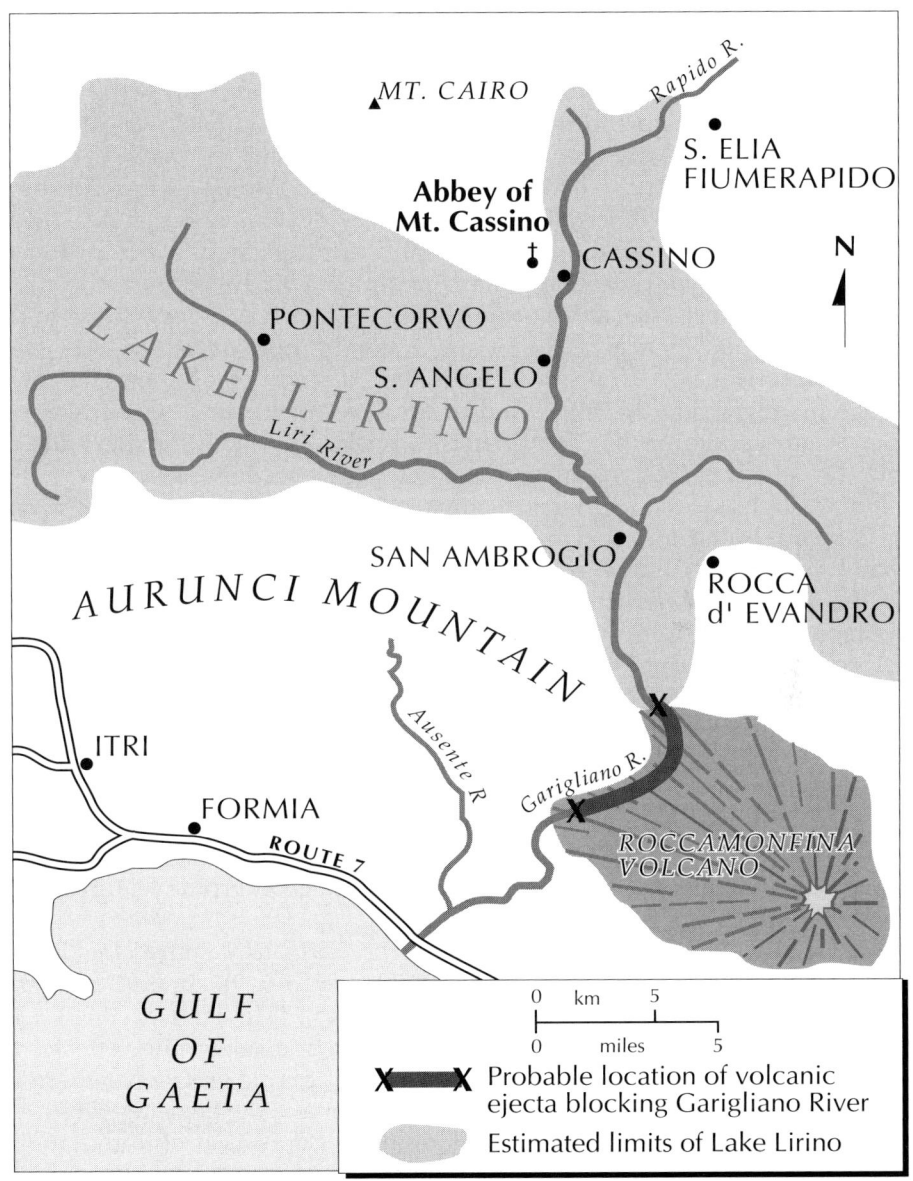

Figure 5. Map showing the limits of Pleistocene Lake Lirino, along with the approximate location of the dam of volcanic ejecta that block the drainage out of the Liri Valley. (Modified from: Carta Geologica d'Italia, Scale 1:100,000)

muddy during this time of the Italian winter, was now extraordinarily marshy. The German defenders made cross-country movement even more difficult by diverting portions of the Rapido waters onto its flood plain. The soldiers, who had already carried their equipment (including large rubber boats to be used for the river crossing) a great distance, had to walk and crawl across the now-submerged flood plain.

The final obstacle, of course, was the Rapido River itself. It had to be crossed without the assistance of heavy equipment, using only what the men could carry. Aside from the military obstacles such as booby traps, barbed wire, mines, machine guns and so on, the soldiers had to contend with the stream itself. The difficulties imposed by the Rapido were threefold. Firstly, the stream itself is not very wide, varying from about eight to 20 m (Majdalany, 1957). Secondly, the stream is appropriately named, for the Italian word *rapido* is translated as fast or rapid. Finally, because the stream flows through clays, silts and volcanic ash, its banks are very steep and in some places nearly vertical. They vary in height from or to 3 m. All of the above characteristics, plus the swirling currents made crossing grave reservations regarding the likelihood of a successful crossing. General Fred L Walker, Commander of the 36th Infantry Division, wrote in his diary the day before the attempt: 'We might succeed, but I don't see how we can ... I don't know of a single case in military history where an attempt to cross a river that is incorporated into the main line of resistance has succeeded. So I am prepared for defeat' (Blumenson, 1970, p. 80).

In the 48 hours of fighting at the Rapido near the village of San Angelo, the Americans lost 1,681 men (Majdalany, 1957). The lacustrine muds of Pleistocene Lake Lirino, the wide flood plain of the Rapido River, the channel characteristics of the stream itself, and the higher ground of the travertine deposits, occupied by the defenders, all conspired to wreak as much havoc as the enemy guns. The crossing of this comparatively insignificant stream was a failure. The Americans withdrew and waited a few days to try again. This time they would try north of the town of Cassino, where the river appeared to be fordable (Figure 8). Here again, the German defenders took excellent advantage of the lacustrine muds and the geomorphic features of the Rapido Valley.

In order to prevent the northern crossing, the Germans dammed the Rapido at the point where it is crossed by Route 6 (Figures 8 & 9). In addition, they blew up a dam in the upper reaches of the valley (Majdalany, 1957). The combination of these two actions had produced exactly the effects the Germans wanted. It inundated the Rapido's flood plain, precisely where the American 34th Infantry Division was assigned to make the crossing (Figure 8). Here too, the soft lacustrine clays and flood plain deposits prohibited the use of heavy, tracked vehicles. The use of tanks was highly restricted, for they could only move in single-file on steel matting laid down by engineering companies (Smith, 1989). So the Germans had only to knock out the lead tank and the others became helpless, unable to manoeuvre if they strayed from the tracks.

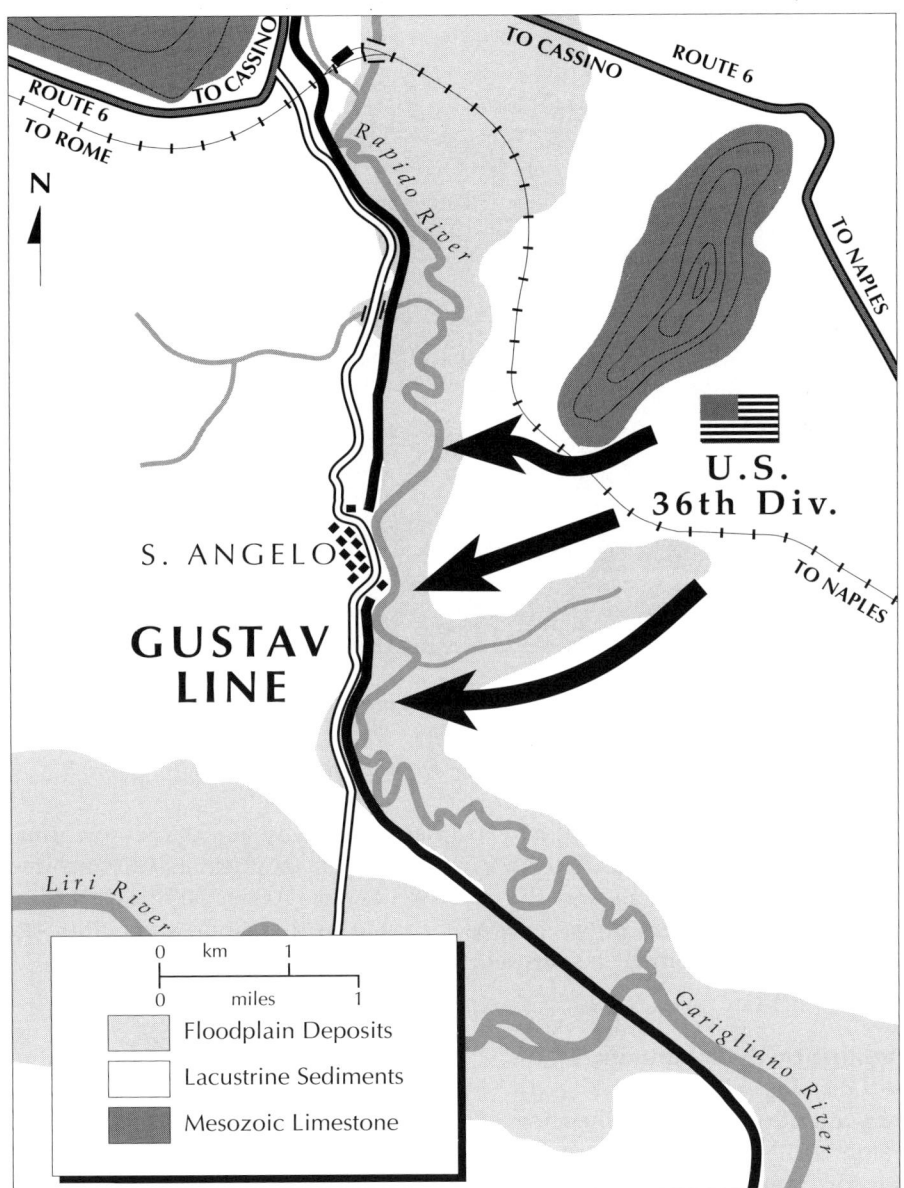

Figure 6. Generalised geological map showing the distribution of lacustrine sediments, floodplain deposits, and exposures of Mesozoic carbonate bedrock. Note the thrust by the US 36th Infantry Division was made entirely on soft, unconsolidated sediments. (Modified from: Smith, 1989 & Carta Geologica d'Italia, Scale 1:100,000)

Figure 7. A view towards the north showing the village of San Angelo situated on a rise above the Rapido River. The advance of the 36th Infantry Division was made from right to left

Once again, the men on foot had to splash their way across the now marsh-like flood plain toward the Rapido River to the base of Monte Cassino itself. The icy, wintry conditions and the mud made the 700 or 800 m trek almost unbearable. The stream however was fordable in this area and, after great cost in men and equipment, a successful crossing was eventually made.

6. Fighting on the mountains

Once the Allied troops gained tenuous footholds across the Rapido River on the slopes of Monte Cassino, their immediate objective was to continue to keep pressure on the Gustav Line, break through it into the Liri Valley and, from there, move on Rome. However, sitting directly atop Monte Cassino, 550 m above the valley floor, is the ancient monastery built in 529 AD by Benedict of Norcia, St. Benedict (Leccisotti, 1987; Figure 12). For centuries this abbey had been a haven and a magnet for pilgrims, tourists and refugees, as well as a centre of education, art, religion and a symbol of the church in this part of Italy. It had seen armies come and go and had been destroyed and rebuilt four

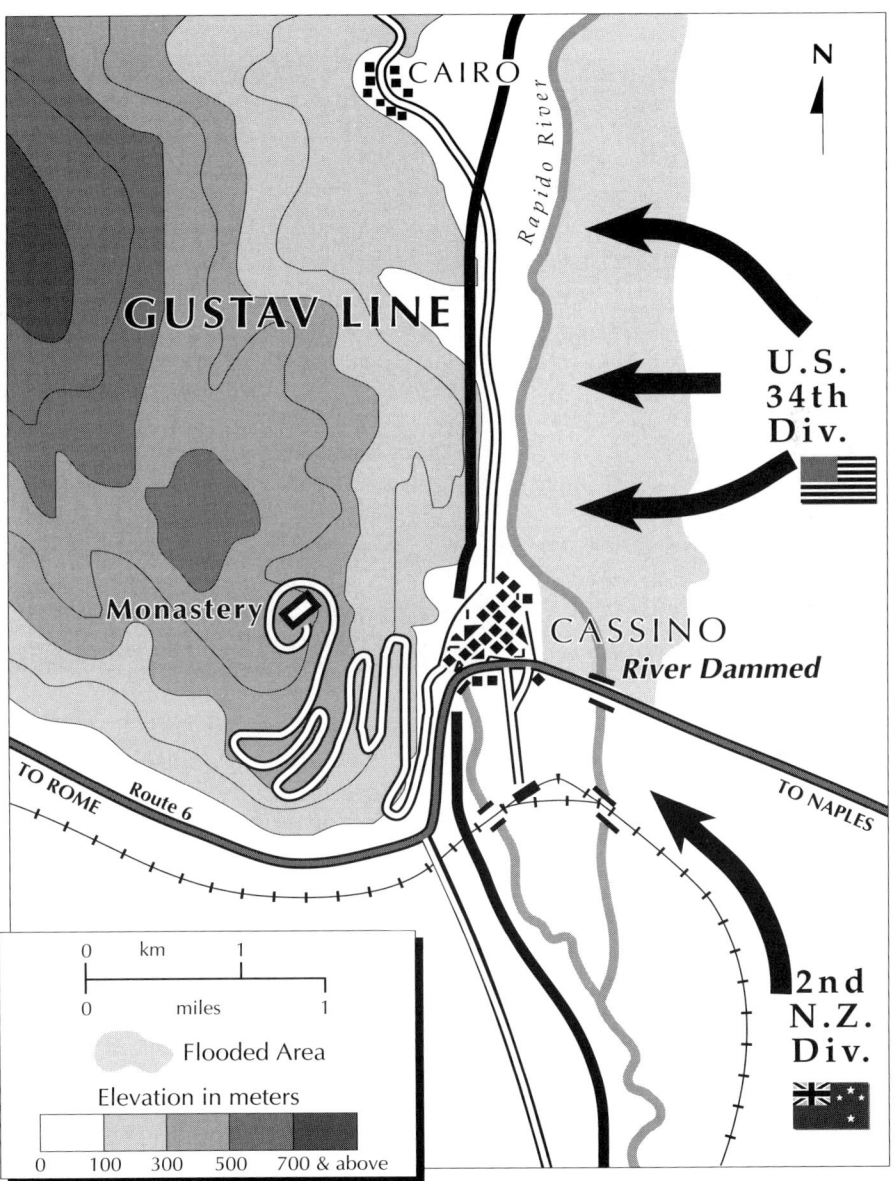

Figure 8. Crossing points north of the town of Cassino, where the US 34th Infantry Division was required to wade across the partially submerged flood plain. The 2nd New Zealand Division attempted to advance along the railroad bed, which was built on embankments to keep it above the soft clays. They occupied the railroad station for a time and then withdrew. (Modified from: Majdalany, 1973)

Figure 9. View of the Rapido River where it is crossed by Rt. 6, the point where it was dammed by the Germans. Note dam location on Figure 8

times in its long history (Leccisotti, 1987). Now in 1944, its three-metre-thick walls shook with a major battle involving troops from seventeen nations (Hapgood & Richardson, 1984).

As the allies began their final assaults on the mountain and the abbey, the building already lay in ruins as a result of intense bombing by planes of the 15th Air Force on the morning of 15th February (Evans, 1988). It was thought that the Germans were using the building (and later its ruins) as observation posts from which to gather intelligence on Allied movements and to direct artillery fire. It is easy to understand why the troops thought this, for every soldier who sees Monte Cassino at once recognises it to be about the finest natural observation post ever encountered (Majdalany, 1957). Though there now seems to be general agreement among students of history that the Germans did not use the abbey itself, they certainly did utilise the slopes up to the base of the building. They also skilfully utilised the rock rubble created by the bombing (Piekalkiewicz, 1980).

The troops who scaled the slopes of Monte Cassino, and ultimately converged on the abbey itself, had to deal with even more geological obstacles. First, the hard Mesozoic limestones and dolomites of the mountain formed very steep, rocky slopes. Slope angles of about 30°-50° are common and

Figure 10. Panoramic view of Monte Cassino as it appears today after its reconstruction at the end of the war. (Reproduced from: Montecassino, 1989)

the surface is profusely littered with cobbles and boulders. This is probably due to the intense faulting and jointing of the brittle carbonates. The rock fragments range in size from 5 or 6 cm to well over a metre across. Slides are common, and cross-country movement on foot is nearly impossible. Even surefooted mules, used to carrying supplies, would often slip and tumble down the steep slopes (D. Mann, pers comm.). The hardships of cross-country movement hindered not only the combat troops but the supporting elements as well. Especially difficult was the task facing stretcher bearers carrying the wounded down the slopes. The stretchers would constantly tilt or suddenly be put down or dropped if one of the bearers stumbled or fell (Majdalany, 1957).

Figure 11 shows that, for the most part, the troops moved on the abbey via routes that kept the gradients of their traverses to a minimum. The Polish and Indian thrusts took a longer but lower gradient route to the objective, whilst the Americans and New Zealanders took a more direct, but somewhat steeper, route. Both routes were dangerous, as the defenders took advantage of the rubble as cover for personnel and weapons.

The second hardship imposed by the geology on the combatants of both sides was the proximity to the surface to the bedrock. There is very little

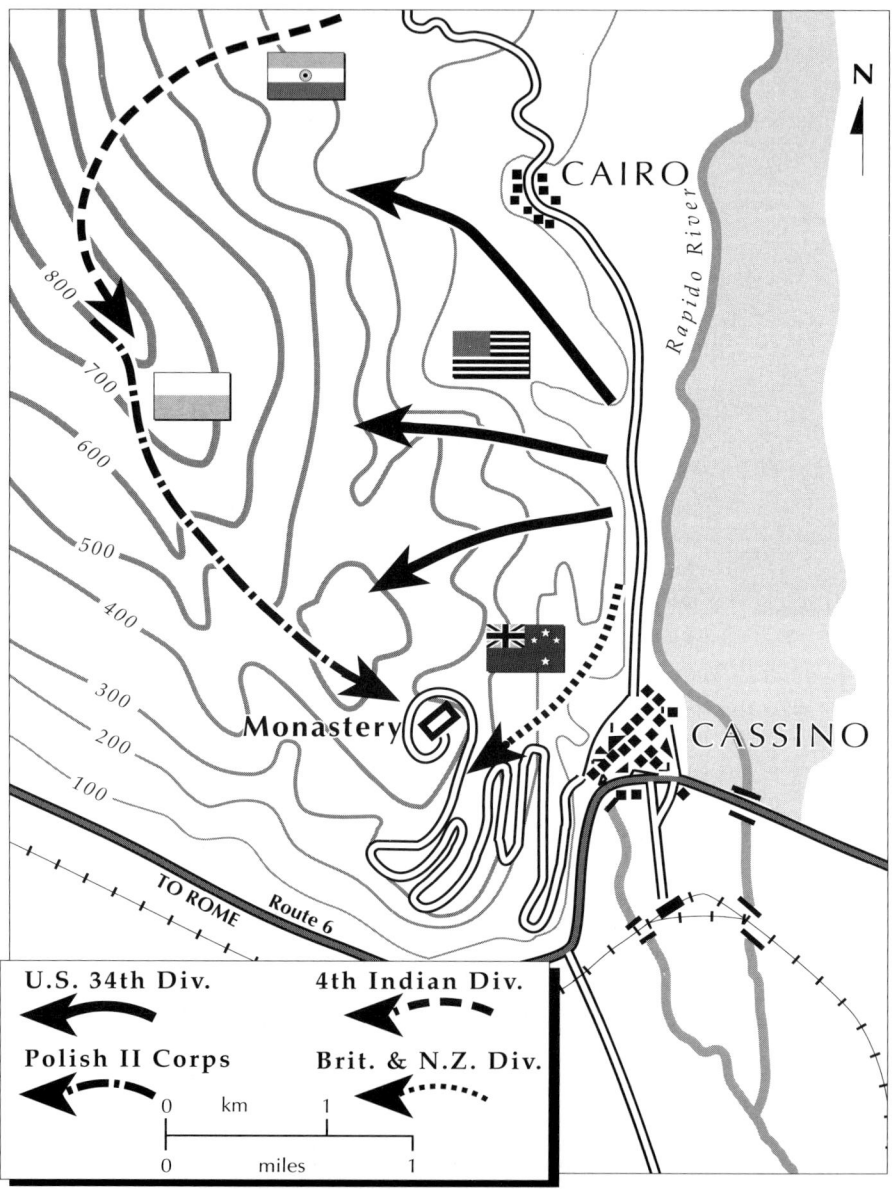

Figure 11. Route taken by troops making the final assault on the abbey. The Polish II Corps was the unit that captured the monastery and ended the battle of Monte Cassino (Modified from: Smith, 1989 & Pianta della Città, Scale 1:1,800)

topsoil and bedrock is abundantly exposed where it is not veneered with the course, rocky colluvium. More than twenty centuries of human activity (deforestation and primitive agricultural practices), combined with natural episodes of vegetation disruption, have denuded the slopes of topsoil (Martini & Wightman, 1987). With the bedrock exposed, or very close to the surface, it was very difficult, and often impossible, for the troops to dig-in to protect themselves from constant enemy fire and from the elements. Daytime rain often turned to snow at night as temperatures fell below freezing, forcing the soldiers to spend nights in the open. This was difficult enough for fresh troops but, for those weakened by numbing cold, wet clothing, fatigue, stress and the lack of warm food and fresh water, it was physically and psychologically devastating. For the seriously wounded, who needed immediate medical attention, the combination was disastrous.

The exposed bedrock caused yet another, and more acute, problem. Because the carbonate rock is very hard and brittle, it shattered in glass-like fashion when hit by exploding artillery shells. The explosions sent shards of rock flying in all directions, inflicting a high percentage of head, face and eye injuries. This direct link between the geology and the combatants can be summarised by these trenchant words: 'It happened that Cassino produced a much higher percentage of head and eye injuries than usual ... The proportion was exceptional enough for two hospitals to have to make special arrangements to deal with these problems. The 92nd General Hospital became partly an eye hospital, the 65th was reserved for head, facial and neurosurgical cases ' (Majdalany, 1957, p. 2337-238).

Figure 12. A close-up view of one of the many shatter-cone-like feature that can be seen on the slopes of Monte Cassino

To this day, over 50 years after the battle, many battle scars remain on the rocky slopes of Monte Cassino. Figure 12 shows one of the numerous examples of the conical, striated-fracture patterns displayed on the exposed rock faces. They are similar to shatter cones found at meteoritic impact sites and are believed to form in response to high-pressure, high-velocity shock waves. On the slopes leading to the abbey, exploding shells and other explosions were so numerous that today these shatter cones are easily spotted, even by the casual observer.

7. Conclusions

It might be argued that nearly every human enterprise is linked and controlled, to some extend, by landscape and, ultimately, geology. This is certainly true for land-based military operations. Any number of battles could have been chosen to illustrate the relationship between geology and military actions. In most others however, the direct influence of the physical aspects of the terrain would not have been as important as they were as Cassino in 1944.

It has been said that the success of mechanised armies, like those facing each other at Cassino, lies in their ability of manoeuvre quickly and with force. The geology, topography and climate of the Cassino area all conspired to hinder mobility. This battle showed how geology can render machines useless. It showed how huge armies of thousands of tanks, trucks, planes and assorted other vehicles were abandoned in favour of primitive muleback transport. At Cassino, in the early months of 1944, one soldier said a single pack mule was worth half-a-dozen tanks. Those conditions generated almost a static siege-like struggle that prompted Hitler to say: 'this is a battle of the First World War fought with weapons of the Second' (Smith, 1989, p. 76).

Military commanders can never really conduct their operations free from the influence of geology and geography. Their plans must be formulated in harmony with the physical characteristics of the area. Ultimately, an understanding of the geology and the events leading to the modern landscape are a prerequisite to success. As the battles at Cassino can attest, the lack of such knowledge, or the inability to use that knowledge, can lead to even greater suffering and waste than would otherwise be the case.

References

Ahlgren, F.F. 1950. *The attack on the Gustav Line in Italy by the 5th Army in WWII.* Unpublished M.A. thesis, University of Oklahoma.

Assessorato al Turismo. 1980. *Pianta della Cittt, Cassino,* Scale 1:1800. Regione Lazio, Assessorato al Turismo.

Bennett, R. 1989. *Ultra and the Mediterranean Strategy.* Morrow & Co., New York.

Blumenson, M. 1970. *Bloody River, The Real Tragedy of the Rapido.* Houghton Mifflin Co., Boston.
Bohmler, R. 1964. *Monte Cassino.* Cassell & Co. Ltd., London.
Devoto, G. 1965. *Lacustrine Pleistocene in the Lower Liri Valley.* Istituto di Geologica e Paleontologia dell'Universita di Roma, IV.
Evans, B.A. 1988. *The Bombing of Monte Cassino.* Pubblicazioni Cassinesi, Montecassino.
Goodenough, S. 1982. *War Maps.* St Martin's Press, New York.
Hapgood, D. & Richardson, D. 1984. *Monte Cassino.* Congdon & Weed, New York.
Leccisotti, T. 1987. *Monte Cassino.* Pubblicazioni Cassinesi, Montecassino.
Majdalany, F. 1957. *The Battle of Cassino.* Houghton Mifflin, Boston.
Mann, D. 1992. Personal interview, telephone, (Captain, Co. B, 142nd. Inf. Regiment 36th Inf. Division, WWII).
Martini, I.P. & Wightman, E.M. 1987. Geomorphology and ancient settlements of the Southern Flank of Mt. Cairo. *Journal of Geoarchaeology* 2, 131-147.
Mauldin, W. 1945. *Up Front.* Henry Holt, New York.
Montecassino. 1989. *Monte Cassino.* Pubblicazioni Cassinesi.
Piekalkiewicz, J. 1980. *The Battle for Cassino.* Bobbs-Merrill, Indianapolis.
Servizio Geologico d'Italia. 1960. *Carta Geologica d'Italia, Cassino Foglio*, No. 160, Scale 1:100,000. Servizio Geologico d'Italia, Rome.
Smith, E.D. 1989. *The Battles for Cassino.* David and Charles, Devon.
Touring Club Italiano. 1973. *Carta Fisico-Politica d'Italia* Scale 1:1,000,000. Touring Club Italiano, Milano.

Notes

1. This paper first appeared as 'The Geology of the Battle of Monte Cassino' in *Journal of Geological Education* 44, 32-42, (1994). Reproduced here in an edited form by permission of the editors of the *Journal of Geological Education.*

John A. Ciciarelli
Pennsylvania State University-Beaver Campus
100 University Drive
Monaca
Pennsylvania, 15061

Terrain as a Factor in the Battle of Normandy, 1944

Stephen Badsey

> **ABSTRACT:** The battle of Normandy lasted from 6th June to 25th August 1944, and is one of the best documented battles of the Second World War, with an extensive primary and secondary literature. It resulted in an overwhelming victory for the Allied forces over the defending Germans. Despite this, the historiography of the battle has been dominated by two controversies. The first of these concerns the failure of the Allied forces to achieve important designated objectives in their amphibious and airborne landing on D-Day, 6th June. The second concerns the time and human cost taken to defeat the defending German forces. Historiography attributes these failings principally to disagreements and poor performances on the part of senior Allied commanders, and/or to superior military skills possessed by German commanders and their forces. This paper addresses these two controversies, and argues that in both cases a principal reason for the battle developing as it did was the nature of the terrain of the battlefield, a factor to which insufficient attention has been paid in the making of historical judgements.

1. Introduction

The Battle of Normandy was one of the most important, and decisive, battles of the Second World War (1939-1945). Following an airborne and amphibious landing on 6th June 1944 ('D-Day'), an alliance of predominantly Anglo-American forces[1] defeated German forces occupying the Calvados region of Lower Normandy. The battle is conventionally regarded as ending with the closing of the Falaise pocket on 22nd August and the liberation of Paris on 25th August. The battle was an overwhelming Allied victory, ending with the virtual destruction of German forces west of the River Seine.

Historiography
The Battle of Normandy is extremely well documented, with primary archival sources in many countries, most of which still await proper analysis. On the Allied side official histories have been published by the major belligerents (Harrison, 1951; Stacey, 1960; Blumenson, 1961; Ellis, 1962). There have also been numerous popular histories of the battle, often published to coincide with significant anniversaries. For the sake of brevity, and because of the ready availability of such narratives, a general knowledge of the battle and its context is assumed in this paper.

Controversies

Oddly for a battle that was such an overwhelming Allied victory, from the earliest accounts by journalists (Moorehead, 1946; Wilmot, 1952), the historiography of the Battle of Normandy has been dominated by two related issues concerning Allied deficiencies. The view of Rommel, that once the Germans failed to prevent the D-Day landings an Allied victory was inevitable (Ryan 1960) has been uncritically and almost universally accepted, and there has been little criticism of the defeated German commanders, some of who also contributed to the general post-war criticism of the Allied performance (Liddell Hart, 1951). One popular explanation for the slow Allied progress in Normandy is a generally superior German military system at all levels, a view heavily endorsed by Dupuy (1977), van Creveld (1983), and Hastings (1984) although partly challenged by Baxter (1995). The first controversial issue of the battle is the failure of the Allied forces to achieve important designated objectives on 6th June or soon after, in particular the early capture of Caen. The second is the slow Allied progress up until the end of July compared to expectations, and the manner in which the Allies closed the Falaise pocket at the end of the battle.

The root of all these issues is the assertion by Montgomery in his memoirs (1946) that 'The outstanding point about the battle of Normandy is that it was fought exactly as planned before the invasion'. This position received little support even at the time, and argument has since concentrated on the 'phase line controversy' and the extent that Montgomery's plans influenced the outcome of the battle. Writers with opinions as diverse as Barnett (1960) and Hamilton (1983), agree that Montgomery's assertion that he could plan a battle in detail beforehand was an essential part of his wartime persona for reasons of troop morale; and that his chief failure was in refusing to admit this after the war. All these issues are described in a critical manner by d'Este (1983).

Attention given to these controversies has largely prevented a scholarly and balanced analysis of the Battle of Normandy being undertaken. In particular, while virtually all accounts of the battle note the importance of terrain, together with the related issue of weather, none has given this factor its full value in explaining the events of the battle. It will be argued here that terrain was a principal reason for the battle developing as it did. Following a brief description of the major terrain features of the Normandy battlefield, it is intended to examine the influence of terrain on several episodes of the battle, starting with the D-Day landings.

2. The Normandy Battlefield

In pursuing this research I have visited and explored the battlefield more than twice a year for almost a decade, often in company with veterans of the battle, military historians, or serving members of various countries' armed forces. This has led to various publications concerning the battle (Badsey,

1990; Badsey, 1993; Nalty, 1993; Chandler & Collins, 1994). The statements concerning the terrain of the battlefield in this paper are based on: personal observation; on photographs or film taken either at the time of the battle, or later on the battlefield; on the testimony of veterans; and on an extensive reading of archival and published sources. However, as an historian, I have relied heavily on acknowledged authorities, particularly Dr E.P.F. Rose, for statements regarding the geology of Normandy.

The terrain of the battlefield
The Calvados region was approximately the maximum southern extremity of the European ice cap during the Pleistocene Epoch. The Cotentin peninsula, a low-lying plateau which becomes progressively more rugged towards the north and west, was for some time a separate island, and its southern landmass is largely reclaimed marsh, with a complex drainage system (Figure 1). One notable result of the raising of the land relative to the sea at the end of the Pleistocene is that parts of the north-facing Calvados coast are also reclaimed low-lying marshland, with a characteristic inshore ridgeline of more solid ground representing the original cliff-face. Attempts were made by the Germans to re-flood all these marshes in 1943 as part of the creation of their Atlantic Wall defences, turning minor roads into causeways running through the marsh. This was particularly a feature of the westernmost Allied landing area, Utah Beach at the base of the Cotentin. It was also a feature, to a lesser extent, of the three almost contiguous British landing beaches, Gold, Juno and Sword, where the inland marsh formed a correspondingly lesser obstacle (Figure 1).

Rivers on the battlefield were minor obstacles in themselves, but again were flooded by German military engineering. In the western or American sector of the battlefield, the base of the Cotentin is joined to the Calvados coast by an extensive river and canal system leading out of the reclaimed marsh into a wide multiple estuary that separates Utah Beach from the other Allied beaches. The principal river marking the western side of the estuary system is the Douve, with its tributaries flowing eastward out of the Cotentin itself and through the small town of Carentan before becoming part of the estuary. The ground rises so markedly north-west of Carentan out of the reclaimed marsh that the next small village is called Saint Côme du Mont. Although the actual height of the ridge at the village above sea level is only just over 30 m, the dramatic nature of this dominating position is best observed and understood by watching from ground level as dawn breaks, just as the terrain was seen by many American paratroopers on D-Day.

The eastern side of the estuary is marked by the River Vire, which flows northwards from beyond the town of Vire itself and through Saint Lô, then through Isigny-sur-Mer before becoming part of the estuary. The French designation 'sur Mer' does not always mean on the coast, but usually at the highest navigable point of a canal or river in mediaeval or early modern times, as with Isigny which is about 4 km from the estuary and almost twice that from the open sea. (Some French towns or villages designated 'sur Mer'

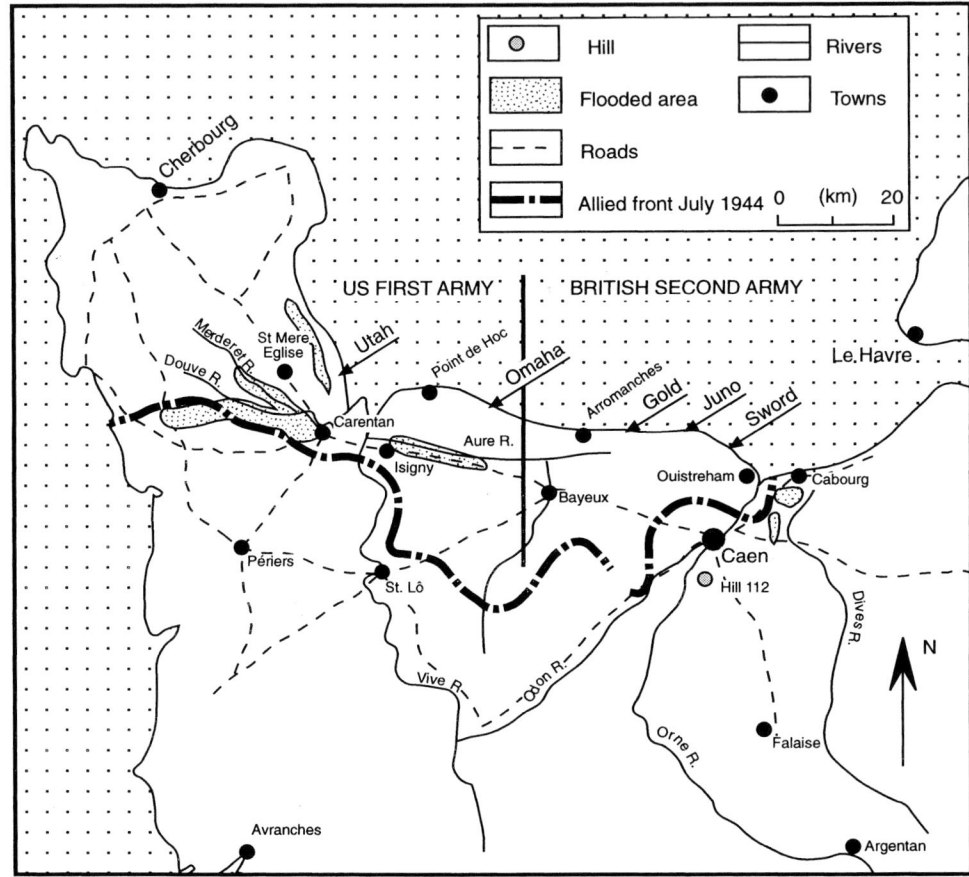

Figure 1. Map of the Normandy battlefield, and its salient topographical points

have no river, but were set back from the sea for safety from pirate raids, or as the first source of fresh, as opposed to saline, water).

In the eastern or British sector of the battlefield the principal river is the Orne (Figure 2), which rises from beyond Argentan, and flows northwards through the centre of Caen, and then on to the sea at Ouistreham, adjacent to Sword Beach. The area of the British landing beaches, with its small coastal villages, was the basis for a pre-war tourist industry, with a casino at Ouistreham. From the centre of Caen northwards the River Orne is canalised, and flows in parallel with its artificial tributary, the Canal de Caen. Other minor rivers in this sector were significant obstacles not in themselves but because of the sometimes sheer gorges that they have cut through the rock. This was particularly true of the River Odon, a tributary of the Orne which flows into the larger river just south west of Caen, across which the British mounted 'Operation Epsom' in June, The ground rises steeply on either side of the valley of the Odon, in the sector where 'Epsom' was mounted, reaching to over 100 m in height within 1000 m of the river. The flat topped 'Hill 112' (m; today measured at 111 m), about 2000 m east of the Odon and 5 km south west of Caen, was a principal British objective for 'Epsom,' and in the early stages of the Battle of Normandy dominated the surrounding area. On the extreme eastern flank of the British landing area, the river Dives was flooded by the Germans from its estuary at the town of Cabourg inland, so effectively sealing off the eastern flank of the battlefield (Figure 2).

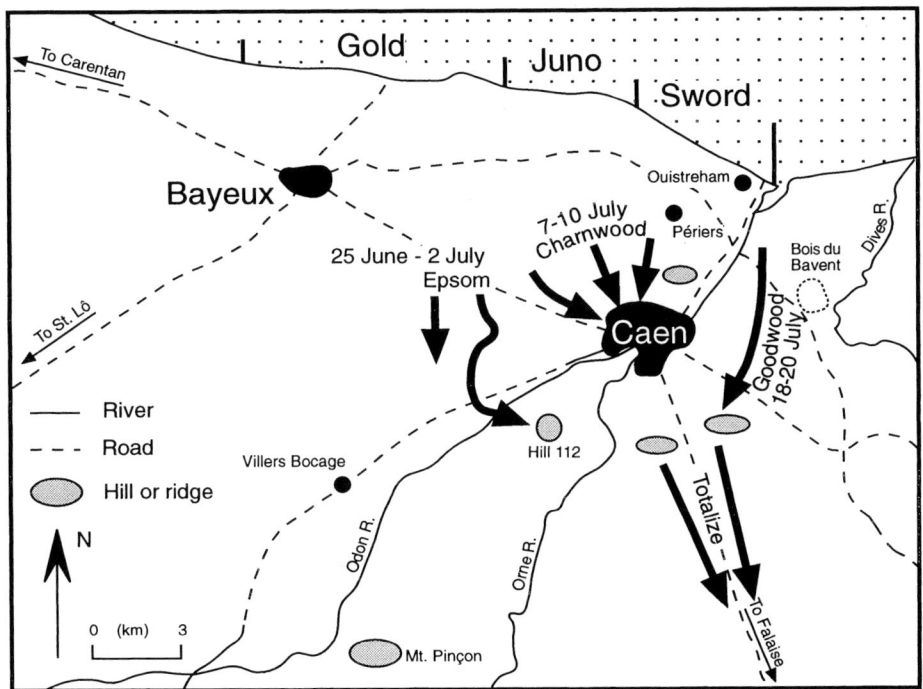

Figure 2 Area around Caen and the British operations in 1944

Away from the beaches and marsh areas, Lower Normandy is chiefly rolling countryside, the yellowish soil of which is identified by geologists (Rose & Pareyn, 1995, 1998; Rose & Nathanail, 2000) as Pleistocene loess over Jurassic limestone. This loess was laid down by winds during the Pleistocene, and has very distinctive properties including a proclivity to break into a powdery dust when dried out and subject to stresses. This was an often-observed hazard for both ground troops and aircraft during the battle, as dust gave away the positions of moving vehicles or fouled engines. There is a repeated anecdote (traceable to Liddell Hart's private papers) that before the British 'Operation Goodwood' attack east of Caen on 18th July a senior German officer put his ear to the ground and heard the sound of massing British tanks amplified through the cavernous limestone rock. Stone quarries to the east and south of Caen were also a feature of the battle, sometimes as German defensive strongpoints, as during the Canadian 'Operation Totalize' offensive of August (Figure 2).

The fertility of the land produces a rich agriculture based on seafood, dairy cattle, and apple orchards: Isigny is the site of a creamery, while Calvados has given its name to an apple brandy. This has produced over much of the battlefield a settlement pattern of mediaeval stone farmhouses and small villages built for defence, often against marauding English in the Hundred Years' War, with thick walls and deep cellars. Village names including 'mesnil', cognate with the English mediaeval 'desmene', are not unusual. The terrain east of the River Orne was and is rather more open, with a different farm field pattern.

Moving inland and further south, the field and farming patterns change as the nature of the soil and underlying rocks also change. From the perspective of the geologist or geographer this alteration is fundamental. But to the military observer of 1944 (and also to the modern observer), the transition appears as one of degree, with a gradual increase in the rolling nature of the ground from the immediate coastal region inland. The local name for this pattern of farming is 'bocage' (as in place-names such as Tilly-Bocage), which in English is a recognised variant of 'boscage' meaning a thicket or grove (Figure 1). The bocage country was and is characterised by irregular fields, woods and orchards averaging about 100 m^2, with embanked sides topped by hedgerows, and connected by narrow dirt roads. Coupled with stone farmhouses, villages and churches, this provided a natural defensive system of considerable strength, which German senior commanders and their staffs had almost a year to study before the battle. A map in Blumenson (1961) pointedly showing the bocage (or 'hedgerow country' to the Americans) as not extending into the British sector of the battlefield as it was on 2nd July 1944 is severely misleading; but it is true that the bocage became progressively more complex and rugged to the west and south west of Caen.

South of Caen and Saint Lô, the terrain of the region is markedly more broken than to the north. The area of the Falaise pocket is part of the 'Suisse Normande', named locally from a supposed resemblance to Swiss mountain valley pastures, heavily marked by cliffs, gullies and forests as well as dense bocage. The highest point to feature prominently in the battle, Mont Pinçon, which is about 5 km south of the village of Aunay-sur-Odon, is 365 m high. Many positions marked by medieval fortifications or their remains have the characteristics of steep cliffs. These include scenes of some of the heaviest fighting in August, such as the American defensive position at Hill 317 (m) near Mortain against the German 'Operation Lüttich', and the château at Falaise itself, captured by the Canadians after heavy fighting in 'Operation Tractable' Indeed, the word 'falaise' itself means cliff, in this case formed by the passage of the minor River Ante through the town.

The battlefield today

As in 1944, Normandy is a living community chiefly associated with agriculture, tourism and maritime pursuits, and continues to resemble its wartime self quite closely. The area extending on both sides of the Vire/Douve river estuary system is designed as the 'Regional National Park of the Marshes ('Marais') of the Cotentin and of Bessin', and adjacent to Utah Beach is a wild bird sanctuary. The heart of the Suisse Normande is also parkland, 'The Regional Park of Normandy-Maine'.

As in 1944, the eastern part of the region is more settled and cultivated than the west, and also shows more marked changes in human habitation patterns. At the time of writing (55 years after the battle) changes in farming practice have led to the reduction or destruction of many of the hedgerows, particularly in the area to the north and west of Caen, but many farms remain or have been rebuilt largely as they were before the battle. Urbanisation has increased, and suburbs are visibly replacing previously isolated villages; while some villages have grown to the point at which they are effectively contiguous. Caen, which in 1944 had about 50,000 inhabitants, has roughly doubled in area; and Lebisy Wood, an important feature that on D-Day was some 2,000 m north of the city, has now virtually vanished into its suburbs. Most of the buildings of Caen were destroyed in the course of the battle, together with almost all of the smaller towns of the area, and they have been rebuilt since in more modern styles. The exception is Bayeux, which was captured intact by the British on 7th June, the inner part of which still retains many of its older characteristics. Metalled roads have replaced dirt tracks, some railways have been removed or replaced by roads, and the main A-13 motorway ('autoroute'), now runs roughly east west across the battlefield through Caen and past Bayeux to Cherbourg, largely replacing an older metalled road.

Some military engineering in the course of the battle has produced permanent changes to the landscape: notably the western part of the ring-road ('peripetique') around Bayeux, built by British troops in 1944 as a way of by-passing the city centre; the remains of the 'Mulberry' harbour of Port Winston at Arromanches-les-Bains; and the present road layouts of Falaise and Saint Lô, both rebuilt after the destruction involved in their capture. In recent years, the economy has benefited from increased 'battlefield tourism'. Particularly in the coastal region, important battlefield sites are well preserved, and there is an abundance of battlefield museums and places of interest. The battle has also been the subject of two major anniversary commemorations, in 1984 and 1994. 'Omaha Beach' (in English) has become the recognised and signposted name for that location.

3. Terrain and the D-Day Landings

Almost every account of D-Day refers to the fact that the original landing date was 5th June, that a delay of 24 hours was imposed through bad weather, and that the weather on 6th June was windy and rainy with a high sea-state, at best marginal conditions for the undertaking. Coupled with German defensive use of certain characteristics of the Normandy terrain, notably the partial re-flooding of marshland, this factor is in itself almost enough to account for several important episodes on D-Day. Certainly, the nature of the terrain, particularly the reflooded marshland, together with the bad weather, was a major factor in accounting for the successes and failings of the Allied landings.

The airborne landings
The American airborne landing as a preliminary to the landing on Utah Beach has been authoritatively described by Marshall (1962). American terrain appreciation was generally good. It was understood that German flooding had turned the road exits from Utah Beach into causeways through sea marsh for up to three km, at which point the ground rises to a flat-topped ridgeline running approximately north-south, with low-lying ground on either side. The main metalled road from Carentan (still mostly extant as the A-13) ran along the highest point of the ridge crest, from Saint Côme-du-Mont through Sainte Mère Eglise and on eventually to Cherbourg. Control of this ridgeline, the road, and Sainte Mère Eglise itself, although still no more than 30 m in elevation, would sever German communications in the Cotentin. Securing the exit causeway roads and Sainte Mère Eglise therefore became the principal mission of the American airborne forces.

A planning deficiency was the failure to appreciate the extent of the inland marshes that lay in some places 1000 m or closer on the immediate western side of the road and ridge crest, caused by German flooding of the Douve and its tributary the Merderet. This tributary, the name of which may be translated as 'little open sewer', was photographed by the author in flood in 1999, in similar conditions to those existing in 1944. The Germans deliberately recreated the conditions of flooding which occurred naturally in bad weather, and which before the war led local farmers to use boats for transport during winter.

This problem was compounded by a bad scattering of the paratroopers as they dropped into the dark. It has been asserted by Keegan (1982) that this scattering was due to the deficiencies of the transport aircrews, who were regarded as inferior to fighter or bomber crews. The unconsidered factor in most assessments is that the prevailing weather system for D-Day developed from west to east, meaning that the bad weather and heavy cloud was significantly worse over the Cotentin than further east. The American transport aircraft, which habitually flew in formation for easier navigation, approached their drop zones by overflying the Cotentin from the west, in order to minimise the chance of dropping their paratroopers into the sea.

Running into heavy cloud at night caused the aircraft to break formation in order to avoid mid-air collisions. After this, a scattered drop into the flooded marsh was inevitable, and from this it was equally inevitable that the Americans would not secure all their D-Day objectives from Utah Beach, or be able to break out easily across the Cotentin or towards Carentan for almost a week.

As described by Crookenden (1976), the weather over the British drop zones was altogether different, with variations that led some battalions to drop in clear and dry air, and others arriving a few minutes later to be scattered by wind and rain. The British were also dropping onto very different terrain. There was no flooding from the canalised Orne and the Canal de Caen. British drop zones lay largely between the canal and the heavily flooded River Dives, about 10 km to the east. The British intended on D-Day to use the flooded Dives as an obstacle to the Germans by destroying its bridges, in which they were successful. Between the two water obstacles of the Canal and the Dives, most of the British drop zones lay in a prominent area of relatively high ground about 4 or 5 km square known as the Ranville ridge, marked to the south by the equally prominent Bavent Wood (Bois du Bavent), which was clearly visible from the air at night, and considered the single most easily distinguished feature in Normandy. Again, the highest point of the Ranville ridge is barely 50 m in elevation, but it dominates the surrounding area, and is clearly and threateningly visible on the approach to Sword Beach (once more this is best appreciated at dawn; the modern cross-channel ferry into Ouistreham has a night crossing that arrives at a time coincident with that of the British landings). After D-Day the German determination to hold ground was greatly assisted in this sector by both the wood and the flooding of the Dives, which made mounting a British offensive due eastward impractical. The town of Cabourg, a D-Day objective if all had succeeded for the Allies, was not actually captured until August, long after the breakout.

Omaha Beach
Montgomery's change to the original COSSAC landing plan to include Utah Beach is well documented, and discussed admiringly by Hamilton (1983). From the Vire/Douve estuary for more than 50 km eastward until the British landing beaches are reached, the Calvados coast is almost entirely sheer cliff, so rugged and broken in places that the local name for the central part is 'Le Chaos'. The only major stretch of sand beach on this entire littoral (other than a few local pockets) is Omaha Beach, which lay considerable distances from the other landings. By direct line across the Vire/Douve estuary it is only about 20 km from Utah Beach, but it is more than double that distance by road through Carentan. Omaha Beach is also over 30 km west from Gold Beach, the westernmost of the British beaches.

Omaha appeared highly suitable as a landing beach, with a gentle shelving slope that in good weather gave some 300 m of beach at low tide. But the corollary of this was that, given the shape of the Cotentin and the estuary system to the west, strong sea currents run in a prevailing easterly direction along the line of the beach. Its existence at all is due to a slight rising of the land relative to the sea, producing instead of a cliff the 'Omaha bluff', a feature rising 30-55 m high about 30 m inland from the shingle of the high tide mark, steep and overgrown but climbable. This bluff in turn is split along the length of the beach by four deep valleys, the 'Omaha draws', which were the obvious exits and heavily defended. Moving inland from the top of the bluff, the transition to the easily defensible bocage came almost at once, within no more than a kilometre, a feature that was remarked upon by more than one American present.

The near-failure at Omaha was a consequence of these factors, together with others; notably the bad weather which degraded the effects of Allied bombing and naval gunnery, and led to swamping of landing craft and vehicles, plus unexpectedly strong German resistance. While no commentator has dared criticise the troops, the failure to secure more than a precarious foothold at Omaha on D-Day severely distorted the Allied plans for the battle. In particular, it left the flank of the British advance towards Bayeux hanging in the air. Joining up the beaches into a continuous front, a notional objective for D-Day itself or shortly afterwards, took until about 12th June.

Bradley and US First Army staff have been criticised for not giving the special nature of Omaha Beach more attention in planning the D-Day landings; and in particular for refusing offers of advice and specialist equipment from the British and from their own subordinate formation commanders with experience of amphibious landings. The Americans might have supported the Omaha Beach landings with airborne forces attacking from inland onto the rear of the German defenders, rather than committing all their airborne troops to support Utah Beach. But the nature of the terrain inland from Omaha Beach provides several problems for this suggestion. Barely 8 km inland the Lesser Aure river ('Aure Inferieur') runs almost parallel with the coast at Omaha Beach, with multiple drainage ditches flowing into it; in 1944 this was flooded to form part of the marsh defences. Air currents are notoriously unpredictable near the coast, and this together with the prevailing winds strongly suggests that a parachute drop at night into the bocage — with the marsh on one side and the sea on the other — would have been a considerable risk. A landing at night or even shortly after dawn using gliders that could fly parallel to the coastline is perhaps another matter; but this was not attempted.

It is also noteworthy that the Americans chose to storm the important German coastal battery at the Pointe du Hoc, 6 km west of Omaha Beach (given by error as the 'Pointe du Hoe' in American official accounts and many others), on D-Day by an assault from the sea directly up the cliff-face. In contrast, the British planned to capture the comparable coastal battery at Longues, about 20 km west of Gold Beach and directly in the area of the 'Chaos', from the landward side, and did so on 7th June. Such decisions were a matter of national style as much as sober military calculation.

Caen and the British beaches

Perhaps the single greatest Normandy controversy is the failure of British troops from Sword Beach to capture Caen on D-Day. Strictly, the controversy arose from the German determination to hold Caen after that date, so that its southern extremities were not captured until 18th July. The other D-Day objectives were Bayeux, captured on 7th June, and the intervening main Caen-Bayeux road, which was largely secured over the next week. Chiefly American writers have criticised British troops and commanders for slowness, despite the very obvious point that the British advance inland from their three beaches was considerably greater than that from either of the two American beaches.

What no recent account of the issue makes sufficiently clear is the quite different nature of the terrain facing the British landings to that of Utah and Omaha, (the American beaches were also very different from each other). Offshore shoals and a short headland separated Sword Beach from Gold and Juno, but otherwise the three beaches formed an almost contiguous landing area. At the shoreline the frontage for parts of the landing zone was — and is — a sea wall and promenade for a line of fishing villages and holiday villas. The obstacle of the sea wall was not continuous, but it was a feature of the major initial landing points. Between these landing points the coastline is in some places low cliffs (in particular between Sword and Juno beaches), and at other places only sand dunes. Lateral communications are based on two main roads which run parallel to the coast and each other, providing a built-up area no more than 500 m deep, which had been turned by the Germans into a series of defensive positions, with beach defences and road blocks.

Once clear of the houses, the British were faced with a variable depth of partially re-flooded marshland, perhaps 1000 m in places, and then the rising ridge of the old coastline. This is usually known as the Périers (or Périers-Dan) ridge from the small village of Périers-sur-le-Dan about 5 km inland from the coast towards Caen. The highest point of the ridge, just north east of the village, was only 61 m (marked as 58 m on modern maps), and the tallest buildings of modern Caen, a further 7 km away, are clearly visible when approaching Sword Beach and Ouistreham from the sea. The settlement pattern was of villages built along this ridge each linked to satellite hamlets on the coast, by means of a single road through the drainage area. For example at Sword Beach the hamlet of La Brèche d'Hermanville on the coast is connected by a single road to the village of Hermanville-sur-Mer on the ridge. As its name suggests, La Brèche d'Hermanville represents a 'breach' or stretch of harder ground through the drained marsh down to the sea. The marsh flooding was not as severe as in the American sector, and the open countryside was generally passable for infantry on foot, but very much more difficult for vehicles except in isolated places. The roads and other areas were also extensively mined, and the Périers ridge was marked by a number of German artillery batteries and underground defensive positions.

Normal Allied practice in the Second World War was to avoid as far as possible both intensive street fighting and attacking heavy defences from the seaward side, but in this case there was no alternative. The British therefore landed not infantry formations but brigade groups based on tanks and artillery as well as infantry, to fight a semi-urban battle at the water's edge, something which had never been done before. As with the American beaches, both Gold and Juno Beach were wide enough for two brigade groups to land side by side, but Sword Beach was wide enough only for one brigade group to land at a time, of the five landed on the beach during D-Day.[2] Extending the landing to the eastern side of the Orne estuary near Cabourg was not possible due to extensive mud flats, something not appreciated by the Germans who had defences there, and who continued to expect a further landing in the course of the battle.

It is again well documented that one consequence of the bad weather in which the D-Day landings took place was to heighten the sea-state and shorten the time that the beaches were free of the tide. By mid-morning, the British beaches were no more than 10 m of hard sand between the intermittent sea-wall and the water's edge. The British needed tanks and other specialist vehicles both to overcome the defences of the shore-line, and also to tackle the ridgeline further inland. But getting these vehicles ashore, off the beach, and then out along the exit roads inland was a considerable problem. The resulting delays were the principal factor in the failure to take either Caen or Bayeux on D-Day.

4. Terrain and the battle after D-Day

The phase lines controversy
Historiography on the battle of Normandy has by no means divided completely along national lines. Nevertheless, the nature of the major controversies of the battle other than those of D-Day is well illustrated by reference to the British and American official histories respectively. Ellis (1962) for the British describes the battle from D-Day to the liberation of Paris in a single volume. The narrative is of a single coherent plan executed by Montgomery, whereby repeated attacks by the British in the Caen sector pulled German forces from opposite the Americans in the west, so creating the breakout which began with 'Operation Cobra' on 25th July. The American official narrative is quite different, being divided into two volumes. Harrison (1951) takes the story from D-Day to the fall of Cherbourg on 27th June, as a British-directed battle which became seriously imperilled; then Blumenson (1961) continues with an essentially American victorious breakout, rescuing a battle in which Montgomery is increasingly marginalised, and the last of his offensives around Caen, 'Operation Goodwood', is dismissed as a 'panacea'.

These arguments depend heavily on the use of 'phase lines' in Montgomery's planning for the battle, of which perhaps the earliest discussion is that of his own chief of staff (de Guingand, 1947). These estimates for the battle predicted that for planning purposes the Allies could expect to be on the line of the River Seine on either side of Paris at D-Day plus 90 days, and this was actually achieved by D-Day plus 78 days. The controversy has arisen from the 'inner' timing of the battle. It is documented that Montgomery understood correctly that Rommel would seek to defeat the landings actually on the beaches. Thereafter, Montgomery expected the Germans to give ground at a reasonably steady rate, husbanding their forces for counter-attacks. This was in fact the preferred strategy of the German theatre commander in France, von Rundstedt, who was over-ruled by Hitler (Liddell Hart, 1951). Instead, Hitler issued repeated orders not to retreat, and to counter-attack to retake lost ground. The total Allied territorial gains by 23rd July, just before 'Operation Cobra', were less than those estimated for 23rd June. It is this Allied failure to advance inland that requires explanation.

Cherbourg and the Mulberry harbours
The belief that the Allies needed to capture a major port on or soon after D-Day was an important factor in German defensive planning, hence the importance of the artificial 'Mulberry' harbours to the Allies. The British harbour, 'Port Winston' at Arromanches-les-Bains, functioned well, but the second harbour started off Omaha Beach was abandoned after the Channel gale of 19th June. It is held particularly by Walter (1986)[3] that the Americans took inadequate precautions with their Mulberry, although again this undervalues the problem of the easterly wind and tide due to the shape of the bay at Omaha. It is true that Bradley was reluctant to trust Mulberry; and that after the establishment of a continuous front on 12th June he made the capture of Cherbourg his priority rather than an advance southward to Saint Lô.

Perfect hindsight has shown that both the Americans and the British were wrong in their assessments (Haswell, 1979). The Allies could have supplied their operations for the Battle of Normandy entirely across the beaches without the use of the Mulberry harbours. As for Cherbourg, it had been so thoroughly sabotaged by the Germans that it was useless as a port until after the Battle of Normandy was over (Harrison, 1951). But none of this was known or apparent at the time.

Cherbourg, as a natural harbour at the north of the Cotentin, is dramatically dominated by higher ground immediately inland, in particular the citadel of the Fort du Roule, which commands the port area, and which played a major part in the fight for the town. Cherbourg's shallow bay is overshadowed by the Cap de la Hague to the west of the peninsula, and the Pointe de Barfleur to the east. This, and the narrowness of the Cotentin, which would have been easy for German defenders to seal off, were among the factors which led Allied planners to reject an attempt to capture Cherbourg from the sea. Other factors included the unsuitability of the soil and underlying rock formations for the construction of temporary airfields.

Once the breakout from Utah began, these factors, and the inundations at the base of the Cotentin, all worked in the Americans' favour. Working against their attack from Omaha southwards to Saint Lô was that this involved fighting through some of the more difficult bocage. This included the important spur 'Hill 192' north east of the town, part of which earned the name 'Purple Heart Draw' (after the medal awarded to wounded Americans) in the course of the battle. Bradley's decision to give priority to Cherbourg meant that it was not until early July that American attacks southwards towards Saint Lô were resumed, as against a 'phase line' expectation of its capture by 16th June.

Goodwood and the appearance of open ground
The British offensive 'Operation Goodwood' mounted southwards to the east of Caen on 18th July (together with the simultaneous Canadian 'Operation Atlantic', which has been neglected by historians but which finally cleared the Germans from the city) is probably the single most controversial episode in the historiography of the battle after that of the failure to take Caen on D-Day (Belfield & Essame, 1965; d'Este, 1983). Its nature in respect to terrain has been well analysed by Dick (1982). What is ignored in many accounts of the battle is the extreme smallness and narrowness of the area from which the British launched a major attack. This was no more than a triangle of land measuring 7 km inland from the coast at Ouistreham and 4 km across its southern front, bounded to the west by the Canal de Caen and to the east by the eastern side of the Ranville ridge and Bavent Wood, held by the enemy. This bridgehead east of the Orne was under observation from Caen and the high ground to the south and southeast, penetrated by German patrols, and was by July dangerously unsanitary for the troops. If only in these terms, there were excellent reasons for mounting an offensive to expand this bridgehead.

The ground of the Ranville ridge was more open than the bocage to the west; although it had some field and hedgerow systems, it consisted more of rolling ridges that were the location of more than one horse stud. This, together perhaps with the closeness to a city, produced a mediaeval settlement pattern of small château-like farms and villages dotted at almost regular intervals through the plain. What the Germans had noticed, and the British had failed to do, was that these villages were each between 1000 and 2000 m apart: for example, the village of Cagny, on the former main road towards Caen, is about 1800 m south east of the hamlet of Le Mesnil-Frèmentel. This was lethal range for the tanks and anti-tank guns of the period. Despite the appearance of being open country, the Goodwood battlefield was among the more easily defended sites in Normandy.

Bocage and the issue of German superiority
The weather, when combined with the terrain and its effects, has already been mentioned as a neglected factor in the Battle of Normandy. In general terms the weather in 1944 was not exceptional for north west Europe between June and August; but as it occurred the start of the battle and period until mid-

July was characterised by overcast, heavy rain and storms, whereas later July and August were brighter and more sunny.

An objection to the idea of an innate German military superiority in Normandy is that a proportion of the troops in German uniform in Normandy were from other nationalities (variously estimated as 10-25 % depending on the unit), and that there was a very wide variation in experience, equipment and general military fitness between German units. The German approach to land warfare in this period, as analysed particularly by van Creveld (1983) placed heavy emphasis on low-level combat skills and the primacy of traditional fighting troops such as infantry and tanks. It has been argued (Badsey, 1980) that this constituted less an absolute or general military superiority as a choice of emphasis on the Germans' part. By specialising in one particular area they became severely deficient in others, notably in the less prestigious fields of supply, transport and military intelligence, all of which were marked Allied superiorities in the Battle of Normandy (Ellis, 1990; Hesketh, 1999). It has also been suggested by R. Hart (1996) that the supply difficulties from which the Germans often complained in Normandy were to some extent of their own making, by trying to keep more fighting troops in Normandy than their system could supply; a strategy which briefly checked the Allies, but which could not succeed.

It has further been argued, particularly by Hart (1996, 2000) that the Allied emphasis on the firepower derived from artillery and airpower, a reluctance to take heavy infantry casualties, and a large administrative and support structure when compared to the traditional fighting forces, were entirely appropriate as a fighting method for mid-20th century democracies. The broad effect of these very deep-seated military cultural and doctrinal choices was to create an appearance of German military superiority in frontline engagements, while the battles and the war were won by the superior Allied methods.

A further front-line problem for the Allies in Normandy was the absence of a tank capable of meeting the small number of German 'Panther' and 'Tiger' tanks on equal terms (Belfield & Essame, 1965; Badsey, 1980). In Normandy both the British and Americans relied on artillery and airpower to offset their weakness in tanks, including very heavy bombardments (characterised by Montgomery as a 'colossal crack') at the start of each major offensive, from the Anglo-Canadian 'Operation Charnwood' of 8th July onwards.

As it happened, the weather for the earlier part of the battle broadly favoured the German fighting method, by degrading the effects of Allied airpower and delaying the landing of some Allied heavy equipment. The nature of the bocage itself suited both the German fighting method and their system of defence, which derived from First World War experience (Griffith, 1981), and consisted of holding front lines very lightly in the hope of drawing attacking troops into more strongly held positions further on. In the maze of the bocage the Allies had great difficulty in directing artillery and airpower to where it was required. Further, the Germans could hold a front with very few troops, as the Allies on more than one occasion interpreted this as a weak front line with stronger defences beyond.

A convincing indicator that it was the terrain more than any other factor that gave the Germans their advantage in holding the Allies in later June and July is that, once either the British or Americans gained ground, the Germans were unable to recapture it. This discounts counter-attacks which gained less than 100 m, or the rare occasions when an exposed position was given up voluntarily, as with the British retirement from Hill 112 at the end of 'Operation Epsom' in June in the face of a major threat to their flank, which was subsequently defeated in consequence of the limited withdrawal. Broadly, once the Allies took ground from the Germans, they could not retake it, suffering from the same problems of attacking in terrain that markedly favoured the defence.

The Falaise pocket and the Suisse Normande
As described by Lucas & Barker (1978), and How (1981), the fighting around and in the Falaise pocket was extremely complex. Whereas the brighter weather and cloudless skies now favoured the Allies, their main attacks in August were launched into the most dense of the Suisse Normande bocage. Most of the historiography of the battle has concentrated on disputes regarding command decisions by the Allied commanders. It has been pointed out by Wilmott (1989) that the Allies were attempting — although eventually with success — a battle of encirclement and annihilation for which they had no common agreed doctrine or operational methods, and no previous experience.

It is hard to generalise about a battle in which fighting stretched as far east as the River Seine both north and south of Paris, and as far south as Alençon. But after what has already been presented, it is sufficient to note that the complexity of the Suisse Normande terrain produced the kind of blind and disjointed 'soldiers' battle' that was beyond the ability of any senior commander to control with the military communications technology of 1944. The fragmented and almost fortuitous nature of the American drive north of Argentan, the Canadian defence of Saint Lambert-sur-Dive, or the Polish capture of the 'Mace' position to seal the pocket, are all evidence of this. Broadly, the Allies were attacking into the deepest bocage in which the Germans found it easier to defend than to escape. Restrictions were necessarily placed on Allied airpower through problems of target identification. The confusion of the Falaise pocket has led to the extreme position taken by Irving (1981) and Rohmer (1981), who argue that the closing of the pocket was mishandled due to command rivalries, and that significant German forces escaped the pocket. However, the evidence remains solid that German losses in the pocket were indeed substantial.

5. Conclusions

The Battle of Normandy was a disastrous defeat for the Germans and a major success for the Allies. An understanding of the terrain of Normandy, often of a detailed nature, is a requirement for understanding the battle from D-Day to its conclusion, and for understanding some of the reasons why the Allies won. Sometimes, the effects of the terrain are not immediately apparent, particularly from general maps; but quite local terrain variations made a significant difference to the conduct of the battle. To this should also be added an analysis of the weather, on a daily basis, particularly for its effects on some of the major attacks mounted as part of the battle.

Understanding the effect of terrain on the battle also sheds new light on some old disputes regarding command decisions and military performance in the battle. The importance, on D-Day and shortly afterwards, of the artificially flooded marshes immediately inland from the invasion beaches in influencing the speed and nature of the Allied advance has, in particular, not been properly acknowledged. While it has always been accepted that the bocage was a problem for the Allies, the advantages given to the Germans in defence by this kind of terrain have seldom been given their full weight. Finally, given the advantage derived by the Germans in defending the Normandy terrain, there is little or nothing in their conduct of the battle to support the view that they possessed an inherently superior military system to that of the Allies.

References

Badsey, S. 1980. The American experience of armour 1919-1953. *In*: Harris, J. R. & Toase, F. H. (Eds), *Armoured Warfare*. Batsford, London, 124-144.

Badsey, S. 1990. *Normandy 1944: Allied Landings and Breakout*. Osprey, London.

Badsey, S. 1993. *D-Day: From the Normandy Beaches to the Liberation of France*. CLB, Godalming.

Barnett, C. 1960. *The Desert Generals*. George Allen & Unwin, London.

Baxter, C. F. 1995. Did Nazis fight better than democrats? *Parameters* 25, 113-118.

Belfield, E. & Essame, H. 1965. *The Battle for Normandy*. Batsford, London.

Blumenson, M. 1961. *Breakout and Pursuit*. US Department of the Army, Washington DC.

Chandler, D.G. & Collins, J.L. (Eds) 1994. *The D-Day Encyclopedia*. Simon & Schuster, New York.

Van Creveld, M. 1983. *Fighting Power*. Arms & Armour, London.

Crookenden, N. 1976. *Drop Zone Normandy*. Allen Lane, London.

Dick, C. J. 1982. The Goodwood concept — situating the appreciation. *Journal of the Royal United Services Institution for Defence Studies* 127, 22-27.

Dupuy, T. N. 1977. *A Genius For War*. Macdonald & Janes, London.

d'Este, D'1983. *Decision in Normandy*. William Collins, London.
Eisenhower, D.D. 1948. *Crusade in Europe*. Doubleday, New York.
Ellis, J. 1990. *Brute Force*. Andre Deutche, London.
Ellis, L.F. 1962. *Victory in the West, Volume I*. HMSO, London.
Griffith, P. 1981. *Forward Into Battle*. Anthony Bird, Chichester.
d'Guingand, F. 1947. *Operation Victory*. Hodder & Stoughton, London.
Hamilton, N. 1983. *Monty: Volume 2, Master of the Battlefield 1942-1944*. Hamish Hamilton, London.
Harrison, G.A. 1951. *Cross-Channel Attack*. US Department of the Army, Washington DC.
Hart, R. 1996. Feeding Mars: the role of logistics in the German defeat in Normandy. *War in History* 3, 418-35.
Hart, S. 1996. Montgomery, morale, casualty conservation and colossal cracks: 21st Army Group operational technique in north west Europe 1944-45. *Journal of Strategic Studies* 19, 132-153.
Hart, S. 2000. *Montgomery and 'Colossal Cracks'*. Praeger, London and Westport.
Haswell, J. 1979. *The Intelligence and Deception of the D-Day Landings*. Batsford, London.
Hastings, M. 1984. *Overlord: D-Day and the Battle for Normandy*. Michael Joseph, London.
Hesketh, R. 1999. *Fortitude: The D-Day Deception Campaign*. St. Ermins, London.
How, J. J. 1981. *Normandy — The British Breakout*. Kimber, London.
Irving, D. 1981. *The War Between the Generals*. Allen Lane, London.
Keegan, J. 1982. *Six Armies in Normandy*. Jonathan Cape, London.
Liddell Hart, B.H. 1951 *The Other Side of the Hill*. Revised edition, Cassell, London.
Lucas, J. & Barker, J. 1978. *The Killing Ground*. Batsford, London.
Marshall, S.L.A. 1962. *Night Drop*. MacMillan, London.
Montgomery, B.L. 1946. *Normandy to the Baltic*. Hutchinson, London.
Moorehead, A. 1946. *Montgomery*. Hamish Hamilton, London.
Nalty, B.C. (Ed.) 1993. *D-Day: Operation Overlord from its Planning to the Liberation of Paris*. Salamander, London.
Rohmer, R. 1981. *Patton's Gap*. Arms & Armour, London.
Rose, E.P.F. & Pareyn C. 1995. Geology and the Liberation of Normandy, France, 1944. *Geology Today* 11, 58-63.
Rose, E.P.F. & Pareyn, C. 1998. British applications of military geography for 'Operation Overlord' and the battle of Normandy, France, 1944. *In:* Underwood, J.R. & Guth, P.L. (Eds), *Military Geology in War and Peace*. Reviews in Engineering Geology, XIII Geological Society of America, Boulder, 55-66.
Rose, E.F.P. & Nathanail, C.P. (Eds) 2000. *Geology and Warfare: Examples of the Influence of Terrain and Geologists on Military Operations*. The Geological Society, London.
Ryan, C. 1960. *The Longest Day: June 6, 1944*. Victor Golancz, London.

Stacey, C.P. 1960. *The Victory Campaign*. The Queen's Publisher, Ottawa.
Walter, A.E.M. 1986. A harbour goes to France. *Royal Engineers Journal* 100, 14-30.
Wilmot, C. 1952. *The Struggle for Europe*. Collins, London.
Willmott, H.P. 1989. *The Great Crusade*. Free Press, New York.

Notes

1. It should be emphasised that the majority of troops who landed on 'Juno Beach' on D-Day were Canadian, and that the term 'British' includes forces of the Commonwealth and Empire. Forces of several nationalities fought in the Battle of Normandy.
2. As with any other era and nationalities, the military terminology of the Second World War is complex and confusing. A British 'brigade group' was broadly the equivalent of an American 'regimental combat team' or RCT.
3. Brigadier A.E.M. Walter was 21st Army Group director of ports and inland water transport, and commander of the port construction force for Mulberry B at Port Winston. A copy of his unpublished paper *The Mulberry Harbours: My Final Assessment* is in the author's possession.

Stephen Badsey
Department of War Studies
Royal Military Academy Sandhust
Camberley
Surrey, GU15 4PQ

Airfield Country: Terrain, Land-Use and the Air Defence of Britain, 1939-1945

Ron N. E. Blake

ABSTRACT: While Britain escaped the ravages of 20th century ground warfare, its landscape bears the imprint of a vital element in national defence: airfields. The British countryside is littered with the sites of c. 1250 aerodromes and landing grounds from two world wards. At the end of the Second World War (1939-45) there were no fewer than 475 aerodromes with hard runway systems, an asset from which the Royal Air Force (RAF) and its kindred arms made a leading contribution to the post-war security of Europe. The ideal environment for military airfields was a surface gradient flatter than 1 in 80, well-drained sub-soil, light woodland cover, nucleated rural settlements, road and rail connections and proximity to the continental mainland. These requirements led to intense competition with agriculture and a variety of engineering solutions designed to balance operational need with a living countryside. In 1945 almost 40% of the United Kingdom lay within c. 8 km of an airfield runway. Over large tracts of eastern, midland and central-southern England airfields were spaced closer than 15 km, creating a continuous impact zone which aviation historians now recognise as 'airfield country'. The best natural foundations for airfields were provided by the Middle Jurassic limestone and Lower Cretaceous Chalk escarpments, but Quaternary tills, sands and gravels were also extensively occupied due to wartime expansion. In southern England high erosion surfaces were typically crowned with airfields while the hydrologically constrained Fens were conspicuously avoided. In Wales and northern Britain coastal sites with inferior ground conditions were necessarily utilised. Adjacent depressions in the landscape were widely used for the siting of encampments as a precaution against blast. To camouflage operational areas special mowing, ploughing, resurfacing and painting techniques were applied.

1. Introduction

Terrain has influenced the course of battle in all three environments of war — land, sea and air. Many of the world's historic land battles were directly shaped by geology (Holmes, 1996; Doyle & Bennett, 1999), while the siting of castles, garrisons and other defence infrastructure was frequently a response to physiographic opportunities at points of expected invasion (Childs, 1998).

Sea battles depended indirectly on dockyards and harbours with safe approach channels, implying a knowledge of submarine geomorphology. Air battles, seemingly unconstrained by landform, were ultimately dependent on level ground for basing aircraft together with an awareness of relief amplitude, meteorology and emergency landing places (Doylerush, 1985).

This paper is concerned with the key ground environment of air defence. Specifically, it examines airfields as exploiters of terrain and as modifiers of the landscape. For practical reasons the study is confined to the United Kingdom (UK) during the Second World War (1939-45), although it is acknowledged that the terrain characteristics of airfields have differed significantly between countries and over time (Neal, 1998). While Britain may not be wholly representative in this regard, its high density of wartime airfields suggest that the present investigation may have international application.

Britain's airfields, as well as being great feats of engineering, are of growing environmental interest (Blake, 1969; Bagley, 1972; Francis, 1996). Since the mid-1950s the Ordnance Survey (OS) has indicated all airfields on its maps, apart from small landing grounds with minimal infrastructure and a handful of larger disused aerodromes, e.g. Aldermaston, Berkshire [SU595635], adapted for top secret storage or research (Oliver, 1991). About twenty years ago an Airfield Research Group (ARG) was formed to establish a nation-wide data-base and has since expanded its remit to include heritage and environmental concerns.

Definitions of 'airfield country', 'terrain' and other key concepts are supplied at appropriate points. The discussion begins with a quantification of the airfield build-up to 1945 when Britain had an extraordinary concentration of air capability. A brief description of airfield dimensions is then given as a foretaste of environmental impact. Next, the concept of 'airfield country' is analysed in terms of its physiographic and land use composition, and finally changes to the landscape attributable to airfield development are summarised.

'Britain', 'British' and 'United Kingdom' are used interchangeably to describe the study area which includes Northern Ireland and the Isle of Man (but excludes the Channel Islands). 'Airfield', 'aerodrome', 'airbase' and 'air (force) station' can be taken as synonymous unless the context suggests otherwise. Ordnance Survey (OS) grid references to a resolution of 100 m are supplied for all individually cited airfields.

2. Airfield development

During the Second World War the United Kingdom contained an estimated 856 military airfields, comprising fully-fledged aerodromes, ancillary landing grounds, flying-boat stations, and marine aircraft anchorages (Figure 1). About 250 of the land-based aerodromes were inherited from the inter-war period and some of the larger establishments, e.g. Farnborough, Hampshire

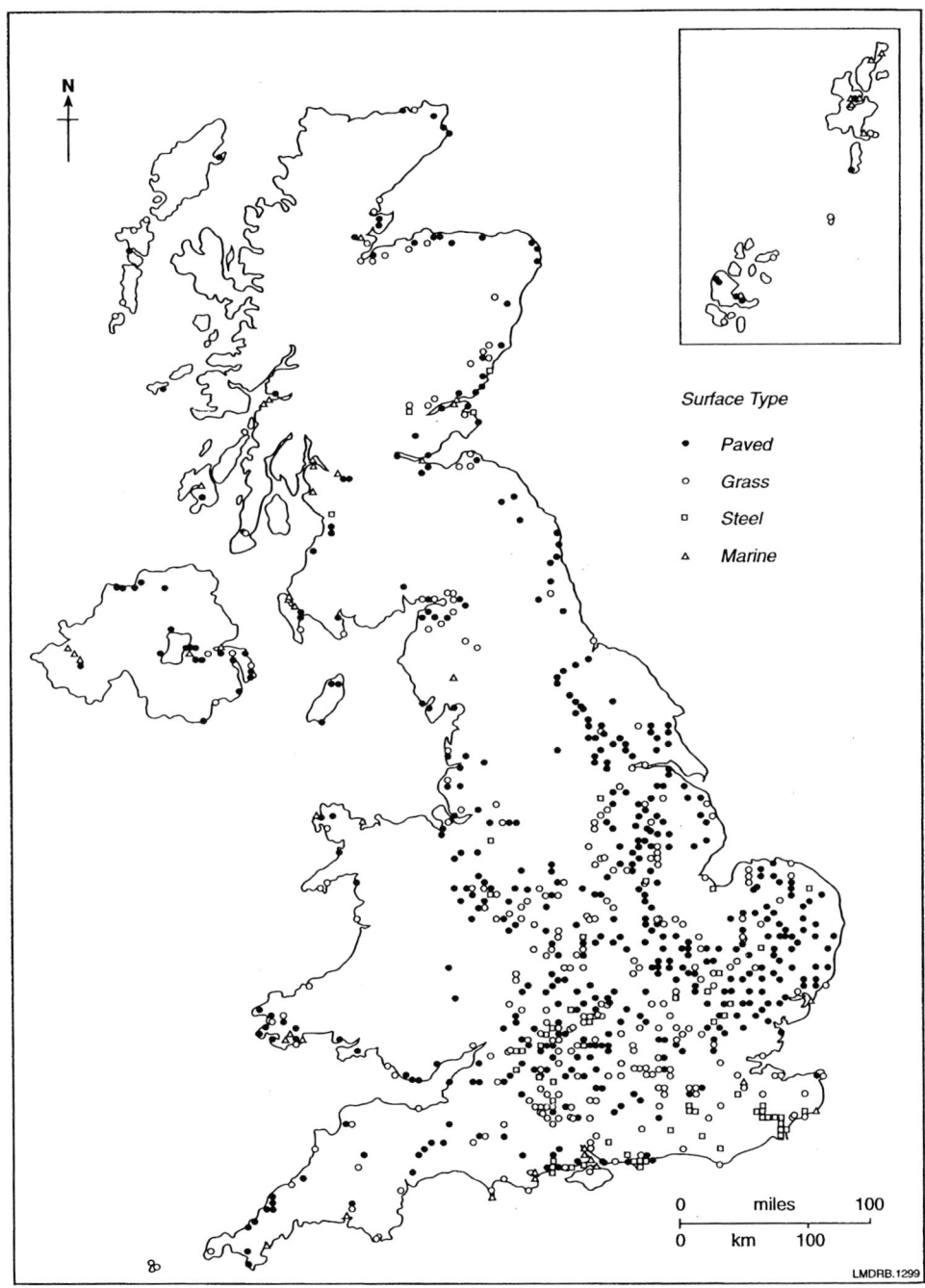

Figure 1. Distribution of airfields in the United Kingdom during the Second World War, 1939-1945

[SU860540] dated back to before the First World War. In 1942 frontline stations were being opened at a rate of one every three days (Smith, 1989).

It has been calculated that Britain's wartime airfields contained enough concrete for a highway from London to Peking (Smith, 1989). By 1945 there were 475 aerodromes (55%) with hard runways, while many of the 275 grass-based airfields possessed a hard perimeter tract. Another 63 had steel-clad strips to sustain all-year use (Betts, 1995).

A total of $c.$ 162,000 ha were covered by airfields during the campaign, equivalent to the County of Hertfordshire (0.7% of the UK's land area) and overwhelmingly at the expense of agriculture. Over much of eastern England the direct airfield footprint was greater than 1% and in the counties of Norfolk, Suffolk and Oxfordshire it exceeded 2%. East Anglian folklore tells of airfields being located every seven miles, a spacing with echoes of historic market towns.

After the war airfield numbers fell dramatically as flying training procedures were streamlined and air power concentrated into a jet-delivered nuclear deterrent. By the mid-1960s there were only 100 still active (Blake, 1969) and since the ending of the Cold War that number has been halved (Marriott, 1997).

3. Airfield design

Before 1939 all UK airfields were grass-based, including large bomber stations, e.g. Marham, Norfolk [TF725085] and London's then chief civil airport at Croydon, Surrey [TQ305635]. Paved runways, commonplace in Europe and North America, were regarded as a luxury in a colonial nation with giant rigid airships, commercial flying-boats and record-breaking trains. The pre-war practice of simultaneous take-offs from grass also explains the late decision to lay concrete and bitumen. Once the war began the needs of twin-engine bombers such as the Wellington made paving inevitable. Conversion to the four-engined Lancaster, Halifax, Flying Fortress and Liberator bombers saw hundreds of new aerodromes constructed with integral paved systems as well as the up-grading of pre-war bomber stations such as Waddington, Lincolnshire [SK985645].

In 1939 the average size of an RAF airfield was 135 ha and state-of-the-art bomber stations could cover $c.$ 200 ha. By 1945 the average size for a paved aerodrome had risen to 272 ha, while the mean for all-grass and steel-strip types was $c.$ 100 ha. Post-war air capability has been mainly concentrated at the larger wartime bases, involving extensions to runways and residential areas which have pushed the average size up to $c.$ 350 ha.

A typical aerodrome of the late-1930s expansion period was roughly square, with a circular landing area bordered by a crescent of brick hangars (Francis, 1996). A compact built-up area contained all the technical, administrative and residential facilities. During the war new hangars and camp areas were constructed in a dispersed pattern to reduce damage from

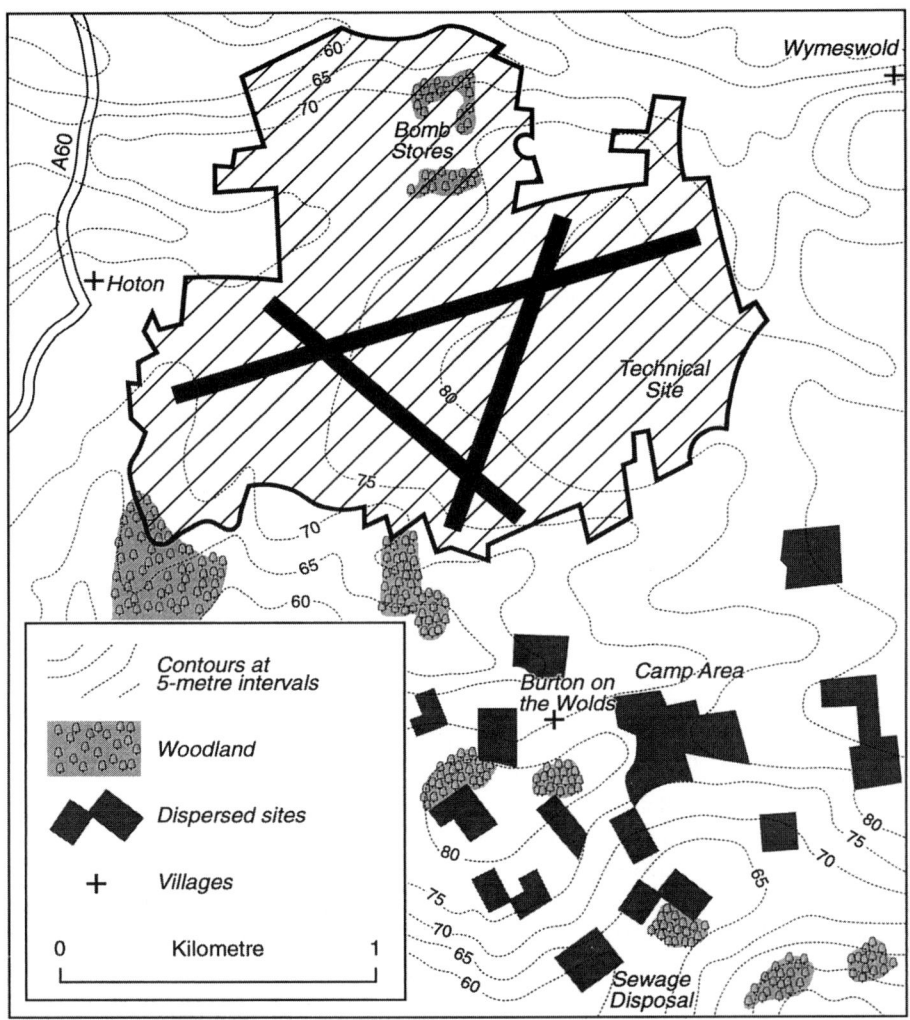

Figure 2. RAF Wymeswold, Leicestershire [SK585225], showing the relationship of the airfield and dispersed camp areas to physiography, land cover and main highway

enemy staffing. Jagged layouts are characteristic of airfield evolution under wartime emergency conditions, as exemplified by RAF Wymeswold (Figure 2).

Most paved aerodromes had three runways mutually disposed at 60° to maintain operations should the main runway be cratered or obstructed (Figure 2). A sizeable minority had two runways laid in a 'cross-bow' configuration and several manufacturers' aerodromes had a single hard strip. Royal Navy airfields typically had four or five shorter, narrower strips to simulate operation from aircraft carriers in adverse winds. After the war single runways were added to a number of wartime all-grass stations, e.g. Duxford, Cambridgeshire [TL465460].

All-grass airfields generally had fewer buildings, although those with a pre-war history as RAF stations or civil airports made a distinctive mark on the skyline by virtue of their hangars, masts, and other infrastructure.

4. Airfield camouflage and concealment

In today's defence environment of electronic surveillance and countermeasures it seems quaint that seventy years ago strenuous efforts were made to deny the existence of airfields. During the pre-war Expansion censorship was even applied to OS-based land utilisation maps, as illustrated by the partial deletion of RAF College Cranwell, Lincolnshire [TF010510] (Board, 1993). For the most secret landing grounds guardrooms were disguised as bungalows (Smith, 1989).

Britain's airfields were vulnerable to attack because of their close spacing, proximity to the coast and discordant layout in a finely-textured landscape (Hoskins, 1995). It was inherently difficult to 'lose' a triangular runway system, curved perimeter track and radiating dispersals within a rectilinear patchwork of small fields. Newly laid concrete was liable to 'shine' and raids by the Luftwaffe were common during the construction phase. At Goxhill, Lincolnshire [TA115215], the exposed chalk sub-soil had to be toned down with soot (Smith, 1989).

Coloured cements, slag, rubber chippings, tannery waste and various other materials were applied to runways but most were eventually rejected because of abrasion to aircraft (Halpenny, 1984). To suggest hedgerows, emulsion and silicate paints were sprayed across grassed areas while differential mowing, seeding and fertilisation techniques were applied. At RAF Northolt, Middlesex [TQ105850], attempts were made to suggest rows of houses (Smith, 1989).

Hangars, being grossly out of scale with traditional rural buildings, were either netted or painted in rustic tones, particularly the brick-built expansion-era type. Administrative and domestic quarters were also toned down in this way, an example of which is still visible at Horsham St Faith, Norfolk, now Norwich Airport [TG220135].

In the early stages of the war many 'spoof' airfields were laid out with dummy aircraft assembled in film studios and special circuit lights which could be switched on at night. Special apparatus was erected to simulate aircraft landing and taxying off presumed runways. To be plausible these decoys were sited in among the genuine airfield clusters, adding to the volume and complexity of engineering works in 'airfield country' (Dobinson, 2000).

By the closing of stages of the war most concealment measures had been abandoned, but artefacts and landscape modifications resulting from their implementation survive as further aspects of the airfield heritage.

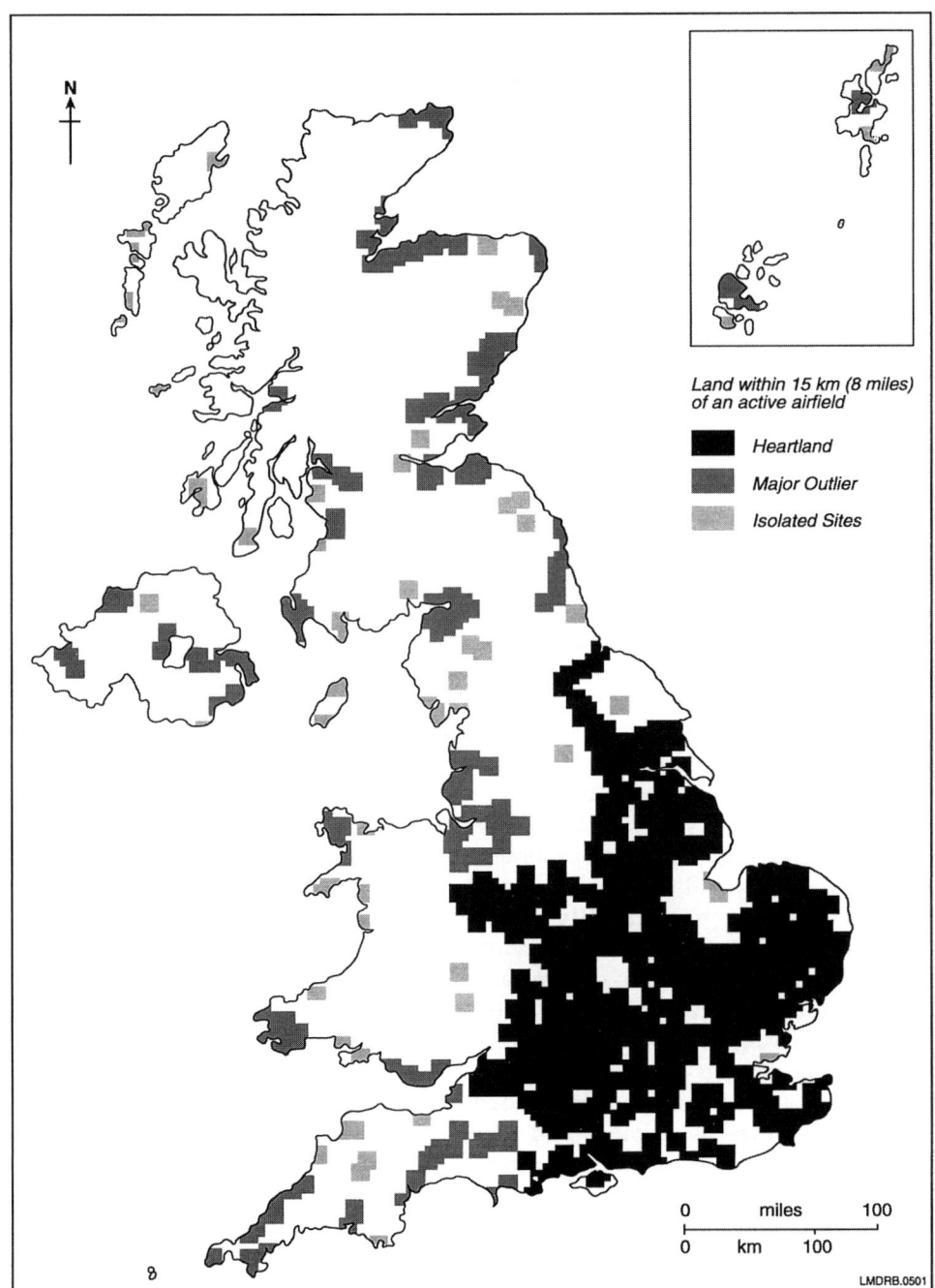

Figure 3. Airfield country, 1945

5. Trafficability

Normal assessment of military terrain in terms of 'the going' has partial relevance for airfields. It is clear that Air Ministry surveyors took a range of environmental factors into account when selecting sites, although no records of terrain classification specifically for home aerodrome construction have yet come to light. The closest parallel are the Army's 'Cross Country Movement (Goings)' maps of northern France and the Low Countries which define 'excellent going' as 'usually open arable', 'fair going' as having 'occasional hedges' and 'very bad going' as characterised by 'obstacles, usually wood and marshy ground' (British Second Army, 1944).

But unlike army operations, where bold sweeps and covert infiltrations are mounted to secure territory, air operations require robust infrastructure behind the battle front where aircraft can be armed, manned and repaired. What effectively is needed therefore are local tracts of high-trafficability country (for take-off and landing) separated by tracts of comparatively low-trafficability country where complementary infrastructure (storage, repair and residential areas) can be effectively concealed.

In a protracted and moving land battle, such as the sequel to 'Operation Overlord' in Normandy in 1994 (Rosenbaum, 1990; Rose & Pareyn, 1995), airfields were intentionally short-lived as the Allied forces progressed eastwards. In Britain, which had withstood the threat of German invasion and mounted a sustained and ultimately successful counter-offensive, airfields were relatively permanent installations deeply embossed in the landscape.

British home airfields required substantial sites of low-gradient and easily engineered ground. In practice, where such sites were plentiful the intervening countryside was also receptive to the ancillary requirements of air defence. An ideal aerodrome stood on a low and mildly convex plateau between villages, involving minimal demolition of property or disruption to local traffic, while enjoying reasonable access to nearby main roads, branch railways, utilities and community services (Nock, 1971).

Two levels of trafficability were therefore pertinent to home-based airfields. In the runway and taxying zone it was essential to have near-perfect ground conditions to minimise wear and tear on flying machines which are characteristically more delicate than tanks or armoured personnel carriers. Swift movement of ambulances and fire tenders was a further requirement. In the outer zone of the airfield it was desirable to have a network of free-draining access roads, with moderate gradients, plus a generous cover of woodland and hedgerows to shelter camp sites from inclement weather, explosions and air raids.

A productive and living countryside was thus generally advantageous. This helps explain the speed with which airfields were constructed and the mixture of affection and resentment felt in many areas towards the wartime air force presence.

6. Defining 'airfield country'

The central premise of this paper is that selected areas of Britain were so densely packed with wartime airfields that their character can in part be defined by the airfield legacy. Classic works on landscape change (e.g. Hoskins, 1955; Nairn, 1955; Brett, 1965) have tended to concentrate on disfigurements, a viewpoint which may inadvertently have delayed the advent of serious airfield studies. Fleeting references to the positive impact of wartime airfields (e.g. Balchin, 1983; Rowley, 1972; Steane, 1974) helped keep the subject alive during the ensuing decades, pending the recent upsurge of interest in the conservation of aeronautical buildings (e.g. Lowry, 1995; Francis, 1996; Higham, 1998).

There are several approaches to a fuller understanding of what 'airfield country' actually means. Figure 1 shows that extensive areas of eastern and southern England had exceptional airfield densities, with significant outlying clusters in lowland pockets such as Pembrokeshire, north Cumbria and the Moray Firth coastlands. This does not necessarily indicate a uniform impact on the landscape however, for the dot distribution conceals subtle differences of size and role. Design knowledge must be invoked to calculate percentage land-take at the district and country levels.

In essence 'airfield country' may be defined as any continuous territory where airfield spheres-of-influence were contiguous or overlapping in wartime and where the landscape still carries tangible aeronautical references. The latter range from the core infrastructure of runways and technical areas, to dispersed camp sites, service roads, lane widenings, water towers, radio beacons, woodland and hedgerow clearances, sand and gravel pits, air memorials, heritage centres, and thematic place names (e.g. Lancaster Farm).

South Kesteven District Council of Lincolnshire has published an 'Airfield Trail' which exploits the rich air war heritage of the area. Low-flying aircraft, crowded skies, crash sites, busy country roads and numerous resident air-force personnel are key ingredients of the nostalgia attached to 'airfield country'.

To summarise the concept, 'airfield country' is a fusion of individual airfield 'shadow zones', within which area it would have been difficult for anyone in wartime to ignore the air force presence and where there are still visible reminders of the period.

7. Mapping 'airfield country'

Systematic recording of all off-site impacts is a laborious task because airfields varied greatly in individual complexity and inevitably time has erased some of the evidence. To simplify the present exercise, airfield shadow zones were assumed to have a standard radius of about 8 km (5 miles) reflecting the circuit for the heavier wartime aircraft (similar to a modern

Military Aerodrome Traffic Zone). To avoid awkward plotting of circles based on exact airfield centroids, the mapping routine was systematised by generalising all locations within a specially devised OS grid.

First, each airfield was plotted on the basis of the 5 km grid-square in which it lay (e.g. TQ46SW for Biggin Hill). Then, eight surrounding grid-squares were added to create a block of country measuring 225 km^2 with a minimum centre-to-periphery distance of 7.5 km, a centre-to-corner distance of a $c.$ 10.5 km and a mean radius of $c.$ 8 km. In areas of medium and high airfield density these blocks were found to touch or overlap, establishing a consistent criterion for the mapping of 'airfield country'.

Figure 3 exhibits the shadowing effect of Britain's airfields when all sites are 'extended' with a local zone of influence. By this method, as much as 38% of the UK land surface was implicitly affected by low-flying aircraft and airfield-orientated ground activity in wartime. Two-thirds of the shadowed area formed a continuous 'heartland' stretching across eastern, southern and central England, with predictable 'windows' caused by the Fens and Greater London.

The most coherent areas within 'heartland' were in East Anglia, south Lincolnshire and the Cotswolds. Suffolk was 95% 'airfield country', while Shropshire, with a substantial lowland portion, had a surprising 54% in shadow. An intriguing gap in south Northamptonshire is possibly explained by the presence of major radio transmitters, canals and the express railway corridor.

Only 9% of the UK consisted of 'major outliers' (each comprising a minimum of three fused airfield circuits), including the coastlands of Cornwall, Northumberland and Tayside. 'Isolated sites' (single, or pairs of adjacent, airfields) accounted for less than 4% of the nation's airfield shadow.

8. Terrain and land-use constraints

In areas characterised by dense natural vegetation, intensive agriculture, urban spread, or any combination of these primary ground-covers, the assessment of terrain for any military purpose necessarily includes an appreciation of the land-use mix. Airfield development in Britain exemplifies par excellence the joint influence upon military operations of landform, surface mantle and pre-existing infrastructure.

This section identifies three measurable variables which clearly acted as constraints on airfield development, but whose influence is best regarded as indirect. In each case the OS 5 km grid was filled with numerically banded data derived from approximately ten thousand readings per variable. From the resultant data-base a picture of the whole UK land surface was assembled for comparison with the several hundred grid-squares containing one or more wartime airfields.

The basis adopted for comparison was the 'location quotient' (LQ), being the relationship between the percentage area occupied by a given terrain type and the percentage of airfields located in that terrain type. In this methodology an LQ value of 1 indicates an airfield occurrence equal to the national density, while a value of 2 signifies an occurrence of twice the average density and 0.5 an occurrence of half the average density.

Terrain characteristics on a national basis were derived from readings at 5-km spacing, while each airfield was 'scored' by a single reading taken for each variable at the centre of its main runway or grass landing area.

Elevation above mean sea level is summarised in Table 1. The location quotients reveal that the land surface between the 8 m and 122 m contours had an airfield density consistently greater than the national average, with distinct thinnings occurring towards the flood-prone lower areas and the more rugged higher terrain. For meteorological reasons hill-top airfields were most common in southern England where level erosion surfaces are wider than elsewhere. In northern Britain airfields were characteristically sited in low coastal pockets because the higher land was rarely level (Appleton, 1962).

Table 1: Elevation basis of Second World War airfields in the United Kingdom

Elevation above sea level (m)	Land area (%)	Airfields No.	Airfields %	Location Quotient
0-8	5.7	60	7.4	1.29
8-30	10.1	209	25.7	2.54
30-61	13.3	160	19.7	1.48
61-122	24.9	251	30.9	1.24
122-183	14.8	100	12.3	0.83
183-244	8.9	29	3.6	0.4
244-304	6.2	4	<0.4	0.06
>304	16	-	-	-

Geology was also an important contributory factor, although its influence is difficult to de-couple from the resultant landforms (Beaver, 1931). Blake (2000) records high location quotients for Middle Jurassic limestone (3.5), Upper Cretaceous Chalk (2.0), Quaternary sands and gravels (3.0) and certain other drift materials including raised beach deposits (6.0). By far the largest number of paved-runway airfields were sited on Quaternary till (Institute of Geological Sciences, 1977), but this association is more likely to have reflected the prevailing geomorphology rather than rock type per se (Gray, 1996, 1997b).

All-grass airfields were especially clustered on Mesozoic chalk and limestone escarpments, but were also found, albeit in small numbers, on Quaternary alluvium because of the lighter aircraft involved.

Agricultural land quantity is perhaps the most curious of all airfield siting factors as it can be interpreted as both an opportunity and a constraint. While attempts were made to minimise land-take through economical design and sensitive land-acquisition procedures, it is ironical that the best ground for aerodromes in Britain coincides, except in the Fens, with the most productive farmland. Large arable fields without banks or walls were ideal for rapid runway construction under emergency conditions.

Despite attempts to keep the Fleet Air Arm away from Cheshire dairy farms (Smith, 1983), nothing could arrest the tide of major aerodrome development across Britain's 'good general purpose farmland', which in wartime was at a premium (Stamp, 1962). Post-war Agricultural Land Classification maps (Ministry of Agriculture, Fisheries and Food, 1968-74) confirm that wartime airfields were disproportionately sited on grades 1 and 2 (the most versatile and scarce soils). Surviving airbases, e.g. Topcliffe, Yorkshire [SE400790] are polarised on these higher grades (Blake, 1989).

9. A 'sieve map' of opportunity

This section examines three further variables which lend themselves better to cartographic analysis (Blake, 1981). These are: undulation, non-agricultural ground cover, and accessibility. Figure 4 shows the results of the 'sieving' analysis.

An impression of undulation, or average gradient, was gained by counting the number of contour lines intersecting the western and southern margins of each 5-km grid-square. This proved faster than gauging amplitude of relief (the difference between the highest and lowest land within a grid-square) and yielded more information pertinent to airfield siting. Data were obtained principally from OS 1:63,360 sheets, as these are roughly contemporary with the airfields themselves but in cases where a whole grid-square stood below the 50-foot (16 m) contour, the relevant 1:25,000 scale map was consulted to distinguish alluvial areas at or near to sea level (≤ 8 m) from those containing low river terraces (8-16m).

Using terminology coined by Dickinson (1979) squares with an undulation score of ≥ 25 were classed as 'hilly' and assumed to be problematical for airfield development. Results confirm that airfield density fell rapidly above this threshold, although several long-term RAF stations, e.g. Little Rissington, Gloucestershire [SP210190], were perched on narrow interfluves with steep adjacent gradients. Flood-prone land similarly had a below-average airfield density, being particularly unsuited to paved runways because of compressible soils and wide drainage channels (Edwards, 1954).

Areas of extensive woodland and urban settlement also had a constraining influence, although this is sometimes difficult to separate from underlying physiographic controls. Grid-squares with a cover of more than a 20% by either land-use type were identified from the New Popular Edition of the OS One Inch map (c. 1945), representing the period of maximum airfield

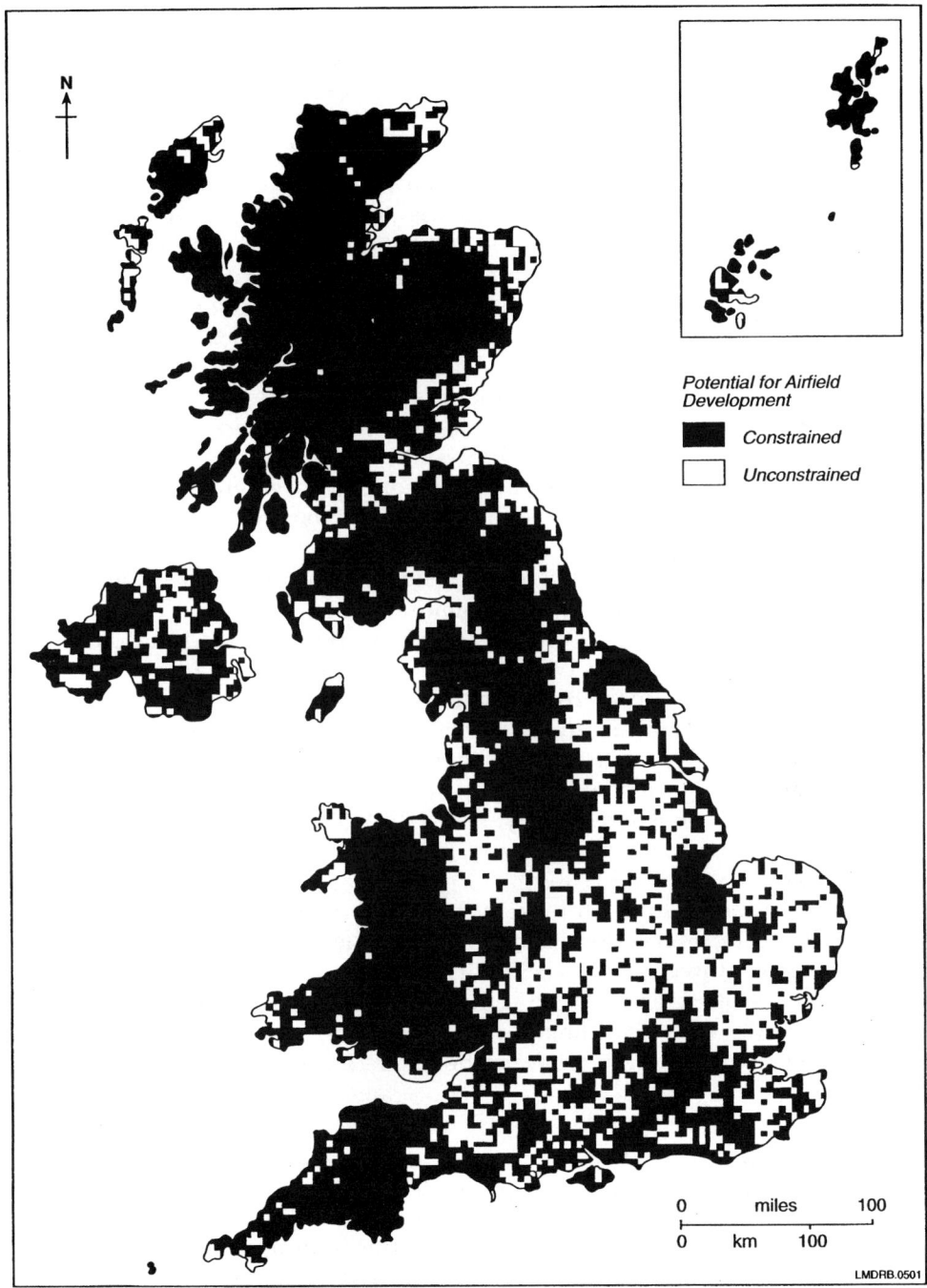

Figure 4. Potential for airfield development in terms of constrained and unconstrained land at the time of maximum expansion

development. Only 6% of the UK was then forested, yet gaps in the airfield distribution are discernible in areas such as Breckland. In northern Britain forest tended to coincide with hilly terrain, but certain areas of southern England, notably the New Forest, had clusters of unusually secluded fighter airfields. Rackham (1986), describing Essex, notes that small woodlands provided essential cover for airfield camps and bomb dumps.

Under 10% of Britain was built-up when war broke out (Stamp, 1962) and continuous urban spread in lowland areas is only discernible on the land use maps for Greater London and other conurbations. Many small aerodromes requisitioned for training and aircraft repair stood on the edge of suburbia and thereby appear in 'constrained' areas according to this analysis. In rural areas a countryside rich in villages and market towns was beneficial because of the existing facilities. Air Ministry planners deserve credit for minimising the need to demolish property and re-house local people.

Surface communications were vital to the efficient operation of airfields, the widely cited attribute of remoteness being applicable only at parish level. Accessibility was measured here according to whether a grid-square contained a class-A road, a serviceable railway, one of those facilities, or neither. Choice of sites alongside Roman roads was common before 1939, e.g. RAF Kemble on the Gloucestershire-Wiltshire border [ST960965], where pilots were trained in navigation by visual reference.

For the wartime transport of minerals, fuel, munitions and personnel the railways were vital (Nock, 1977). In the least accessible areas of Britain (those lacking both a main road and railway) there was usually a strong physiographic constraint in any case, but the analysis confirms that even in the lowlands there was a spatial association between relatively good accessibility and high airfield density. Inevitably a minority e.g. Binbrook, Lincolnshire [TF190960] and Brawdy, Dyfed [SM850250] lay in comparatively inaccessible terrain, and significantly neither station has survived as an active airbase.

When overlaid (Figure 4), these last three major constraints produce the 'sieve' map comprising 'constrained' (black) and 'unconstrained' (white) squares. Almost three-quarters (73%) of the UK had at least one development constraint (the black area), with a high degree of contiguity in the hilly north and west. Correspondingly, the white area accords reasonably well with popular expectation of where airfields are located.

10. Quantifying the 'fit'

From the patterns presented in Figures 1, 3 and 4 it is clear that the distribution of airfields in wartime Britain was broadly consonant with the theoretically permissive environment identified as being free from major development constraints. While manual overlaying of distributions confirms the general 'fit', a statistical matching adds credence to the principal influences described earlier in the discussion. Two further analyses, focused respectively on the deterrent effect of multiple constraints, and the constraints profile of the independently derived 'airfield country', are summarised below.

Table 2: Relationship of Second World War land airfields (terrestrial) to the number of environmental constraints on development

Number of constraints	% area of the UK	Airfields No.	Airfields %	Location Quotient
None	27.2	498	61.3	2.25
1	43.2	259	31.8	0.74
2	27.4	53	6.5	0.24
3-4	2.2	3	0.4	0.18
	100	813	100	-

Table 2 indicates that while slightly over a quarter (27%) of the UK's land area was free from any of the three key development constraints, approaching two-thirds (61%) of the airfields lay in 'white' grid squares. A location quotient of 2.25 for the latter signals an airfield density two-and-a-quarter times the national average. In areas with one constraint, LQ fell to one-third of the value for unconstrained land (0.74). In areas with two or more constraints, airfield density plummeted to less than a quarter of the national average (0.24), or one-tenth of the unconstrained value. Fewer than 7% of wartime airfields were sited in areas of multiple (i.e. 2 or more) constraints, being predominantly all-grass landing grounds tolerant of rolling, forested, remote, or suburbanised terrain.

Table 3 takes a more sweeping approach by examining the degree of environmental constraint within the three categories of 'airfield country' distinguished in Figure 3, and also includes the unshadowed ('airfield-free') remainder of the UK as a comparator. Predictably, more than half (57%) of the grid-squares comprising 'airfield heartland' were free from any key development constraint. The proportion dropped to about two-fifths (42%) in the 'major outliers' and below a quarter (24%) around 'isolated sites' (Doylerush, 1985).

'Airfield country' as a whole was almost half free from any constraint, whereas the more extensive 'non-airfield country' was little over a quarter free.

11. The landscape legacy

A final aspect of Britain's wartime airfields meriting comment is their long-term imprint on the landscape (Blake, 1969). The three principal legacies are: (1) enlargement of fields and eradication of topographic detail such as ponds, ditches, hedgerows and isolated cottages; (2) geometric superimpositions due to the reinstatement of public roads along stretches of abandoned runway and perimeter track; and (3) the retention of hangars as core features of industrial estates, institutions and recreational complexes.

While nature has restored biomass through the regeneration of woodland and scrubland around derelict and re-used buildings, the former operational areas are still characterised by either an absence of traditional tree cover or the presence of regimented rows of poplars and elongated coniferous plantations. Incomplete demolition has left piles of rubble in an otherwise rural environment, inviting squatters, scrap metal dealers and fly-tipping. The fact that there is no specific government policy for comprehensive landscape restoration has allowed the ambience of 'airfield country' to persist (Gray, 1997a).

Table 3: Composition of 'airfield country' in terms of environmental constraint on development

Type of airfield country	% area of the UK	% area free of development constraint
Heartland	24.9	56.9
Major Outlier	9.1	42.2
Isolated Sites	3.6	24
All airfield country	37.6	50.3
Non-airfield country	62.4	27.2

The key attributes of openness, ample concrete, cheap buildings and interstitial rough land account for the remarkable range of after-use encountered on airfields (Blake 1986; 1995). Institutions such as Silverstone motor-racing circuit, Northamptonshire [SP675420] and the Fire Services College at Moreton-in-Marsh, Gloucestershire [SP220330] are unmistakable on OS maps because the original runway layouts have been retained and exploited. Poultry breeding enterprises, prisons and vehicle test-tracks are equally discernible, but sundry activities such as learner-driving, land-yachting and casual use by private aircraft have gone largely unrecorded on maps. From visual clues, the passer-by is nonetheless reminded on a daily basis that these are zones of special land-use character and history.

All manner of artefacts, including dilapidated bomb-aiming towers, guardrooms and remnants of security fencing, are now being conserved privately or under the auspices of the many museums which have been based in old control towers and communal buildings (Innes, 1995).

12. Conclusions

Britain's wartime airfields were remarkable in their density, land-take and resource utilisation, and are probably unique both globally and historically. Land acquisition and construction were played out in a productive, finely textured and well settled landscape as there were few level wildernesses available for development in the Nation's primary zone of air defence.

A geomorphological coincidence, whereby the most easily engineered ground was ranged on the enemy side of the nation's urban-industrial complexes, accounts for the packing of airfields into East Anglia, the East Midlands and Wessex. The current OS 1:50,000 sheet 121 Lincoln & Surrounding Area contains as many as nineteen Second World War airfields (Blake et al., 1984). Relative to the best agricultural areas the wartime land requirement for aerodromes was significantly greater than the c.1% headline figure cited earlier in this review.

Analysis of terrain and land use through the 'sieve mapping' technique has revealed that slightly over a third of the UK was theoretically suitable for airfields. Plotting of local airfield hinterlands confirms a similar proportion of the landscape to have been 'shadowed' by aircraft operations. There is strong spatial association between the two distributions, suggesting a topographical logic to airfield siting.

The fact that constrained landscape was roughly twice as extensive as unconstrained landscape highlights 'airfield country' as a spatially discrete and definable environment exploited for air defensive purposes. A broad dichotomy of 'airfield-rich' and 'airfield-poor' areas emerges from the data analysed, inviting deeper investigations into cause and effect as further official records pass into the public domain.

From an historic perspective, the recognition of 'airfield country' provides fresh insights into Britain's island defences and its wider geopolitical role in European conflicts. From an applied perspective, a structured appreciation of airfields in terms of environmental carrying capacity, landscape change and land-use controls is of potential interest to planners, developers, conservationists and guardians of the nation's heritage.

References

Appleton, J.H. 1962. *The Geography of Communications in Great Britain*. Oxford University Press, Oxford.

Bagley, J.A. 1972. A gazetteer of Hampshire aerodromes. *Proceedings of the Hampshire Field Club and Archaeological Society* 24, 93-108.

Balchin, W.G.V. 1983. *The Cornish Landscape*. Hodder & Stoughton, London.

Beaver, S. H. 1931. The Jurassic scarplands. *Geography* 12, 298-307.

Betts, A. 1995. *Royal Air Force Airfield Construction Service 1939-46*. Airfield Research Publishing, Ware.

Blake, R.N.E. 1969. The impact of airfields on the British landscape. *Geographical Journal* 135, 508-28.

Blake, R.N.E. 1981. The changing distribution of military airfields in the East Midlands, 1914-80. *East Midland Geographer* 7, 286-302.

Blake, R.N.E. 1986. Old airfields take off: new uses for aerodromes. *Geographical Magazine* 58, 272-274.

Blake, R.N.E. 1989. *The development of military and civil airfields in the United Kingdom since 1909, with special reference to land use.* Unpublished PhD Thesis, University of London.

Blake, R.N.E. 1995. Alternative strategies for the restoration and re-use of abandoned airfields. In: Coulson, M. & Bladwin, H. (Eds), *Pilot Study on Defence environmental Expectations.* Proceedings of the International Symposium on the Environment and Defence (University of Wales, Swansea, 13-15 September 1995), North Atlantic Treaty Organisation, Committee on the Challenges to Modern Society, Report No. 211, 321-332.

Blake, R.N.E. 2000. Geological influences on the siting of military airfields in the United Kingdom, In: Rose, E. P. F. & Nathanail, C. P. (Eds), *Geology and Warfare: Examples of the Influence of Terrain and Geologists on Military Operations.* The Geological Society, London, 276-312.

Blake, R.N.E., Hodgson, M. & Taylor, W.J. 1984. *The Airfields of Lincolnshire since 1912.* Midland Counties Publications, Leicester.

Board, C. 1993. Falsification and security. In: Barber, P. & Board, C. (Eds), *Tales from the Map Room: Fact and fiction about Maps and their Makers.* BBC Books, London, 106-107.

Brett, L. 1965. *Landscape in Distress.* Architectural Press, London.

British Second Army 1944. *Cross Country Movement (Goings).* Scale 1:25,000 GSGS 4347, Military Survey.

Childs, J. 1998. *The Military Use of Land: A History of the Defence Estate.* Peter Lang, Bern.

Dickinson, G.C. 1979. *Maps and Air Photographs: Images of the Earth.* Edward Arnold, London.

Dobinson, C. 2000. *Fields of Deception: Britain's Bombing Decoys of World War II.* Methuen, London.

Doyle, P. & Bennett, M.R. 1999. Military geography: the influence of terrain in the outcome of the Gallipoli Campaign, 1915. *Geographical Journal* 165, 12-36.

Doylerush, E. 1985. *No Landing Place.* Midland Counties Publications, Leicester.

Edwards, K.C. 1954. Changing geographical patterns in Lincolnshire. *Geography* 39, 78-90.

Francis, P. 1996. *British Military Airfield Architecture: From Airships to the Jet Age.* Patrick Stephens, Cambridge.

Gray, J.M. 1996. The containment properties of glacial tills: a case study from Hardwick Airfield, Norfolk (UK), In: Bentley, S. P. (Ed.), *Engineering Geology of Waste Disposal.* Geological Society, Engineering Special Publication No. 11, London, 299-307.

Gray, J.M. 1997a. Planning and landform: geomorphological authenticity or incongruity in the countryside? *Area* 29, 312-324.

Gray, J.M. 1997b. Environment, policy and municipal waste management in the UK. *Transactions of the Institute of British Geographers* 22, 69-90.

Halpenney, B. B. 1984. *Action Stations 8: Military Airfields of Greater London.* Patrick Stephens, Cambridge.

Higham, R. 1998. *Bases of Air Strategy: Building Airfields for the RAF 1914-1945.* Airlife Publishing Ltd, Shrewsbury.

Holmes, R. 1996. *War Walks: From Agincourt to Normandy.* BBC Books, London.

Hoskins, W. G. 1955. *The Making of the English Landscape.* Hodder & Stoughton, London.

Innes, E.B. 1995. *British Airfield Buildings of the Second World War.* (Aviation Pocket Guide No. 1), Midland Publishing, Leicester.

Institute of Geological Sciences. 1977. *Quaternary Map of the United Kingdom. Scale 1:625,000, 2 sheets (North & South).* Ordnance Survey, Southampton.

Lowry, B. (Ed.) 1995. *Twentieth Century Defences in Britain* (Practical Handbooks in Archaeology No. 12), Council for British Archaeology, York.

Marriott, H.L. 1997. *British Military Airfields: Then and Now.* Ian Allan, Shepperton.

Nairn, I. 1955. *Outrage.* Architectural Press, London.

Neal, I.L. 1998. Playas in military operations. *In:* Underwood, J. R. and Guth, P. L. (Eds), *Military Geology in War and Peace.* Geological Society of America. Reviews in Engineering Geology 13, 165-172.

Nock, O. S. 1971. *Britain's Railways at War 1939-45.* Ian Allen, London.

Oliver, R. R. 1991. *An Introduction to the Ordnance Survey One-inch Seventh Series Map: with a list of editions.* Ordnance Survey, Southampton.

Rackham, O. 1986. *History of the Countryside.* I M Dent, London.

Rose, E.P.F. & Pareyn, C. 1995. Geology and the liberation of Normandy, France, 1944. *Geology Today* 11, 58-63.

Rosenbaum, M.S. 1990. Geologists at war: the D-Day operations and subsequent advance. *Proceedings of the Geologists' Association* 101, 163-166.

Rowley, T. 1972. *The Shropshire Landscape.* Hodder & Stoughton, London

Smith, D.J. 1989. *Britain's Military Airfields 1939-45.* Patrick Stephens, Cambridge.

Stamp, L.D. 1960. *Britain's Structure and Scenery.* Fontana, London.

Stamp, L.D. 1962. *The Land of Britain: Its use and misuse.* 3rd Edition. Longmans, London.

Steane, J. 1974. *The Northamptonshire Landscape.* Hodder & Stoughton, London.

Ron N. E. Blake
Department of Building Studies & Environmental Health
Nottingham Trent University, Nottingham, NG1 4BU

Subject Index

Battles and engagements are given by names and dates.

Air defence of Britain (1939-45), 365-383
Airfields, 365-383
 - camoflage, 370-372
 - 'country', 366, 371, 373, 381
 - design of, 368-370
 - development, 366-368
 - distribution of, 367
 - and geology, 375
 - and landuse, 374-376, 381
 - and landscape, 379-380, 381
 - mapping of, 373-374, 376-379
 - trafficability of, 372
Air photographs, 172, 177, 184-185, 190, 194
American Civil War (1861-65), 4, 63-115
 - battles of, 64-68
 - causes of, 64-66
 - and terrain, 63-115
Aldbourne (1643), 34
Alton (1643), 22
Amiens (1918), 180
Anglo-Scots Border Wars, 20, 21, 26
 - battles of, 19-31
 - and geology, 19-31
Anglo-Zulu War (1897), 117-135
 - archaeology of, 117
 - battles of, 120-123
 - causes of, 118, 135
Arras (1917), 5, 178
Artillery
 - maps, 176-177
 - survey, 183, 192
Atlantic Wall, 265, 278-295
Aubers Ridge (1915), 177

Babylon Hill (1642), 22
Battlefield
 - archaeology, 5, 117
 - geometry, 171-175
Bishop's War (1640), 20, 23
 - battles of, 19-31
 - and geology, 19-31
Boer War (1899-1902), 4
 - and General Buller, 137-141
 - and military maps, 4, 138, 140-141
Bosworth (1485), 20
Broodseinde (1917), 243
Brunanburh (937), 11

Cambrai (1917), 178, 180, 240
Cheriton (1644), 34, 35, 45-46
Chester (613), 12, 13
Chicamauga (1863), 99, 113
 - battlefield of, 108
Colenso (1899), 137, 138, 140

Defence-in-depth, 220, 223, 241
Digital Elevation Models (DEM)
 - of Fredricksberg, 64, 76, 78, 83, 85, 95-97
 - of Messines, 205, 214, 221, 222, 223
 - of Navarre, 59-60
Dunkirk (1940), 278

Ebro (1938), 6, 144, 145, 148, 257-264
 - battlefield, 260-262
 - as a didactic resource, 257-264
 - memorials of, 145, 258-259
English Civil Wars, 20, 23, 33-50
 - battles of, 7-8, 19-31
 - and geology, 7-8, 19-31, 33-50

Fall of France (1940), 278
Festubert (1915), 177
First World War (1914-18), 4, 5, 141-144, 145, 171-255, 342
 - Gallipoli, 4, 5, 149-169
 - Messines, 178, 179, 216-218, 253
 - and military maps, 3, 4, 149, 160, 162-163, 171-204, 205-224
 - and military mining, 4-5, 172, 216-218, 225-236
 - Third Ypres, 5, 178, 180, 237-255
 - Western Front, 171-204, 205-206, 207
Flash-spotting, 172, 192
Flodden (1513), 20, 22
Fort Sumter (1861), 66
Fredricksberg (1862), 63-97
 - battlefield terrain, 68-72, 80, 90-92

Gallipoli (1915), 4, 5, 149-169
 - and geology, 151-153
 - and hydrology (water supply), 153-155, 156, 162, 163, 166-167
 - landing beaches, 162, 165-166
 - and military maps, 4, 149, 160, 162-163
 - terrain classification of, 155-159
 - and topography, 151-155
Geographical information systems (GIS), 95-97, 211-214
Geology and warfare, 19-31, 33-50, 110-111
George Booth's Rising (1659), 20
German Offensives (1918), 197-198, 199-200, 211, 219
Gettysberg (1863), 258
'Going' surfaces, 4, 19, 29, 161
Great War (see First World War)
Ground conditions, 4-5, 161

Hamel (1918), 180

Hastings (1066), 12
Hazard Mitigation, 5
Hooge (1915), 227

Isandlwana (1879), 120-122
    - archaeological potential of, 131-132
    - battlefield of, 124-128
    - monumentalisation of, 128-131

Jacobite Uprising (1745-46), 22
Jersey fortifications, 265-309
    - German, 280-290
    - and geology, 266-270, 271, 295-302
    - history of, 273-278
    - and labour, 291-292
    - and resources, 290-295
    - similarity to Gibraltar, 303, 305
    - and topography, 270-273

Landscape, 3-6, 123-124
    - of battle, 3-5, 123-124
    - iconic, 5, 145-147, 257-262
    - symbolic, 3, 128-131
Landsystems, 155-159
Lansdown (1643), 34, 35
Little Bighorn (1876), 131
Loos (1915), 177-178
Ludford Bridge (1459), 22

Maldon (991), 12, 14
Manassas
    - First Battle of (1861), 66
    - Second Battle of (1862), 67
Marston Moor (1644), 20
Mediaeval wars, 9-17
    - and battle location, 10-12, 13-16
    - and terrain, 9
    - and warfare type, 12-13
Memorialisation, 128-131, 144-145
Messines (1917), 178, 179, 216-218, 253
    - and geology, 219-220
    - and military mining, 216-218, 225-236
Military maps
    - and the American Civil War, 110-113
    - and the Boer War, 4, 138, 140-141
    - and Gallipoli, 4, 149, 160, 162-163
    - and the Napoleonic Wars, 4, 54
    - and the Western Front, 3, 171-204, 205-224
Military mining, 4-5, 172, 216-218, 225-236
    - craters, 208, 218, 232-234, 236
    - explosions, 231-232
    - explosives, 225-227, 235
    - detonations, 227-229, 235-236
    - mine chambers, 227
    - 'tamping', 230
    - tunnels, 290, 292, 296, 298, 300
Mobility, 4
Monte Cassino (1944), 4, 325-343
    - difficulties of exposed rock, 339-341
    - fortification of, 331
    - and geology, 325-343

    - ancient history of, 336
    - and the Rapido River, 334-336
    - slopes of, 336-339
    - and topography, 327-329, 331-336
Murfreesboro (1862), 104

Napoleonic Wars, 4, 51-62, 278
Menin Road (1917), 241
Neuve Chappelle (1915), 177
Newburn Ford (1640), 20
Newbury
    - First Battle of (1643), 34, 35-39
    - Second Battle of (1644), 34, 35, 39-44
Nivelle Offensive (1917), 188-189
Normandy (1944), 4, 145, 257, 345-363
    - battlefield of, 347-349, 351
    - bocage, 350, 358-360
    - and German 'superiority', 346, 359, 361
    - historiography of, 345-346
    - landing beaches, 347, 351, 353-356
    - Mulberry Harbours, 357-358, 363
    - 'phase-line' controversy, 346, 356-357
    - terrain, 347-351, 352-356, 361

Paris Gun (1918), 198-199
Passchendaele (1917), 243-246
Peninsula War (1808-1814), 51-62
    - and guerrilla warfare in Navarre, 51-62
Phenomenological archaeology, 123
Photogrammetry, 174-175, 184-185, 189-190, 193-195
'Pillboxes' (MEBU), 220, 223, 241
Plan directeurs, 176, 181, 182-183, 185-189
Poelcappelle (1917), 238, 243-246
Polygon Wood (1917), 241
Position, 4
Powick Bridge (1642), 20
Predicted shooting, 174

Refuge, 159-161
Resource Provision, 5, 29, 104, 290-295
Rorke's Drift (1879), 120, 122, 127
Roundway Down (1643), 34, 35, 46-48
Russo-Japanese War (1902), 172

Schlieffen Plan (1914), 189
Second World War (1939-45), 4, 144, 262, 264, 342
    - air defence of Britain (1939-45), 365-383
    - fortification of Jersey, 265-309
    - Monte Cassino, 4, 325-343
    - Normandy, 4, 145, 257, 345-363
    - Western Desert, 5, 311-323
Sedgemoor (1685), 22
Sevenoaks (1450), 20
Solway Moss (1542), 20
Somme (1916), 144, 146, 178, 179, 240, 253, 257
Sound-ranging, 172, 192
Sowerby (1644), 22
Spanish Civil War (1936-1939), 144, 145, 257-264
Spioenkop (1900), 137

SUBJECT INDEX

Stoke Field (1487), 20
Survey, 171-204, 191

Tadcaster (1643), 22
Terrain, 1-7, 63-64, 365-366
- analysis, 55-58
- assessment, 4-5, 159-164
- classification, 155-159
- definitions of, 1, 2, 366
- effective use of, 90-92
- and guerrilla warfare, 51-62
- intelligence, 3-4, 54, 110-111, 137, 138
- and strategy, 1, 29
- and tactics, 1

Third Battle of Champagne (1915), 186
Third Ypres (1917), 5, 178, 180, 237-255
- combat logistico-engineering (roads) of, 237-255

'Thirty years war' (1914-1945), 143-148
Topography
- and tactics, 29, 139-141
Torrington (1646), 22
Towton (1461), 5, 20
Trafficability, 372 (see 'going surfaces')
Traversability (see 'going' surfaces)
Trench
- maps, 176-201, 209-213, 215-217
- position in relation to geology, 219-220
Turnham Green (1642), 22
Tullahoma (1863), 99-115
- battlefield, 99-102, 110-111
- and General Rosencrans, 102-113

Ulundi (1879), 123

Valkraans (1900), 137
Vantage, 159-161
Verdun (1916), 145, 186-188, 195-196, 257
Vimy Ridge (1917), 179, 253
'Virtual battlefield', 95-97

Wars of the Roses (1450-1487), 5, 20, 21
- battles of, 3, 6-7
West Wall (see Atlantic Wall)
Western Desert (1940-1943), 5, 311-323
- and aircraft, 320-321
- and Ralph A. Bagnold, 313-315
- and the Long Range Desert Group, 311-323
- topography of, 311-313, 319, 320
- and vehicles, 316, 317
Western Front (1914-1918), 171-204, 205-206, 207
- 'learning curve' on, 252
- and military engineering, 237-255
- and military maps, 3, 171-204, 205-224
- and military mining, 4-5, 172, 216-218, 225-236
- and trench maps, 176-201, 209-213, 215-217
Worcester (1651), 20

Ypres Salient (1914-1918), 4, 205-224, 225, 238-241

Zulu War (see Anglo-Zulu War)

The GeoJournal Library

38. J.A.A. Jones, C. Liu, M-K. Woo and H-T. Kung (eds.): *Regional Hydrological Response to Climate Change.* 1996　　ISBN 0-7923-4329-8
39. R. Lloyd: *Spatial Cognition.* Geographic Environments. 1997　　ISBN 0-7923-4375-1
40. I. Lyons Murphy: *The Danube: A River Basin in Transition.* 1997 ISBN 0-7923-4558-4
41. H.J. Bruins and H. Lithwick (eds.): *The Arid Frontier.* Interactive Management of Environment and Development. 1998　　ISBN 0-7923-4227-5
42. G. Lipshitz: *Country on the Move: Migration to and within Israel, 1948–1995.* 1998
ISBN 0-7923-4850-8
43. S. Musterd, W. Ostendorf and M. Breebaart: *Multi-Ethnic Metropolis: Patterns and Policies.* 1998　　ISBN 0-7923-4854-0
44. B.K. Maloney (ed.): *Human Activities and the Tropical Rainforest.* Past, Present and Possible Future. 1998　　ISBN 0-7923-4858-3
45. H. van der Wusten (ed.): *The Urban University and its Identity.* Roots, Location, Roles. 1998　　ISBN 0-7923-4870-2
46. J. Kalvoda and C.L. Rosenfeld (eds.): *Geomorphological Hazards in High Mountain Areas.* 1998　　ISBN 0-7923-4961-X
47. N. Lichfield, A. Barbanente, D. Borri, A. Khakee and A. Prat (eds.): *Evaluation in Planning.* Facing the Challenge of Complexity. 1998　　ISBN 0-7923-4870-2
48. A. Buttimer and L. Wallin (eds.): *Nature and Identity in Cross-Cultural Perspective.* 1999　　ISBN 0-7923-5651-9
49. A. Vallega: *Fundamentals of Integrated Coastal Management.* 1999
ISBN 0-7923-5875-9
50. D. Rumley: *The Geopolitics of Australia's Regional Relations.* 1999
ISBN 0-7923-5916-X
51. H. Stevens: *The Institutional Position of Seaports.* An International Comparison. 1999
ISBN 0-7923-5979-8
52. H. Lithwick and Y. Gradus (eds.): *Developing Frontier Cities.* Global Perspectives – Regional Contexts. 2000　　ISBN 0-7923-6061-3
53. H. Knippenberg and J. Markusse (eds.): *Nationalising and Denationalising European Border Regions, 1800–2000.* Views from Geography and History. 2000
ISBN 0-7923-6066-4
54. R. Gerber and G.K. Chuan (eds.): *Fieldwork in Geography: Reflections, Perspectives and Actions.* 2000　　ISBN 0-7923-6329-9
55. M. Dobry (ed.): *Democratic and Capitalist Transitions in Eastern Europe.* Lessons for the Social Sciences. 2000　　ISBN 0-7923-6331-0
56. Y. Murayama: *Japanese Urban System.* 2000　　ISBN 0-7923-6600-X
57. D. Zheng, Q. Zhang and S. Wu (eds.): *Mountain Geoecology and Sustainable Development of the Tibetan Plateau.* 2000　　ISBN 0-7923-6688-3

The GeoJournal Library

58. A.J. Conacher (ed.): *Land Degradation.* Papers selected from Contributions to the Sixth Meeting of the International Geographical Union's Commission on Land Degradation and Desertification, Perth, Western Australia, 20–28 September 1999. 2001
ISBN 0-7923-6770-7
59. S. Conti and P. Giaccaria: *Local Development and Competitiveness.* 2001
ISBN 0-7923-6829-0
60. P. Miao (ed.): *Public Places in Asia Pacific Cities.* Current Issues and Strategies. 2001
ISBN 0-7923-7083-X
61. N. Maiellaro (ed.): *Towards Sustainable Buiding.* 2001   ISBN 1-4020-0012-X
62. G.S. Dunbar (ed.): *Geography: Discipline, Profession and Subject since 1870.* An International Survey. 2001   ISBN 1-4020-0019-7

KLUWER ACADEMIC PUBLISHERS – DORDRECHT / BOSTON / LONDON